Rolls Royce
SILVER CLOUD
and Bentley S series
Gold Portfolio
1955~1965

Compiled by
R.M. Clarke

ISBN 1 85520 1593

Brooklands Books Ltd.
PO Box 146, Cobham, KT11 1LG
Surrey, England

Printed in Hong Kong

BROOKLANDS BOOKS

BROOKLANDS ROAD TEST SERIES

AC Ace & Aceca 1953-1983
Alfa Romeo Alfasud 1972-1984
Alfa Romeo Alfetta Coupes GT. GTV. GTV6 1974-1987
Alfa Romeo Giulia Berlinas 1962-1976
Alfa Romeo Giulia Coupes Gold Portfolio 1963-1976
Alfa Romeo Giulia Coupes 1963-1976
Alfa Romeo Giulietta Gold Portfolio 1954-1965
Alfa Romeo Spider Gold Portfolio 1966-1991
Alfa Romeo Spider 1966-1990
Allard Gold Portfolio 1937-1959
Alvis Gold Portfolio 1919-1967
American Motors Muscle Cars 1966-1970
Armstrong Siddeley Gold Portfolio 1945-1960
Aston Martin Gold Portfolio 1972-1985
Austin Seven 1922-1982
Austin A30 & A35 1951-1962
Austin Healey 100 & 100/6 Gold Portfolio 1952-1959
Austin Healey 3000 Gold Portfolio 1959-1967
Austin Healey Sprite 1958-1971
Avanti 1962-1990
BMW Six Cylinder Coupes 1969-1975
BMW 1600 Col. 1 1966-1981
BMW 2002 1968-1976
BMW 316, 318, 320 Gold Portfolio 1975-1990
BMW 320, 323, 325 Gold Portfolio 1977-1990
Buick Automobiles 1947-1960
Buick Muscle Cars 1965-1970
Buick Riviera 1963-1978
Cadillac Automobiles 1949-1959
Cadillac Automobiles 1960-1969
Cadillac Eldorado 1967-1978
High Performance Capris Gold Portfolio 1969-1987
Chevrolet Camaro SS & Z28 1966-1973
Chevrolet Camaro & Z-28 1973-1981
High Performance Camaros 1982-1988
Camaro Muscle Portfolio 1967-1973
Chevrolet 1955-1957
Chevrolet Corvair 1959-1969
Chevrolet Impala & SS 1958-1971
Chevrolet Muscle Cars 1966-1971
Chevelle and SS 1964-1972
Chevy Blazer 1969-1981
Chevy EL Camino & SS 1959-1987
Chevy II Nova & SS 1962-1973
Chrysler 300 Gold Portfolio 1955-1970
Citroen Traction Avant Gold Portfolio 1934-1957
Citroen DS & ID 1955-1975
Citroen SM 1970-1975
Citroen 2CV 1949-1988
Shelby Cobra Gold Portfolio 1962-1969
Cobras and Cobra Replicas Gold Portfolio 1962-1989
Cobras & Replicas 1962-1983
Chevrolet Corvette Gold Portfolio 1953 1962
Corvette Stingray Gold Portfolio 1963-1967
Chevrolet Corvettee Gold Portfolio 1968-1977
High Performance Corvettes 1983-1989
Daimler SP250 Sport & V-8250 Saloon Gold Portfolio 1959-1969
Datsun 240Z 1970-1973
Datsun 280Z & ZX 1975-1983
De Tomaso Collection No.1 1962-1981
Dodge Charger 1966-1974
Dodge Muscle Cars 1967-1970
Excalibur Collection No.1 1952-1981
Facel Vega 1954-1964
Ferrari Cars 1946-1956
Ferrari Dino 1965-1974
Ferrari Dino 308 1974-1979
Ferrari 308 & Mondial 1980-1984
Ferrari Collection No.1 1960-1970
Fiat-Bertone X1/9 1973-1988
Fiat Pininfarina 124 + 2000 Spider 1968-1985
Ford Automobiles 1949-1959
Ford Bronco 1966-1977
Ford Bronco 1978-1988
Ford Consul. Zephyr Zodiac MkI & II 1950-1962
Ford Cortina 1600E & GT 1967-1970
Ford Fairlane 1955-1970
Ford Falcon 1960-1970
Ford GT40 Gold Portfolio 1964-1987
Ford RS Escorts 1968-1980
Ford Zephyr Zodiac Executive MkIII & MkIV 1962-1971
High Performance Capris Gold Portfolio 1969-1987
High Performance Escorts Mk1 1968-1974
High Performance Escorts Mk II 1975-1980
High Performance Escorts 1980-1985
High Performance Escorts 1985-1990
High Performance Fiestas 1979-1991
High Performance Mustangs 1982-1988
Holden 1948-1962
Honda CRX 1983-1987
Hudson & Railton 1936-1940
Jaguar and SS Gold Portfolio 1931-1951
Jaguar XK120 XK140 XK150 Gold Portfolio 1948-1960
Jaguar MkVII VIII IX X 420 Gold Portfolio 1950-1970
Jaguar Cars 1961-1964
Jaguar Mk2 1959-1969
Jaguar E-Type Gold Portfolio 1961-1971
Jaguar E-Type 1966-1971
Jaguar E-Type V-12 1971-1975
Jaguar XJ12 XJ5.3 V12 Glold Portfolio 1972-1990
Jaguar XJ6 Series II 1973-1979
Jaguar XJ6 Series III 1979-1986
Jaguar XJS Gold Portfolio 1975-1990
Jeep CJ5 & CJ6 1960-1976
Jeep CJ5 & CJ7 1976-1986
Jensen Cars 1946-1967
Jensen Cars 1967-1979
Jensen Interceptor Gold Portfolio 1966-1986
Jensen Healey 1972-1976
Lamborghini Cars 1964-1970
Lamborghini Countach Col No.1 1971-1982
Lamborghini Countach & Urraco 1974-1980
Lamborghini Countach & Jalpa 1980-1985
Lancia Stratos 1972-1985
Land Rover Series I 1948-1958
Land Rover Series II & IIa 1958-1971
Land Rover Series III 1971-1985
Land Rover 90 & 110 1983-1989
Lincoln Gold Portfolio 1949-1960
Lincoln Continental 1961-1969
Lincoln Continental 1969-1976
Lotus and Caterham Seven Gold Portfolio 1957-1989
Lotus Cortina Gold Portfolio 1963-1970
Lotus Elan Gold Portfolio 1962-1974
Lotus Elan Collection No.2 1963-1972
Lotus Elite 1957-1964
Lotus Elite & Eclat 1974-1982
Lotus Turbo Esprit 1980-1986
Lotus Europa Gold Portfolio 1966-1975
Marcos Cars 1960-1988
Maserati 1965-1970
Maserati 1970-1975
Mazda RX-7 Collection No.1 1978-1981
Mercedes 190 & 300SL 1954-1963
Mercedes 230/250/280SL 1963-1971
Mercedes Benz SLs & SLCs Gold Portfolio 1971-1989
Mercedes Benz Cars 1949-1954
Mercedes Benz Cars 1954-1957
Mercedes Benz Cars 1957-1961
Mercedes Benz Competion Cars 1950-1957
Mercury Muscle Cars 1966-1971
Metropolitan 1954-1962
MG TC 1945-1949
MG TD 1949-1953
MG TF 1953-1955
MG Cars 1959-1962
MGA & Twin Cam Gold Portfolio 1955-1962
MGB MGC & V8 Gold Portfolio 1962-1980
MGB Roadsters 1962-1980
MGB GT 1965-1980
MG Midget 1961-1980
Mini Cooper Gold Portfolio 1961-1971
Mini Moke 1964-1989
Mini Muscle Cars 1961-1979
Mopar Muscle Cars 1964-1967
Morgan Three-Wheeler Gold Portfolio 1910-1952
Morgan Cars 1960-1970
Morgan Cars Gold Portfolio 1968-1989
Morris Minor Collection No.1
Mustang Muscle Cars 1967-1971
Oldsmobile Automobiles 1955-1963
Old's Cutlass & 4-4-2 1964-1972
Oldsmobile Muscle Cars 1964-1971
Oldsmobile Toronado 1966-1978
Opel GT 1968-1973
Packard Gold Portfolio 1946-1958
Pantera Gold Portfolio 1970-1989
Panther Gold Portfolio 1972-1990
Plymouth Barracuda 1964-1974
Plymouth Muscle Cars 1966-1971
Pontiac Tempest & GTO 1961-1965
Pontiac Firebird and Trans-Am 1973-1981
High Performance Firebirds 1982-1988
Pontiac Fiero 1984-1988
Pontiac Muscle Cars 1966-1972
Porsche 356 1952-1965
Porsche Cars in the 60's
Porsche Cars 1960-1964
Porsche Cars 1964-1968
Porsche Cars 1968-1972
Porsche Cars 1972-1975
Porsche Turbo Collection No.1 1975-1980
Porsche 911 1965-1969
Porsche 911 1970-1972
Porsche 911 1973-1977
Porsche 911 Carrera 1973-1977
Porsche 911 Turbo 1975-1984
Porsche 911 SC 1978-1983
Porsche 914 Gold Portfolio 1969-1976
Porsche 914 Collection No.1 1969-1983
Porsche 924 Gold Portfolio 1975-1988
Porsche 928 1977-1989
Porsche 944 1981-1985
Range Rover Gold Portfolio 1970-1988
Reliant Scimitar 1964-1986
Riley 11/2 & 21/2 Litre Gold Portfolio 1945-1955
Rolls Royce Silver Cloud Gold Portfolio 1955-1965
Rolls Royce Silver Shadow 1965-1981
Rover P4 1949-1959
Rover P4 1955-1964
Rover 3 & 3.5 Litre Gold Portfolio 1958-1973
Rover 2000 + 2200 1963-1977
Rover 3500 1968-1977
Rover 3500 & Vitesse 1976-1986
Saab Sonett Collection No.1 1966-1974
Saab Turbo 1976-1983
Shelby Mustang Muscle Portfolio 1965-1970
Stubebaker Gold Portfolio 1947-1966
Stubebaker Hawks & Larks 1956-1963
Sunbeam Tiger & Alpine Gold Portfolio 1959-1967
Thunderbird 1955-1957
Thunderbird 1958-1963
Thunderbird 1964-1976
Toyota Land Cruiser 1956-1984
Toyota MR2 1984-1988
Triumph 2000. 2.5. 2500 1963-1977
Triumph GT6 1966-1974
Triumph Spitfire Gold Portfolio 1962-1980
Triumph Stag 1970-1980
Triumph Stag Collection No.1 1970-1984
Triumph TR2 & TR3 1952-60
Triumph TR4-TR5-TR250 1961-1968
Triumph TR6 Gold Portfolio 1969-1976
Triumph TR7 & TR8 1975-1982
Triumph Herald 1959-1971
Triumph Vitesse 1962-1971
TVR Gold Portfolio 1959-1990
Valiant 1960-1962
VW Beetle Collection No.1 1970-1982
VW Golf GTi 1976-1986
VW Karmann Ghia 1955-1982
VW Kubelwagen 1940-1975
VW Scirocco 1974-1981
VW Bus. Camper. Van 1954-1967
VW Bus. Camper. Van 1968-1979
VW Bus. Camper. Van 1979-1989
Volvo 120 1956-1970
Volvo 1800 Gold Portfolio 1960-1973

BROOKLANDS ROAD & TRACK SERIES

Road & Track on Alfa Romeo 1949-1963
Road & Track on Alfa Romeo 1964-1970
Road & Track on Alfa Romeo 1971-1976
Road & Track on Alfa Romeo 1977-1989
Road & Track on Aston Martin 1962-1990
Road & Track on Auburn Cord and Duesenburg 1952-1984
Road & Track on Audi & Auto Union 1952-1980
Road & Track on Audi 1980-1986
Road & Track on Austin Healey 1953-1970
Road & Track on BMW Cars 1966-1974
Road & Track on BMW Cars 1975-1978
Road & Track on BMW Cars 1979-1983
Road & Track on Cobra, Shelby & GT40 1962-1983
Road & Track on Corvette 1953-1967
Road & Track on Corvette 1968-1982
Road & Track on Corvette 1982-1986
Road & Track on Corvette 1986-1990
Road & Track on Datsun Z 1970-1983
Road & Track on Ferrari 1950-1968
Road & Track on Ferrari 1968-1974
Road & Track on Ferrari 1975-1981
Road & Track on Ferrari 1981-1984
Road & Track on Ferrari 1984-1988
Road & Track on Fiat Sports Cars 1968-1987
Road & Track on Jaguar 1950-1960
Road & Track on Jaguar 1961-1968
Road & Track on Jaguar 1968-1974
Road & Track on Jaguar 1974-1982
Road & Track on Jaguar 1983-1989
Road & Track on Lamborghini 1964-1985
Road & Track on Lotus 1972-1981
Road & Track on Maserati 1952-1974
Road & Track on Maserati 1975-1983
Road & Track on Mazda RX7 1978-1986
Road & Track on Mazda RX7 & MX5 Miata 1986-1991
Road & Track on Mercedes 1952-1962
Road & Track on Mercedes 1963-1970
Road & Track on Mercedes 1971-1979
Road & Track on Mercedes 1980-1987
Road & Track on MG Sports Cars 1949-1961
Road & Track on MG Sprots Cars 1962-1980
Road & Track on Mustang 1964-1977
Road & Track on Nissan 300-ZX & Turbo 1984-1989
Road & Track on Peugeot 1955-1986
Road & Track on Pontiac 1960-1983
Road & Track on Porsche 1961-1967
Road & Track on Porsche 1968-1971
Road & Track on Porsche 1972-1975
Road & Track on Porsche 1975-1978
Road & Track on Porsche 1979-1982
Road & Track on Porsche 1982-1985
Road & Track on Porsche 1985-1988
Road & Track on Rolls Royce & B'ley 1950-1965
Road & Track on Rolls Royce & B'ley 1966-1984
Road & Track on Saab 1955-1985
Road & Track on Toyota Sports & GT Cars 1966-1984
Road & Track on Triumph Sports Cars 1953-1967
Road & Track on Triumph Sports Cars 1967-1974
Road & Track on Triumph Sports Cars 1974-1982
Road & Track on Volkswagen 1951-1968
Road & Track on Volkswagen 1968-1978
Road & Track on Volkswagen 1978-1985
Road & Track on Volvo 1957-1974
Road & Track on Volvo 1975-1985
Road & Track - Henry Manney at Large and Abroad

BROOKLANDS CAR AND DRIVER SERIES

Car and Driver on BMW 1955-1977
Car and Driver on BMW 1977-1985
Car and Driver on Cobra, Shelby & Ford GT 40 1963-1984
Car and Driver on Corvette 1956-1967
Car and Driver on Corvette 1968-1977
Car and Driver on Corvette 1978-1982
Car and Driver on Corvette 1983-1988
Car and Driver on Datsun Z 1600 & 2000 1966-1984
Car and Driver on Ferrari 1955-1962
Car and Driver on Ferrari 1963-1975
Car and Driver on Ferrari 1976-1983
Car and Driver on Mopar 1956-1967
Car and Driver on Mopar 1968-1975
Car and Driver on Mustang 1964-1972
Car and Driver on Pontiac 1961-1975
Car and Driver on Porsche 1955-1962
Car and Driver on Porsche 1963-1970
Car and Driver on Porsche 1970-1976
Car and Driver on Porsche 1977-1981
Car and Driver on Porsche 1982-1986
Car and Driver on Saab 1956-1985
Car and Driver on Volvo 1955-1986

BROOKLANDS PRACTICAL CLASSICS SERIES

PC on Austin A40 Restoration
PC on Land Rover Restoration
PC on Metalworking in Restoration
PC on Midget/Sprite Restoration
PC on Mini Cooper Restoration
PC on MGB Restoration
PC on Morris Minor Restoration
PC on Sunbeam Rapier Restoration
PC on Triumph Herald/Vitesse
PC on Triumph Spitfire Restoration
PC on VW Beetle Restoration
PC on 1930s Car Restoration

BROOKLANDS HOT ROD 'MUSCLECAR & HI-PO ENGINE SERIES

Chevy 265 & 283
Chevy 302 & 327
Chevy 348 & 409
Chevy 350 & 400
Chevy 396 & 427
Chevy 454 thru 512
Chrysler Hemi
Chrysler 273, 318, 340 & 360
Chrysler 361, 383, 400, 413, 426, 440
Ford 289, 302, Boss 302 & 351W
Ford 351C & Boss 351
Ford Big Block

BROOKLANDS MILITARY VEHICLES SERIES

Allied Mil. Vehicles No.1 1942-1945
Allied Mil. Vehicles No.2 1941-1946
Dodge Mil. Vehicles Col. 1 1940-1945
Military Jeeps 1941-1945
Off Road Jeeps 1944-1971
Hail to the Jeep
Complete WW2 Military Jeep Manual
US Military Vehicles 1941-1945
US Army Military Vehicles WW2-TM9-2800

BROOKLANDS HOT ROD RESTORATION SERIES

Auto Restoration Tips & Techniques
Basic Bodywork Tips & Techniques
Basic Painting Tips & Techniques
Camaro Restoration Tips & Techniques
Chevrolet High Performance Tips & Techniques
Chevy-GMC Pickup Repair
Custom Painting Tips & Techniques
Engine Swapping Tips & Techniques
Ford Pickup Repair
How to Build a Street Rod
Mustang Restoration Tips & Techniques
Performance Tuning - Chevrolets of the '60s
Performance Tuning - Ford of the '60s
Performance Tuning - Mopars of the '60s
Performance Tuning - Pontiacs of the '60s

CONTENTS

6	Silver Cloud and Series S	Autocar	April 29 1955
12	New Rolls Royce and Bentley Cars	Motor Sport	June 1955
13	The Rolls Royce — British Aristocrat	Car Life	Aug. 1954
16	The First New Rolls in 10 years	Wheels	July 1955
18	Bentley and Rolls-Royce	Motor	Oct. 12 1955
20	Rolls-Royce Silver Cloud Road Test	Motor	Jan. 18 1956
24	Refinements from Crewe	Autocar	Oct. 5 1956
27	Long-wheelbase Silver Cloud Limousine	Motor	Sept. 25 1957
28	Silver Cloud Consumer Analysis	Car Life	November 1957
32	Rolls-Royce Automatic Gearbox	Motor	Dec. 18 1957
34	Still the Best?	Modern Motor	April 1958
36	Rolls-Royce in Texas	Motor	May 14 1958
39	Rolls-Royce Silver Cloud Road Test	Autocar	May 16 1958
43	Again, a Rolls-Royce Drophead	Wheels	December 1958
44	Yes — Still the Best Road Test	Modern Motor	July 1958
49	By Royce to Rome	Motor	Dec. 31 1958
53	Rolls-Royce Silver Cloud Road Test	Motor Life	September 1959
56	Silver Cloud II & Bentley S2	Motor	Sept. 23 1959
60	New V-8 Rolls-Royce Engine	Autosport	Sept. 25 1959
62	Silver Cloud II Touring Test	Wheels	January 1960
66	Silver Cloud II Road Test	Autocar	May 13 1960
70	Silver Cloud II Road Test	Motor	May 18 1960
74	Silver Cloud II Road Test	Road & Track	Sept. 1960
78	New Air Conditioning for Rolls-Royce & Bentley	Autocar	Sept. 25 1959
79	Silver Cloud I Used Car Test	Autocar	Oct. 21 1960
80	Silver Cloud II Road Test	Modern Motor	July 1961
82	Silver Cloud III & Bentley S3	Autocar	Oct. 19 1962
84	Silver Cloud III Road Test	Motor	Aug. 21 1963
90	Full Stop by Rolls-Royce	Motor	Mar. 4 1964
91	Silver Cloud III — Opinion	Motoring Life	1964
94	Rolls Royce Silver Cloud, Bentley Series S Buying Secondhand	Autocar	Dec. 27 1975
98	Silver Cloud III	Autocar	Sept. 25 1964

Part Two

100	Bentley Introduces the "S" and Rolls-Royce "Silver Cloud"	Autosport	April 29 1955
102	Bentley Series S Saloon Road Test	Autocar	Oct. 7 1955
106	Bentley 'Drivescription'	Motor Trend	April 1956
108	Bentley "S" Series	Autosport	Aug. 17 1956
110	Round the Clock with a Bentley "S" Series	Motor Sport	Nov. 1956
113	3-Figure Cruiser	Autocar	Jan. 18 1957
117	Rolls Royce Road Test	Road & Track	May 1958
120	Motoring on a Cloud	Sports Cars Illustrated	June 1958
123	Beautiful "S" Series Bentley Road Impression	Wheels	Oct. 1958
126	The Bentley Continental	Autosport	Aug. 22 1958
128	Fast and Slinky — Its Continental	Sports Cars Illustrated	Oct. 1958
132	Rolls Royce Silver Cloud	Autosport	Nov. 1958
135	Rolls Royce — Bentley	Motor Trend	Jan. 1959
136	Rolls Royce Technical Analysis	Road & Track	Sept. 1960
142	Bentley S2	Autosport	Oct. 7 1960
144	The Most Perfect Car Road Test	Wheels	June 1961
148	Bentley S2 LWB Saloon Road Test	Car — South Africa	Nov. 1961
152	Rolls Royce Silver Cloud Road Test	Sports Car Graphic	Sept. 1962
156	Rolls Royce and Bentley	Motor	Oct. 17 1962
158	The Rolls Royce Silver Cloud	Autosport	Aug. 21 1964
160	In the Grand Manner	Sporting Motorist	Aug. 1964
163	Silver Cloud III	Car — South Africa	Sept. 1964
164	Not so Much a Motor Car More a Way of Life	Motor Sport	Sept. 1965
169	1961 Bentley Continental S2 Used Car Test	Autocar	Aug. 16 1973
172	Triple Silver	Classic and Sportscar	Oct. 1988
176	Crewe or Crude?	Classic and Sportscar	May 1991

ACKNOWLEDGEMENTS

Since we first put out a Brooklands book on the Rolls-Royce Silver Cloud and Bentley S-type models, several years ago, a lot of new stories on these fine motor cars has come to light. When it came time to consider reprinting our first volume, then, our choice had been made for us. Hence this Gold Portfolio, which adds more than 80 new pages of material to that which appeared in our earlier volume and creates a better balance between the Rolls-Royce and Bentley models of what were, after all, the same car.

Without the co-operation of those who own the original copyright to these articles, this book — and the others in the Brooklands Books series — could not exist. So we are pleased to acknowledge the debt we owe for the material in this book to the publishers of Autocar, Autosport, Car Life, Car — South Africa, Classic and Sportscar, Modern Motor, Motor, Motor Sport, Motor Trend, Motoring Life, Road & Track, Sports Car Graphic, Sports Cars Illustrated, Sporting Motorist and Wheels. We are grateful, too, to motoring writer James Taylor for his brief introduction to this book.

R.M. Clarke

When the Silver Cloud and S-type models were introduced in 1955, there was no question in the popular mind whether these cars were or were not the "best cars in the world". By the time the last one was built, some 11 years later, the question had been raised not once but several times, and the answer had not always been in favour of Britain's most famous car manufacturer.

This, however, does not reflect on the quality of the Clouds and S-types; rather does it reflect on the speed with which other manufacturers improved their products in the 1950s and 1960s. The Crewe products in fact started with a disadvantage, being designed with a separate chassis and body at a time when monocoque construction was rapidly gaining ground. By the time they were due for replacement (by Crewe's own first monocoque design), the many advantages of monocoque construction had become known to the public at large, and the Clouds and S-types appeared to be lagging behind lesser manufacturers on the technological front.

Unbelievers only needed a brief acquaintance with one of these cars to become converts, however. Indeed, that is still the case, a quarter of a century after they ceased production. No matter that some of the technology was outmoded; the fact was that everything worked so superbly well, and went on working without failure for so many thousands of miles. The number of Clouds and S-type Bentleys still around today is eloquent testimony of the sheer quality of these luxurious vehicles.

Of course, the days of the truly individual, hand-crafted Rolls-Royce were long gone by the time these cars came on the market. The majority were delivered with "standard steel" coachwork, which was indeed finished by hand but originated in the dies at Pressed Steel's body-pressing plant. Yet this coachwork was so enegantly styled and so beautifully balanced that even today it does not look out of place — it simply stands out from the crowd, as it always did. And the separate chassis of the Clouds and S-types allowed the specialist coachbuilders to create their magic in the way they always had, for those who were fortunate enough to afford something which would stand out even amongst its peers.

This is not the place to discuss the relative merits of the six-cylinder engine in the first-series models and the V8 in the second and third series cars, or of the merits or otherwise of the four-headlamp styling on the Silver Cloud III and Bentley S3. Such things are for the individual to decide, and this book provides an excellent way in to the subject. It is one which no enthusiast of these models will want to be without.

James Taylor

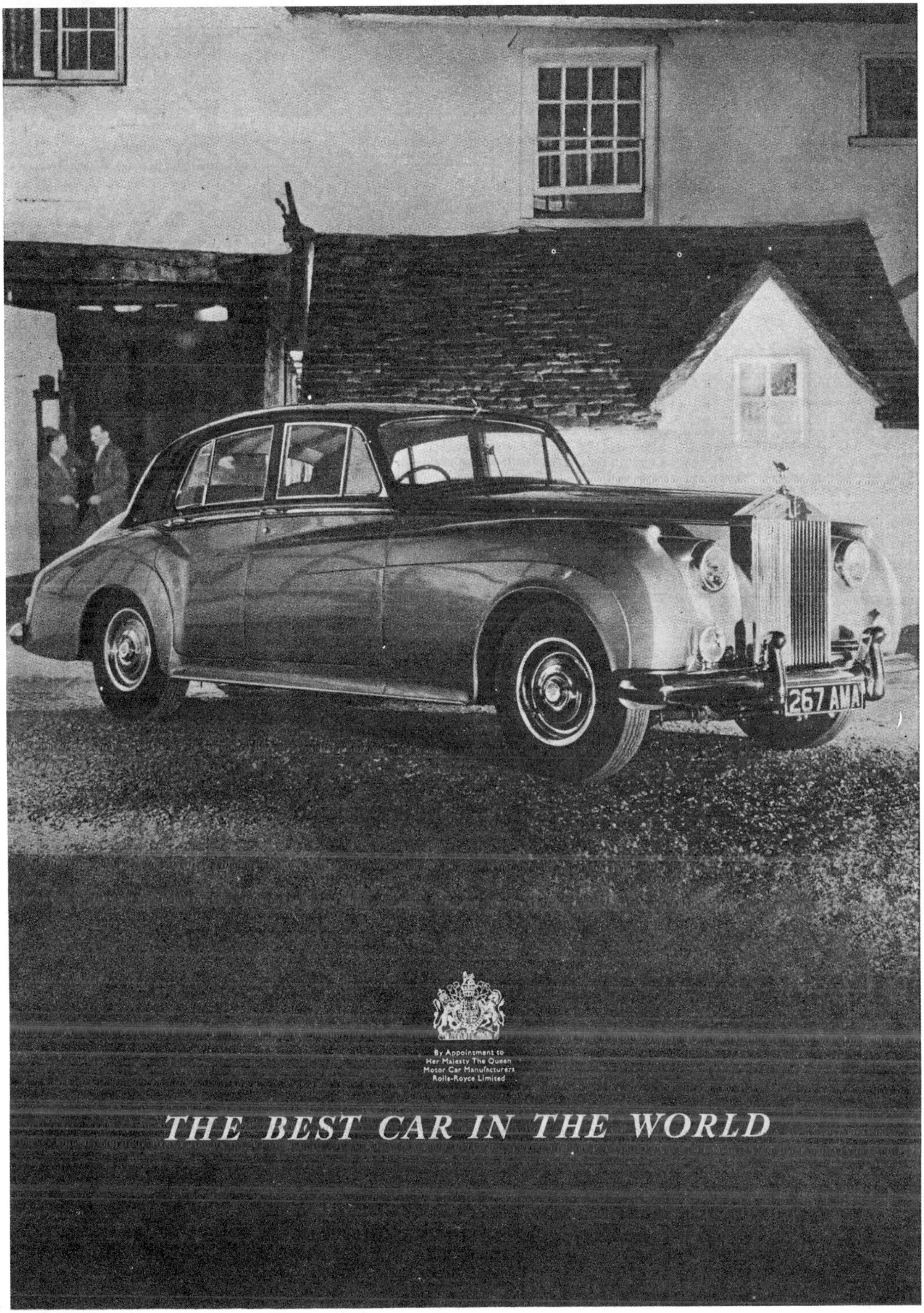

By Appointment to
Her Majesty The Queen
Motor Car Manufacturers
Rolls-Royce Limited

THE BEST CAR IN THE WORLD

NEW CARS DESCRIBED

SILVER

NEW ROLLS-ROYCE AND BENTLEY MODELS

Larger Engine
New Front Suspension
Z-Bar addition to Rear Suspension
New Head and Porting
Automatic Transmission Standardized
Smaller Wheels—Larger Brake-drum Area

WITH the reputation for making the finest car in the world it is no easy task to introduce a new range of cars that will be better than the models they supersede but cost very little more. Yet with the introduction of the new Rolls-Royce Silver Cloud and the Bentley Series S this is just what has been done. That the nearer one approaches the ultimate the greater will be the standardization is evident by the similarity between the Rolls-Royce and the Bentley.

The automatic transmission, now made standard equipment on all models, remains unchanged. Chassis can be supplied for specialist coachwork. The engine is similar in design to that used on the Bentley Continental.

Cylinder bore diameter has been of the journal. From there it is conveyed by another drilling to the crankpin, which is cross-drilled to lubricate the big-end bearing. The connecting rod is drilled throughout its length to lubricate the gudgeon pin, and is cross-drilled on the thrust side, below the piston skirt, for cylinder bore lubrication.

At the rear of the back main bearing an oil thrower is now incorporated in addition to the scroll type of oil return. A flange is formed at the rear of the shaft, and to this is attached the flexible disc flywheel which carries the starter ring, and to which is bolted the housing for the fluid coupling used in conjunction with the automatic transmission. At the front of the shaft there is a torsional vibration damper consisting of a metallic mass attached to a thin flange on the shaft by spring-loaded friction linings. In between this and the front main bearing there is the camshaft drive gear; again to reduce the effects of torsional vibration, a spring drive is incorporated in this. Both this

valve seats. To combat the effects of corrosion and reduce pre-ignition the heads of the exhaust valves are given a bright ray treatment. This coats the valve head with 80 per cent nickel and 20 per cent chromium applied with a welding torch.

The most noticeable change to the power unit has been redesigning the cylinder head, the four-port arrangement being replaced by a new six-port head to

increased from 92 to 95.25mm, so that the capacity is now 4,887 c.c. Unified threads are now used for all nuts and bolts. A single casting forms the cylinder block and crankcase, and this is extended well below the crankshaft centre line, the crankchamber being adequately stiffened by the five webs which house the intermediate bearings for the seven-bearing shaft. End thrust is taken by the centre main bearing, and all main bearings have 2.75in journals. The crankpins are 2in in diameter and have an effective width of 1¾in. Indium-coated lead-bronze steel-backed shells are used for both main and big-end bearings.

The forged crankshaft now has integral balance weights; these are placed on either side of the centre main bearing, on the outsides of numbers 2 and 6 main bearings, and on the insides of numbers 1 and 7 journals. Both main bearing and crankpin journals are hollow, the ends being sealed by bolted-in plugs of stainless steel. On earlier models these plugs were light alloy, but this was changed to prevent adverse effects from leaded fuels. Each main journal has radial drillings which allow lubricant to pass to the hollow centre gear and the torsional vibration damper are mounted inside the engine front cover.

Full-length cylinder liners, pressed into the bores, provide wear-resisting surfaces for the long split-skirt pistons. These pistons have four rings, the top one being chromium plated.

The well-known Rolls-Royce system of overhead inlet and side exhaust valves is retained, the exhaust valves being produced from KE965 material with Stellite improve the breathing and permit increased power output. Cast in aluminium, the cylinder head is attached to the block by 37 ½in set bolts. By using these in place of the normal arrangement of studs and nuts it has been possible to increase the spanner clearance, particularly on the bolts adjacent to the rocker pedestals. The construction of the cylinder head is interesting: half of the manifold is actually formed in the head itself. The other half,

CLOUD AND SERIES S

containing the flanges for carburettor attachment, is also a light alloy casting, the two parts being held together by 22 studs and nuts. This arrangement results in a fairly simple casting as far as the ports are concerned, and simplifies fettling and inspection.

The inlet valves are 1.85in diameter and have a lift of 0.4in. The corresponding figures for the exhaust are 1.625in and 0.375in. The inlet valves are produced from S65, a nickel chrome alloy steel, and a similar material is used for the inlet valve inserts; these are screwed into place and, once they are in position, the spanner lugs are machined off. Particular attention has been paid to the design of the inlet porting to improve flow, and the internal shape of the inserts is curved, to blend with the line of the inlet porting. This improves gas flow round the back of the valve heads.

Brass inserts are screwed into the head to take the sparking plugs, and these are locked in place by additional screwed rings to prevent the possibility of the insert being removed when the sparking plugs are unscrewed.

Both Bentley and Rolls-Royce models have twin horizontal S.U. carburettors in conjunction with the automatic cold starting device described in the September 19, 1952 issue. Both this unit and the carburettors are different from those previously used on Bentley models. The object of the cold-starting device is to eliminate the manual choke control and the need to raise and lower the carburettor jet block to provide mixture enrichment.

On previous carburettors the jet block was held in place by cork washers which enabled it to be adjusted and also provided a fuel seal. The bottom end of the carburettors has now been completely redesigned, and a new assembly is used with the jet block attached to a rubber diaphragm. This diaphragm permits vertical movement of the jet and provides a better seal. Because of the cold starting device variation of the jet position is required only for initial carburettor tuning.

The carburettors also include a new slow-running adjustment which does not rely on the setting of the main throttle valves in order to produce the correct tickover. Instead, the linkage is adjusted so that when the throttle pedal is released the butterfly valves are closed and the tickover speed is then adjusted by a taper-ended screw which controls the flow through a small port which by-passes the main throttle valve.

The cold starting device has been improved. On the previous model the solenoid which closes the choke valve was in circuit with the starter motor switch;

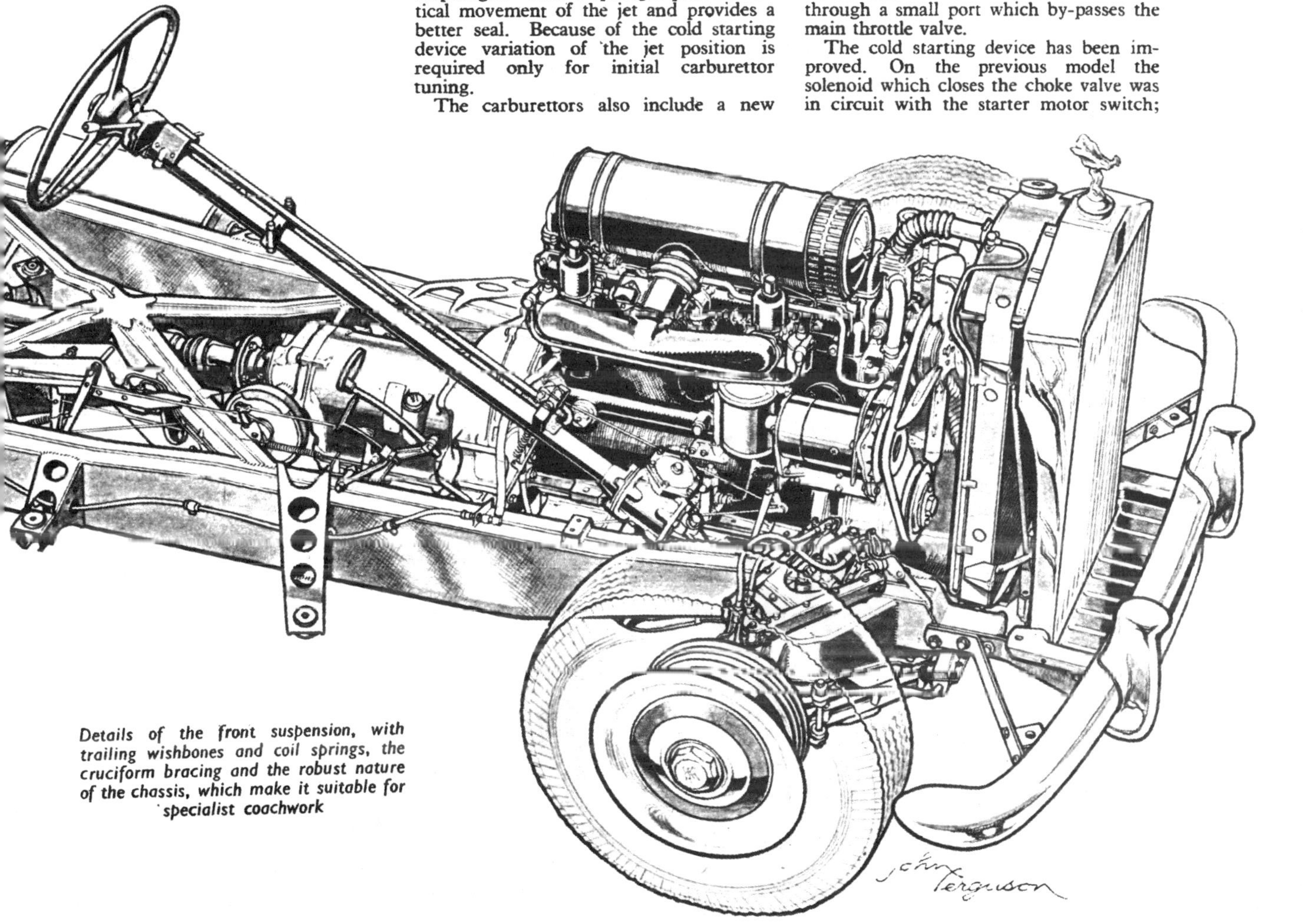

Details of the front suspension, with trailing wishbones and coil springs, the cruciform bracing and the robust nature of the chassis, which make it suitable for specialist coachwork

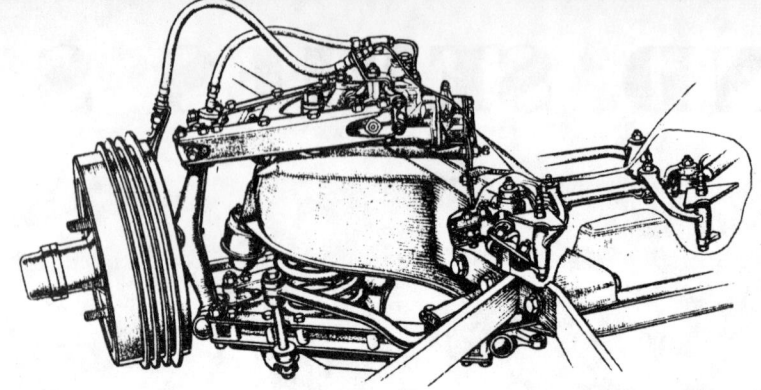

SILVER CLOUD AND SERIES S
.... continued

consequently, as soon as the engine fired and the starter switch was released, the circuit ceased to be energized. The effect could be to stall the engine after a few revs.

To overcome this the solenoid is placed in the ignition circuit; in addition, a pressure-sensitive switch is tapped into the oil gallery which breaks the circuit as soon as the pump produces a given pressure in the oil line. A temperature-sensitive switch is included in the cold-starting circuit to prevent the system from operating at under-bonnet temperatures.

On the new cars automatic transmission, identical with that optionally fitted to the previous model, is a standard feature. The drive from the transmission to the rear axle is by two-piece propeller-shaft with a flexibly mounted centre bearing. The rear section of the shaft consists of conventional Hardy Spicer couplings with a sliding spline just to the rear of the centre bearing. The front section, however, contains a special joint which will slide with the application of only a very light end load, even when it is transmitting full torque. This has been introduced in the quest for complete silence and freedom from vibration.

In designing the braking equipment Rolls-Royce engineers were faced with two problems: first, the performance of

The independent front suspension units consist of half-trailing long and short wishbones, the inner fulcrum bearing for the upper wishbones being formed by the damper unit. The plate welded to the chassis frame and bolted to a plate on the inner end of the damper prevents deflections which could cause a change in castor angle under heavy braking

Although retaining a dignified style, the facia panel has been restyled to give a cleaner appearance

the car was to be considerably improved, with the result that the brakes would have much more work to do; secondly, the demands of the stylist meant that there would be less space available for the brakes inside the wheels as these were reduced in size from 16 to 15in. It was necessary to reduce brake drum diameters from 12¼ to 11¼in; however, the effective width was increased from 2¼ to 3in with the result that there is 22 per cent more lining area than on the previous brakes. To increase stability under severe operating conditions the brake system has been changed at the front from the single wheel cylinder layout to two trailing shoes. At the rear the normal Rolls-Royce layout with single wheel cylinder and compensating link mechanism has been retained. In place of a high-rate spring, special low-rate springs are used on the shake-back stop assembly to enable the friction force to be controlled more accurately, and the friction surfaces for these springs consist of cadmium plating on the web of the brake shoe itself, in conjunction with chromium plated washers. In place of rolled section brake shoes, the linings are riveted to fabricated shoes on the new model.

Previous cars have sometimes been criticized because of servo lag. This has been rectified by increasing the servo speed so that the motor now runs at $\frac{1}{5.6}$ of the propeller-shaft speed compared with $\frac{1}{10.5}$. This modification has also reduced servo noise and the slight judder that was sometimes experienced. In the new system the front brakes depend wholly upon hydraulic operation by the servo, and 60 per cent of the rear braking is so applied; the remaining 40 per cent at the rear is applied mechanically direct from the brake pedal (the old figure was 50 per cent). The total distribution of

A Z-bar pivoted to the chassis frame and the rear axle prevents road spring wind-up caused by torque reaction, and modifies the rear roll stiffness to produce the desired degree of understeer

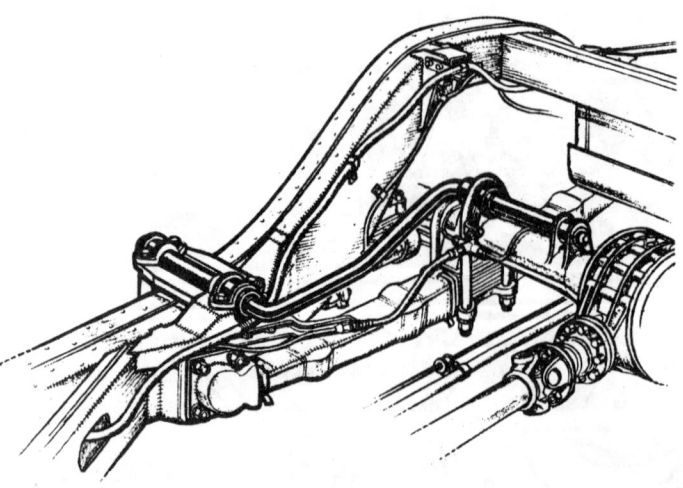

The new S.U. carburettor has auxiliary drillings in conjunction with the needle valve to adjust the slow running, and a diaphragm member which forms the seal for the jet block assembly. A small lever with screw adjustment is used to vary the height of the jet block for carburettor tuning

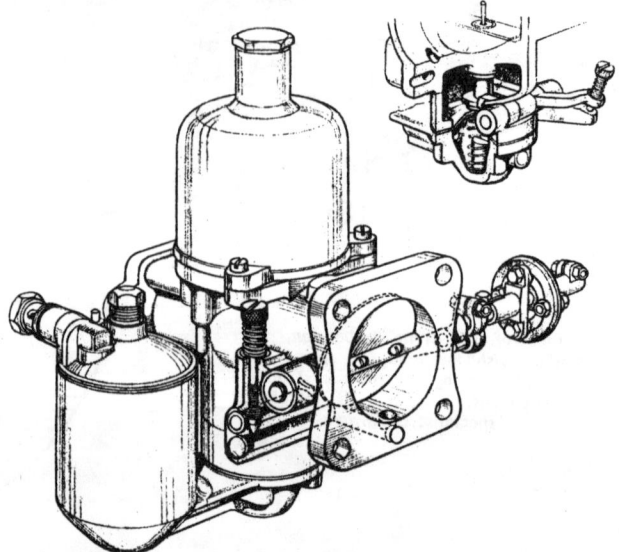

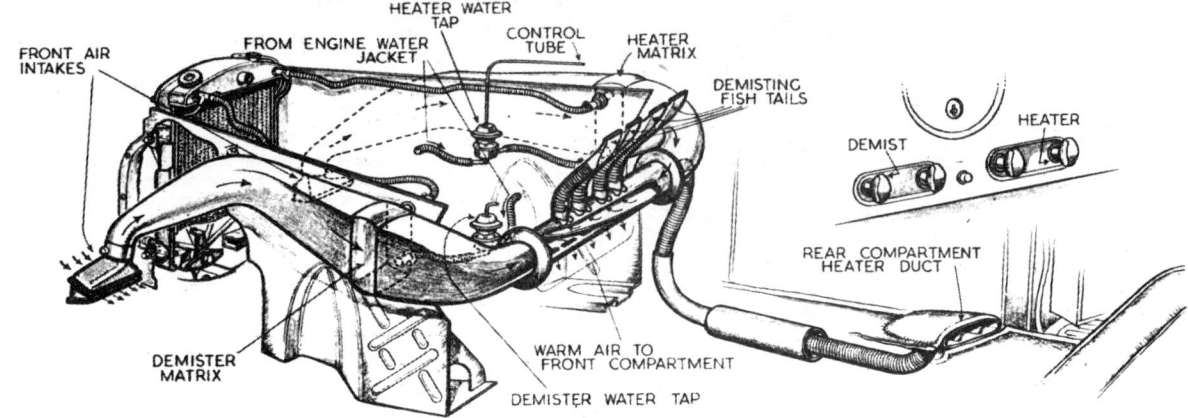

Layout of the heating system. Fans are provided to assist the flow through both heater units, and the speed of these is controlled by multi-position switches

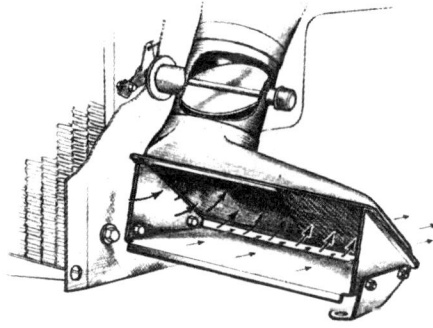

SILVER CLOUD AND SERIES S .... continued

braking between front and rear is 1.23 to 1. Automatic adjustment is provided for the front brakes, and there is a special safety device to prevent the brake shoes from wearing out against the drums and to indicate when relining is necessary. This consists of a high rate spring which, in effect, over-rides the automatic adjustment, with the result that free pedal travel becomes noticeable after a predetermined amount of lining wear has taken place. Manual adjustment is necessary for the rear brakes, and this automatically adjusts the hand brake.

The most noticeable change in the front end is to be observed in the new suspension. A conventional independent arrangement with long and short wishbones and coil springs is now employed. This change has been brought about in order to increase the permissible amount of wheel movement. With the new system 3in of bump and 4in of rebound are provided.

In the new suspension the wishbones are half-trailing, with the inner fulcrum bearings set at an angle of 62½ deg to the longitudinal centre line of the chassis. The fulcrum bearings for the lower wishbones are bolted on to the main front cross member, while those for the upper wishbones are formed by the damper unit. Both upper and lower wishbones are channel section pressings with bolted-on bearing and attachment brackets, and additional bolted-on plates and pressings to form the spring mounting and attachment for the rebound rubber. The abutment for the road spring consists of a welded pressing located below the damper. Although the top end of the road spring is finished flat, the lower end has a pig tail which fits into the pressing bolted to the bottom of the lower wishbone assembly.

Lubricant is piped to all front suspension pivots with the exception of those provided for the front damper, and screwed bushes are used in all suspension joints. Caster angle adjustment is provided by moving the inner ends of the top wishbones along the squared ends of the damper spindle and re-clamping in the correct position, suitably shaped steel blocks being placed between the channel section wishbone members, the centre flange of which is cut back to provide the necessary flexibility. Bolted-on lugs also provide attachment for the anti-roll bar, which is supported by rubber bushes and runs in front of the suspension unit. To prevent vibration the bar has a centre rubber bush bearing.

Steering and track rods are located behind the front wishbones, and the system consists of a three-piece track rod with two slave levers pivoted to the rear of the front cross member. To ensure that accuracy is maintained, lugs attached to the inner wishbone fulcrums provide the pivot points for the slave levers, thereby ensuring that the correct relative distance between the slave lever pivots and the wishbones is maintained. A lug extending back from the left slave lever is connected to a link pivoted to the forward facing lever on the steering box. The steering box itself is located inside the frame member, and has a forward facing lug on its casing so that the frame attachment point is in line with the cross link (viewed from above), thereby preventing slight frame deflections from affecting the steering.

Although the rear suspension is conventional with half-elliptic leaf springs and a half-floating rear axle, a number of changes have been made. To simplify the frame and enable straight side members to be used, the springs are now placed closer together so that they are on the inside of the chassis frame. This arrangement also tends to make for greater interior body space as the frame side members run closer to the wheel arches. The new layout also increases the roll understeer characteristics. During the development stages it was found necessary to reduce the stiffness of the rear springs in order to improve the rear seat ride and this in turn reduced the rear roll stiffness, with the result that the car then had rather too much understeer. Further, it was found necessary to reduce spring wind-up brought about by torque reaction.

These two factors could have been controlled by the use of a rear anti-roll bar and torque arms, or some other form of link mechanism. However, it is very difficult to accommodate an anti-roll bar on the rear suspension (unless it is of a peculiar shape) and to provide, as well, clearance to permit rear axle movement. The Rolls-Royce engineers solved the problem by fitting a Z-bar, one end of which is pivoted to the frame, the other end to the top of the right-hand axle

Half of the inlet manifold is formed in the new cylinder head; the other half is bolted to it.

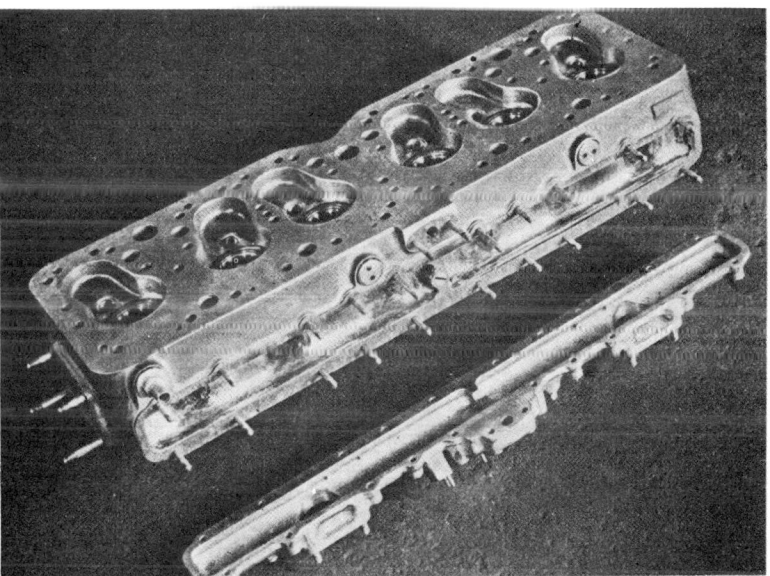

THE AUTOCAR, 29 APRIL 1955

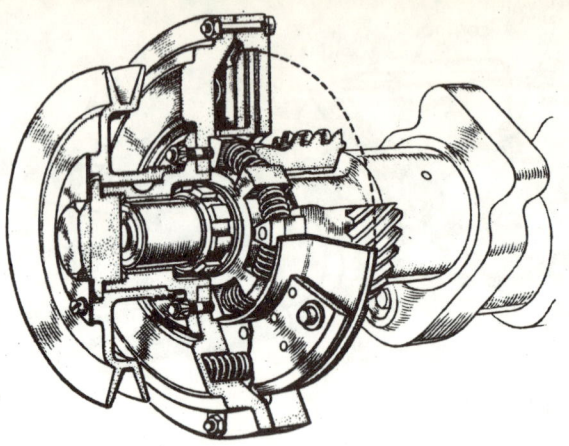

The crankshaft vibration damper and spring drive to the camshaft are located on the front of the crankshaft, and are enclosed by the engine front cover

SILVER CLOUD AND SERIES S
. . . . continued

casing tube. This bar, in effect, trims the car to provide the correct degree of understeer. The addition of the Z-bar caused a considerable increase in the loading on the front spring eye, located almost directly beneath it, under braking. It was, therefore, necessary to increase the size of the rubber bush, which is now housed in a wrapped and welded spring eye.

The leaf springs are double grooved. To reduce road noise rubber interlining is placed between the top four leaves. The leaves are Parkerized to assist retention of lubricant, and the complete spring is packed with Ragosine 204G lubricant containing 20 per cent molybdenum disulphide. The springs are, of course, enclosed in gaiters which greatly reduce wear by excluding grit.

Improved damping is provided on both front and rear suspensions, giving greater fluid flow, so that the whole of the fluid is circulated as opposed to a small quantity of it continuously. The ride control, which modifies the setting of the rear dampers, has been altered. It is now electrically controlled and alters the "slow leak" on the dampers; operating the switch on the steering column gives instantaneous change, and when in the hard position the damping is twice as hard as that provided when in the normal position.

Produced in 16-gauge 20-ton steel, the chassis frame is 50 per cent stiffer torsionally between axles than on the previous models. This increase has been brought about without noticeably increasing the weight, the figures for the old and new frame being 286 lb and 300 lb respectively. The side members are of box section, composed of two "top hat" section pressings, with additional stiffening plates spot welded to the upper and lower flange of each section before the main pressings are welded together to form a box. In the centre of the frame there is a deep box section cruciform, drilled to provide the necessary clearance for the propeller-shaft. In addition to the main front cross member, there are two cross members at the rear of the body behind the rear axle centre line.

On a car of this type smoothness and silence are two extremely important requirements. Particular attention has been paid to the body mounts, and these have

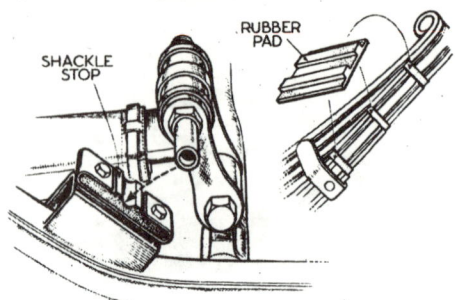

To prevent the rear spring shackle from going "over centre," a shackle stop is built on the top of the frame, and this engages with an extension on the outer end of the shackle bolt

The crankshaft now has integral balance weights. This sub-assembly shows the shaft with a vibration damper and fan pulley at the front, and thin disc fly-wheel and a starter ring at the rear

A forward extension on the steering box brings its frame attachment point close to the ball joint on the steering lever, thereby reducing inaccuracies and vibrations that might occur as a result of frame deflection. Note the pipe around the lower end of the steering box which supplies lubricant to the steering lever joint

been placed at the greatest possible distance from the chassis longitudinal centre line in order to provide the maximum amount of rock control with relatively soft rubbers. Further, it has been found that to reduce noise all body mounts must be equally loaded. To solve this problem the housings for the mountings can be adjusted, and to ensure that all twelve mounts are equally loaded, the body is attached to the front pair of mounts, and air jacks are placed underneath the remainder. The air pressure then applies a uniform loading, and the mountings are then secured. Two additional body mountings are provided on the rear arms of the cruciform to prevent floor vibrations.

In producing the new body the aim has been to provide a car with faster and more modern lines, and also provide extra width in both front and rear compartments. The main body panels are in 20-gauge steel, although 18- and 16-gauge light alloy are used for the doors, luggage locker and bonnet panels. To resist corrosion, zinc-plated steel is used for the sills and bulkhead. To improve visibility and give the car a lighter appearance, thin, polished, stainless frames are used for the top halves of the doors. Like the door

A heritage from the coachbuilder's trade, the razor-edge treatment blends well with the curves and still the "finest car in the world" shuns the flush-sided treatment

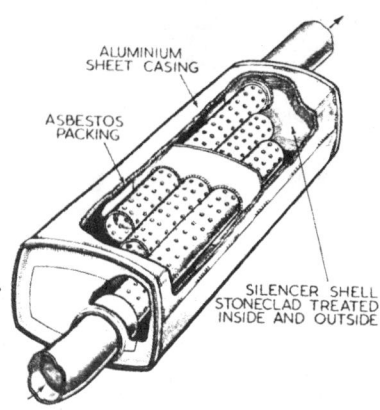

The first silencing chamber consists of a stoneclad treated steel box surrounded by asbestos, with an aluminium casing

panels, the fuel tank is also in light alloy, as are the fuel feed pipes.

Both the Rolls-Royce and Bentley now have completely new heating and demisting systems. In effect, there are two complete heat exchanger units, one on each side of the engine compartment. The combined output varies from 8 kw at 30 m.p.h. to 11 kw at 80 m.p.h. Both heating units have forward facing air intakes mounted low on each side of the main engine radiator. These are provided with special intake ducts which, in addition to the forward opening, have a diagonal grille and a rear outlet slot. The purpose of this is to prevent insects and foreign bodies from being drawn up into the main heater matrix. In addition, if the main forward facing grille becomes blocked with snow, the heater will still function because air can then be drawn through the narrow slot at the rear of the duct. The right-hand unit supplies air to the windscreen, the left-hand heater discharges it along the toe-board in the front compartment and via an additional tube into the rear of the body. This enables the temperature at the bottom and top of the car to be varied, and that very desirable combination of warm air at the bottom of the car and cool, fresh air around the driver's face can easily be provided. As on previous models, the electric element rear window demister is still retained.

In keeping with the remainder of the car, much thought and detail development have been given to the design of the seats. Basically the upholstery consists of a spring case with a Dunlopillo overlay, and in place of individual seats there is a single bench in the front. To increase the comfort and prevent the effects of weight on one side of the cushion being transferred to the other side (from driver to passenger, or vice versa), a measure of damping is provided by a vertical diaphragm fitted midway in the cushion. Another instance of detail development work is the use of a flexible panel under the tray behind the rear seats. This re-

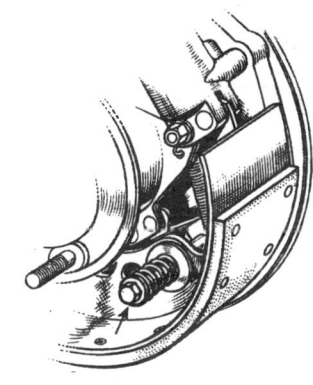

To permit more accurate adjustment a low-rate spring is used for the shake-back stops on the brake shoes

duces boom. The new cars also have greater luggage carrying capacity. No part of the car can be analysed in detail without bringing to light a most interesting development story, and there is little doubt that these fine new cars will carry on the makers' tradition and reputation.

SPECIFICATION

Engine.—6 cyl, 95.25 × 114.3mm, 4,887 c.c. Compression ratio 6.6 to 1. B.h.p.—not quoted. Maximum torque—not quoted. Seven-bearing crankshaft. Overhead inlet valves, side exhaust valves operated by single side four-bearing camshaft.

Transmission.—Fluid coupling and four-speed automatic transmission. Overall ratios: top 3.42 to 1; third 4.96 to 1; second 9.0 to 1; first 13.03 to 1. Reverse 14.7 to 1.

Final Drive.—Half-floating rear axle with hypoid gears; ratio 3.42 to 1 (12 to 41). Four-pinion differential.

Suspension.—Front: independent, wishbones and coil springs; anti-roll bar. Rear: half-elliptic leaf springs with Z bar. Rolls-Royce hydraulic dampers. Suspension rate (at the wheel): front, 92.5 lb per in; rear, 127 lb per in. Static deflection: front, 10in; rear, 9in.

Brakes.—Hydraulically operated two trailing shoe, front; hydro-mechanical interlinked leading and trailing, rear. Mechanical servo assisted. Drums: 11¼in dia, 3in wide, front and rear. Total lining area 240 sq in (120 sq in front).

Steering.—Rolls-Royce cam and roller. Ratio (straight ahead) 20.6 to 1. Five turns from lock to lock.

Wheels and Tyres.—8.20-15in tyres on 6L × 15in rims. Five-stud fixing.

Electrical Equipment.—12 volt, 57 ampère-hour battery. Double-dip head lamps. 60-36 watt bulbs.

Fuel System.—18-gallon tank. Oil capacity 16 pints.

Main Dimensions.—Wheelbase 10ft 3in. Track, front 4ft 10in; rear 5ft 0in. Overall length 17ft 8in. Overall width 6ft 2¾in. Overall height 5ft 4½in. Ground clearance 7in. Frontal area 26.4 sq ft. Turning circle 41ft 8in. Weight (with 5 galls fuel) 4,228 lb. Weight distribution 49.7 per cent front.

Price.—Bentley: Basic £3,295, British purchase tax £1,374 0s 10d, total £4,669 0s 10d. Chassis £2,465. **Rolls-Royce**: Basic £3,385, British purchase tax £1,411 10s 1d. Total £4,796 10s 1d. Chassis £2,555.

Although the bench-type front seat has a one-piece cushion, the squab is divided. Both parts can be adjusted individually for rake, and this adjustment also results in a variation in available leg room

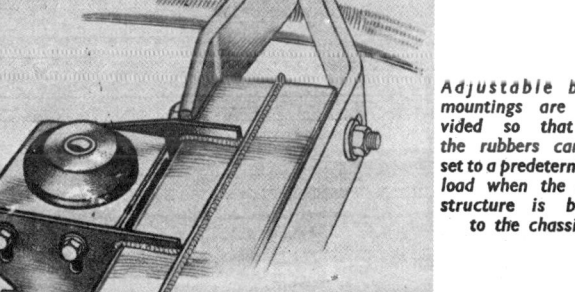

Adjustable body mountings are provided so that all the rubbers can be set to a predetermined load when the body structure is bolted to the chassis

MOTOR SPORT   JUNE, 1955

NEW ROLLS-ROYCE AND BENTLEY CARS

> " My new engine will be as silent and as swift as clouds
> moving before a storm ! "—*Chapter 1.*
> " Like an arrow the Silver Cloud shot forward on its last,
> most desperate race against time."—*Chapter 22.*
> " *The Silver Cloud,*" by Katrin Holland (*Nicholson
> and Watson, 1936*).

LATEST BENTLEY is the Model S, which has a 4.9-litre engine, automatic transmission, and a body by the Pressed Steel Company. The price, after purchase tax has been met, is £4,669 0s. 10d.

THE " magic of a name " is such that an announcement of new Rolls-Royce and Bentley models arouses more than the normal amount of attention. Consequently, interest attaches to the recently introduced Rolls-Royce Silver Cloud and S-series Bentley cars.

Both are virtually the same design, which means that the Rolls-Royce Silver Cloud is a high-performance car. It supersedes the Silver Dawn, which was the 10-ft. wheelbase 4¼-litre Rolls-Royce of last year, although the 11 ft. 1-in. wheelbase 4½-litre Silver Wraith Rolls-Royce remains in production.

The new Rolls-Royce and Bentley have an engine capacity of 4,887 c.c. to ensure ample performance for their considerable bulk, this being the engine developed for the Bentley Continental in 1954, further enlarged by 37 c.c. It retains i.o.e. valve layout but now has a six-port alloy head with two S.U. HD6 carburetters, feeding through redesigned water-jacketed inlet manifolds, and new exhaust manifolding. The engine is of the old long-stroke type (95 by 114 mm.) employing the modest compression ratio of 6.75 to 1. No figures are revealed for b.h.p. or b.m.e.p., as is R.-R. practice, but road speed in top gear at 2,500 ft. per min. piston speed equals 82½ m.p.h.

The chassis is an entirely new one, being a separate welded structure with box-section side and cruciform members, and the wheelbase increased by three inches, to 10 ft. 3 in.

It has been said that the late Sir Henry Royce excelled in perfecting conventional automotive practice rather than in the development of unorthodox designs; that this tradition is being maintained is apparent after studying the specification of the Bentley-S and Rolls-Royce Silver Cloud. The transmission is automatic, the delightful r.h.-change gearbox having gone before the American demand for self-selection, although driver skill can still be employed and four forward ratios of 3.42, 4.96, 9.0 and 13.03 to 1 result. But independent rear suspension, now common on Continental cars, has been eschewed in favour of an old-fashioned back axle sprung on ½-elliptic cart-springs, to which have been added such refinements as electric ride-control, anti-roll bar, rubber-bushed shackles, grooved Neoprene spring inserts, and grease-filled leather gaiters.

Front suspension is independent by unequal-length wishbones and coil-springs, damped by R.-R. hydraulic dampers. The braking system retains the famous R.-R. mechanical servo introduced thirty years ago, but this gearbox-driven brake now rotates twice as fast as before. Hydraulic operation is used at the front and trailing shoes are fitted, with automatic adjustment of almost-zero clearance. The cast-iron brake drums are of 11¼ in. diameter and 3 in. wider than before, giving a lining area of 240 sq. in. (compared with 186 sq. in. of the Bentley B7). The hand-brake is the all-too-familiar pull-out " umbrella handle."

There is a new steering layout with cam-and-roller gear and three-piece transverse track-linkage. The beautiful centre-lock wheels of old have given way to stud-attached pressed-steel discs, the diameter of which has been decreased from 16 in. to 15 in.

Both these new Rolls-Royce and Bentley cars have Pressed Steel Company four-door saloon bodies, with alloy doors, bonnet-top and boot-lid. Luggage space is increased, and there are such refinements as one-shot chassis lubrication, a very comprehensive heating and ventilation system with rear-window demister, separate adjustable back-rests to the bench-type front seat, and dial-type instruments for speedometer, water temperature, oil pressure and sump contents, and ammeter. Both front doors lock and upholstery is in English hide with foam-rubber overlays on spring cases; there are pile carpets and French walnut veneer facia and garnish rails.

Those who seek the best kind of English motor car bound with tradition will find their ideal in these new Bentley and Rolls-Royce cars, the prices of which are, respectively, £4,669 0s. 10d. and £4,796 10s. 1d. inclusive of p.t. Those who think in terms of small high-efficiency engines, all-independent suspension and tubular chassis, etc., may, perhaps, consider that the magic lies mainly in the name.

SILVER CLOUD is the type name of the new Rolls-Royce. The forward mounting of the radiator is retained and the engine is now of 4,887 c.c., like that of the new Bentley-S.

The original Rolls-Royce 'Silver Ghost' of 1906, the 48-horsepower engineering marvel that sent the automobile world of the time into rhapsodies by the smooth, noiseless motion. It is still in good running conditon — 500,000 miles later.

From its inception in 1904 down to the present day, this famous car has represented Quality

The ROLLS-ROYCE
British Aristocrat

By AUSTEN CHANNING and TED BONNER

THE CAR SPLUTTERED to a standstill; from the radiator cap puffs of steam signalled the fact that, once again, it had boiled. The driver sat in thought, arms resting on the wheel, and considered, as he had considered many times before, the variety of things that were wrong with his automobile: the inefficient ignition system, the difficulty of starting, the hellish noise it made and the monotonous regularity with which it over-heated. It was the year 1904 and motoring was in its infancy.

Although Henry Royce was an electrical engineer and one of the directors of a small company in Manchester, England, he was essentially a craftsman and an inventor. He knew little about internal combustion engines, but he was learning fast from the French car he had bought. He had stripped it down, re-built it and stripped it again, seeking some means to correct the many faults in its design and construction. Now, seated by the roadside, he suddenly determined that he would build a motor car to his own design.

So, in a factory designed for the manufacture of cranes and dynamos, he set to work. The surprise and skepticism of his co-directors failed to deter him. In fact, during his early experiments and tests, he resolved to build three cars. Together with a few picked men, working late into the night and personally doing much of the precision work, he built the first 'Royce' car—a 10 hp. two-cylinder open tourer—and had it ready for its trial run in April, 1904.

The workers who had helped him stood by silently as Royce swung the handle; the motor fired at once and, calmly triumphant, the forty-year-old inventor climbed in and drove the fifteen miles to his home without the slightest trouble or breakdown. Pleased though he must have been with this vindication of his theories and his engineering, even Henry Royce could not have foreseen that that short journey was to prove the first step in laying the cornerstone of a legend.

In London at this same time, a young man named Charles Rolls, the third son of Lord Llangattock, had already made a name for himself in the automobile world. Rolls was not content to adopt the leisurely life of his class; he had taken degrees in engineering and had driven cars in many competitions in Europe. He was convinced that the motor car was the means of transportation of the future and he had opened an auto business in London with another enthusiast named Claude Johnson. Together they had built this into a prosperous concern and were already well established. They did not, however, handle a single British car because Rolls did not believe that any of the then existing British cars were good enough.

It was inevitable that these three men—the skilled, imaginative craftsman in Manchester and the two young businessmen in London—would meet. Their introduction was arranged by a friend of Rolls' who had bought the third car Royce had built and who had lost no time in telling Rolls all about it.

Rolls and Royce met in Manchester

CAR LIFE

One of the first three cars built by Henry Royce in 1904, just before the inventor joined forces with Charles Rolls to form the Rolls-Royce company.

about the smoothness and the sensation of effortless and noiseless motion.

While this was all very pleasant for the Rolls-Royce company, Johnson was determined that the Ghost would prove beyond a doubt that it was as efficient as it was beautiful and so he entered it in practically every competition he could find. The Scottish reliability trial, the 1906 T.T., hill-climbing rallies all over England—all these and many more the Rolls swept through triumphantly.

Then Johnson determined to attack the world record for an observed non-stop endurance test which was held by a Siddeley car with a total distance of 7,089 miles. The Ghost covered 14,371 miles non-stop. The car was finally withdrawn when the company felt they had achieved all they had set out to do.

As a further instance of their com-

and took to each other right away. Rolls, however, did not like two-cylinder cars and he was prepared to be disappointed in the one he was then to test. But he had never driven an automobile like the Royce before and in a few miles he was completely sold.

It was decided to form a company; Royce was to be chief designer and was to be guided by Rolls' knowledge of what the customer wanted. This was particularly invaluable since Rolls had been dealing with the carriage trade and he realized that the wealthy aristocracy of England demanded reliability, comfort and silence in their cars; they were not interested in the adventurous aspects and glorious uncertainty of motoring as it then was. So Rolls and Johnson handled the selling and general organization and left Royce free to get on with production.

From the very first Royce was determined that smoothness and silence were to be the keynotes of every car he produced and he pursued those ends with a determination that took little account of meals or sleep. He would work late into the night, with no more sustenance than a few sandwiches consumed absent-mindedly at his bench. His desire for mechanical perfection has probably never been excelled and it was nothing unusual for Royce to reject a part—even a comparatively trivial part—time and time again until he was convinced that it was as near perfect as human skill could make it.

For smoothness, he modified the then current clutch design until he achieved the type with which we are now familiar; the clutches of those days were either 'in' or 'out'—there were no intermediate positions.

The new company produced four models, all of which were of superlative quality. But there is no doubt that they really began to command worldwide attention when, in 1906, they produced an entirely new car which was to purr its way into motoring headlines all over the globe.

It had a 48 hp. engine (at 1,700 rpm.) which was completely square—both bore and stroke were 4½ inches. The six cylinders were arranged in two

The aristocratic Bentley Continental, with a top speed of 120 mph., is now manufactured by the Rolls-Royce firm.

blocks of three, flexibly mounted on the frame at three points; the radiator was also flexibly mounted instead of being rigidly fitted as was the custom of the time; the expansion chambers referred to earlier were incorporated and the smoothness of the motor was enhanced by fitting the chassis with thin, multi-leaved springs and dampers.

By working at an almost superhuman pace, Royce managed to have two of these cars ready for the motor show of 1906 and there they were displayed—one as a stripped chassis and the other as a Pullman; the chassis price was $4,750.

Claude Johnson took one of these and fitted the body with silver-plated accessories; he painted it with an aluminum finish and, in a flash of inspiration, christened it the 'Silver Ghost'. The motoring press of the day ran out of superlatives in describing this new offering and rhapsodized

plete faith in their product, however, they invited the Royal Automobile Club to strip the car at the end of the trial and replace any part that showed the slightest sign of wear. This was done and the bill, for replacement, was not quite nine dollars!

Any other designer would have been content to display, in the motor show of the following year, the car which had shattered the non-stop record, and leave it at that. But Royce put on show the same basic model with improved steering, brakes, springs, carburetor and lubrication system. This action demonstrated quite clearly his philosophy that no machine was ever perfect.

In the incredibly short time of three years, Rolls-Royce had achieved a reputation second to none for quality. The Derby plant was opened in 1908 and from then until 1914 the cars it produced boosted that reputation to the very peak of motoring fame. Their

AUGUST 1954

in 1906 Charles Rolls averaged 39 mph. in this 20-hp. Rolls-Royce to win the Isle of Man Trophy. In 1910, Rolls met his death in an airplane accident.

Sir Frederick Henry Royce, developer of the world famous Rolls-Royce auto.

Hon. Charles Stewart Rolls handled the selling of the famous automobiles.

Silver Ghost had earned the right to be called The Best Car in the World—and Rolls-Royce were only too eager to compete against anyone who doubted its claim to that proud title.

Then tragedy struck. Charles Rolls, who had become deeply interested in aviation, was killed in an air accident in 1910 at the age of 33; he was the first Englishman to be killed in an air crash. Coming as it did after years of driving his mind and body mercilessly, this blow almost finished Henry Royce and he became seriously ill—so seriously, in fact, that specialists gave him no more than three months to live.

But they had reckoned without his tenacity and Johnson's determination. The latter reorganized the company's executive structure, delegated his responsibilities, and took Royce abroad for a long holiday. As a result of this trip Royce recovered, although physically he was never the same man again. It was arranged that he live in France in the winter and in England in the summer and that his own staff of secretaries and draftsmen should accompany him. By this means he was able to go on designing and improving without the added burden of management worries. Johnson ran the plant and supervised production; Royce never visited it again.

In 1910, a car made by the Napier company recorded 76.4 mph. in a speed test at Brooklands after having done a top-gear run of almost a thousand miles, during which its gas consumption averaged 19.33 miles per gallon. When these results were published, Johnson immediately entered a Rolls-Royce for the same test and it turned in a maximum speed of 78.2 mph. and averaged 24.3 miles per gallon. Further, just to see how fast a Rolls-Royce could travel, a specially prepared model with a lightened body was tested at Brook-

Continued on Page 47

The Park Ward Touring Saloon represents a new styling for Rolls-Royce, with fine edged coach work, a high wing line and a curved windshield. Interior fittings include de-misters for the windows, tables recessed into the back of the front seat.

Flush sides to body give more space and luggage room, yet don't date older model cars. Car on top is the R.R. Silver Cloud; one opposite is a Bentley.

I drove the first new Rolls in 10 years

Modern lines, more comfort, more silence than ever and a stinging 104 m.p.h. are talking points about the first new Rolls-Royce in nearly ten years.

By GORDON WILKINS

NOT many people can afford the new Silver Cloud, the first new Rolls to come from the most famous factory in the world since they released their Silver Dawn right after the war. Nevertheless there won't be a motorist anywhere who won't read about it—even though the factory has no high-pressure publicity or sales campaigns.

The motorists will be joined by salesmen, engineers, manufacturers, overseas dealers — for what Rolls-Royce bring out in a new model can be said to be the best of mechanical design and the near-perfection of current auto practice.

The new Rolls is called the Silver Cloud, in the tradition of ghostly names started with the Ghost and running through Phantom, Wraith and Dawn. With the Cloud comes a corresponding Bentley—identical, except for a different radiator grille and a £90 sterling cheaper price ticket in England before tax is added.

The new R.R. looks quiet and dignified—yet its top speed is 104 m.p.h. One may relax and enjoy the sensation of swift, silent and effortless travel, leaving the automatic transmission to do its work unnoticed. Alternatively, one may drive it as one would a sports car for the pleasure of exploiting a responsive machine.

A lever under the steering wheel allows the driver to over-ride the automatic gear change, holding second gear up to 40, and third up to 70 m.p.h.

With four occupants, the car will easily accelerate from a standstill to 50 m.p.h. in less than 10 seconds.

The steering is light and accurate, the brakes are more powerful than ever, there is very little roll and the car can be driven through fast corners like a sports model without squeal from the tyres.

It remains quiet, discreet and unruffled, with acceleration and braking to meet any situation. Forward vision is wide, the deep seats give the right support, and every control works with the ease and precision which has long delighted R.R. owners.

The design of the new car is simple —deceptively simple. When I saw the car for the first time, the Company's Chief Engineer said:

"If there is any detail which seems complicated, it's because we haven't been clever enough to find a simpler way of doing it."

Chassis and body are entirely new. The engine, developed from the previous straight six, is larger and more powerful, with a new alloy cylinder head and new, specially-designed twin S.U. carburettors.

A new exhaust system, designed for quietness as well as power, has two big mufflers with a smaller silencer near the end of the pipe to damp out the last hiss.

The transmission retains the Hydramatic drive, made under licence from General Motors of America. Rolls-Royce have added their own over-ride control so that the driver can either use his skill on the gearbox or motor without gear changing at all.

The conventional gearbox, with four forward speeds, offered as optional on the previous Silver Dawn model, has been dropped altogether.

It is still available on the Silver Wraith, Rolls' special-order limousine.

Suspension at the front is independent by semi-trailing wishbones and coil springs, with R.R. opposed-piston shock absorbers and a torsional anti-roll bar.

Rear suspension is by leaf springs, enclosed by steel gaiters and pumped full of grease. A Z-bar is fitted for extra stability.

To eliminate the slightest steering movements at the front, no rubber bushes are used in the wishbone suspension or steering. However, they are included in the rear to eliminate greasing from certain points.

R.R. say that the new chassis is stiffer than any they have ever built. When tested to destruction on a special rig, it lasted two and a half times as long as the previous model, yet it weighs only 30 lb. more.

The ride control, by which the

WHEELS, July 1955

driver can adjust the ride from soft to hard, is retained, but is now worked electrically instead of by oil pipes.

Improvements have been made to the brakes. The wheels are 15in. in diameter, instead of 16in., so the drums have been made wider to compensate for the reduction in diameter. The lining area is 240 sq. in. instead of 186 sq. in. previously. The friction servo-assistance remains.

The body, despite its conservative shape, has been designed to give less wind resistance and noise than before. There is now only as much wind noise at 70 m.p.h. as there was at 60 m.p.h. in the old model.

Inside, the car is more than ever the perfect gentleman's carriage.

There are two separate heating and ventilation systems, one for each side of the car, the doors have push-button locks, and there is a R.R. master control switch which immobilises the electrical equipment as an anti-thief precaution.

Instruments include a rev. counter as well as a speedometer.

SPECIFICATIONS:
ENGINE: 6-cyl., 95.2 x 114.3 mm., 4,887 c.c. Horsepower not quoted. Overhead inlet and side exhaust valves. Twin S.U. carburettors.
TRANSMISSION: R.R. automatic gearbox (General Motors Licence), with manual over-ride control. Ratios: 13.06, 9.00, 4.96, 3.42 to 1.
SUSPENSION: Front by wishbones and coil springs, anti-roll bar. Semi-elliptics at rear with Z-control bar. Double-acting lever dampers.
STEERING: Marles cam and roller, three-piece track rod.
BRAKES: Girling hydraulic, hydrostatic shoes, 11¼" drums. Friction area, 240 sq. in. Servo-assisted. Additional safety control for rear wheels in case of hydraulic failure.
DIMENSIONS: Wheelbase, 10' 3", front track, 4' 10", rear track, 5' 0", length, 17' 8", width, 6' 5", unladen height, 5' 4", turning circle, 41' 8", drive-away weight, 38 cwt.
PERFORMANCE: Top speed, 104-106 m.p.h. (works figure). Gearing, 25 m.p.h. in top at 1,000 r.p.m.

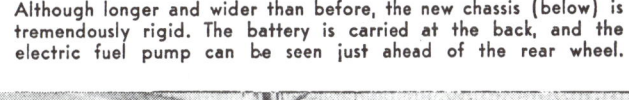

Interior features (above) include walnut facings, bonded to glass fibre mountings to give durability in severe climates. Seats have sponge rubber over springs and first quality hide coverings.

New front suspension (below) has semi-trailing wishbones, coil springs, an anti-roll bar and Rolls-Royce shock absorbers of opposed-piston design.

Although longer and wider than before, the new chassis (below) is tremendously rigid. The battery is carried at the back, and the electric fuel pump can be seen just ahead of the rear wheel.

NEW STANDARDS.—The Silver Cloud Rolls-Royce with 4.9-litre engine and automatic gearbox offers a higher common factor of maximum speed, acceleration, roadworthiness and refinement than any previous model in the distinguished history of the company.

1956 CARS

BENTLEY and

THE Bentley and Rolls-Royce programme for the coming year will be based on three related but differing types of chassis. As the standard model of the range will be the Bentley S series and Rolls-Royce Silver Cloud which were fully described in *The Motor* of April 27 last, it will clarify the picture if the leading features of this type are briefly recapitulated.

Although stemming directly from the R Series Bentleys and Silver Wraith Rolls-Royces introduced shortly after the war the present cars can be considered a new design. The six-cylinder engine with a bore and stroke of 114.3 mm. has a swept volume of 4,887 c.c. and the iron cylinder block and crankcase is surmounted by a light alloy cylinder head which contains inlet ports and inlet valves only, the exhaust valves being placed at the side in the main block. There are separate ports to each valve drawing mixture from a pair of special S.U. carburetters, and though the engine is exceptionally compact seven main bearings give support on each side of each crank throw. This ensures not only long life but also exceptional smoothness of running, and another contribution to this end is the improved manifolding which has provided benefits both in respect of greater power and more stable idling at very low speeds.

Standardized Automaticity

Power is transmitted through a single two-piece hydraulic coupling to a four-speed epicyclic gearbox, the gear trains being automatically engaged or uncoupled by an automatic control responsive to both engine speed and torque. Thus on full throttle openings and at maximum available torque the engine r.p.m. will be maintained at the highest value, whereas on light throttle openings correspondingly low r.p.m. will be enjoyed. It is, however, possible for the driver to override this mechanism, and to ensure, except under extreme conditions, that the car is held in either second or third gear, an arrangement which makes for safety on long mountain descents and for ease of driving in mixed traffic conditions.

The automaticity of the gearbox confers two-pedal control and the use of an improved version of the long-established Rolls-Royce servo mechanism with the plate clutch now being driven at 0.18 times engine speed, ensures stable stopping with a pedal pressure but little greater than that applied to the accelerator. A two-piece propeller shaft takes the drive to the conventional live axle supported on leaf springs, and at the front end a change has been made by adopting normal type unequal length wishbones with open coil springs and a divided type trackrod.

The completely box section frame has a large cruciform centre member and the relation between engine size, position, and wheelbase is such that all the passengers are seated well forward of the rear axle, which component is located not only by the springs but also by a Z-shaped rod which imparts additional roll stiffness to the rear suspension.

Comfort and Quiet

The wheelbase on this model is 10 ft. 3 in. and both the Bentley S series and Rolls-Royce Silver Cloud models have a similar pressed steel four-door saloon body in which modern full-width styling has been combined with gracefully flowing lines, the tail at the same time providing quite exceptional luggage-carrying capacity.

Another feature of these two models which has received especial attention is internal ventilation. Air ducted from each front wing passes through a filter to an axial-type fan. This fan forces the air through a light alloy heat exchanger to four nozzles at the base of the windscreen from which it passes up the screen and beneath the roof. Each duct has its own heat exchanger and air from the right-hand system can be diverted to the floor of the car so as to evenly warm the whole interior. Further detail refinement can be seen in the utmost endeavours that have been made to exclude noise from the passenger compartment, the normal central controls running down the steering column having been removed so as to ensure that they will not act as a telephone system. The result is a car commonly agreed to set new standards of power, speed, roadworthiness and refinement of running.

To meet the needs of those who are prepared to accept some diminution of accommodation in return for a sub-

LIGHTWEIGHT CONVERTIBLE.—Closed or convertible bodies are available on the Bentley Continental chassis which with greater power output and higher gearing than the standard type can achieve road speeds of the order of 120 m.p.h. The all-round performance of the Park Ward convertible shown here is enhanced by a body weighing only 7½ cwt.

ROLLS-ROYCE

Three Basic Chassis Types Cover a Wide Range of Motoring Requirements of the Highest Order

stantial increase in maximum and potential average speed the Bentley is also offered in Continental form with a choice of saloon bodies by H. J. Mulliner or by Park Ward, the latter concern also supplying a drophead coupé. As compared with the standard model just described, the Continental has a braking area increased from 240 sq. in. to 275 sq. in., giving, in conjunction with a lower weight for the complete car, a gain in brake area per unladen ton of over 20%. Both body constructors have given great attention to the elimination of unnecessary weight by the use of light alloy structure and aircraft technique, and the complete weight of the drophead coachwork is now only 7½ cwt., a reduction of no less than 40% over the previous model of this kind.

The engine installed in this chassis is modified to give a higher power output, and the driver-controlled dampers have an exceptionally rapid response to the control gear.

By contrast, for those who seek additional accommodation and bodywork of town carriage type, the long wheelbase (11 ft. 1 in.) Silver Wraith model has been brought into line with the new design. That is to say, it has the new enlarged engine, which gives a power increase of 8½%, and the automatic gearbox. The rear track on this model has been enlarged to 5 ft. 4 in. and the lowest priced model, the Park Ward touring saloon, is now catalogued at £4,695. Other bodies standardized on this E series Rolls-Royce are a Mulliner touring limousine, a James Young saloon, and six seven seater saloons by Park Ward and Hooper. A limousine body is also offered by Messrs. Freestone & Webb.

TOWN CARRIAGE.—With an 11-ft. 1-in. wheelbase the Silver Wraith Rolls-Royce chassis can support elegant and capacious coachwork suitable for both town and country travel. This model now has the 4.9-litre engine and is shown here fitted with a touring limousine by H. J. Mulliner.

The Motor Road Test No. 1/56 (Continental)

Make: Rolls-Royce **Type:** Silver Cloud
Makers: Rolls-Royce Ltd., 14-15 Conduit St., London, W.1

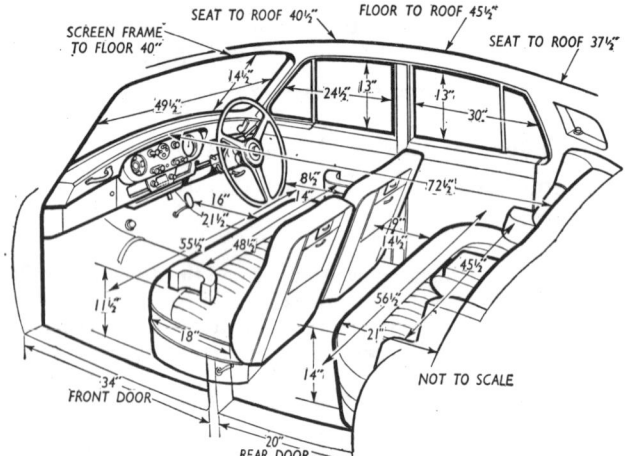

Test Data

CONDITIONS. Damp road; Tyre pressures raised to 28 rear and 23 front.

INSTRUMENTS
Speedometer at 30 m.p.h. .. 1% fast
Speedometer at 60 m.p.h. .. 1% fast
Speedometer at 90 m.p.h. .. accurate
Distance recorder .. .. accurate

MAXIMUM SPEEDS
Flying Quarter Mile
Mean of four opposite runs .. 102.9 m.p.h.
Best time equals .. .. .. 104.6 m.p.h.

Speed in gears
Max. speed in 3rd gear .. 63 m.p.h.
Max. speed in 2nd gear .. 40 m.p.h.

FUEL CONSUMPTION
24.0 m.p.g. at constant 30 m.p.h
22.0 m.p.g. at constant 40 m.p.h.
20.0 m.p.g. at constant 50 m.p.h.
17.5 m.p.g. at constant 60 m.p.h.
15.5 m.p.g. at constant 70 m.p.h.
12.0 m.p.g. at constant 80 m.p.h.
Overall consumption for 1,004 miles, 72 gallons, =14.5 m.p.g. (19.5 litres/100km.)
Fuel tank capacity 18 gallons.

ACCELERATION TIMES Through Gears from Rest
0-30 m.p.h. .. .. 4.3 sec.
0-40 m.p.h. .. .. 6.7 sec.
0-50 m.p.h. .. .. 10.0 sec.
0-60 m.p.h. .. .. 13.5 sec.
0-70 m.p.h. .. .. 19.0 sec.
0-80 m.p.h. .. .. 26.1 sec.
0-90 m.p.h. .. .. 35.1 sec.
Standing Quarter Mile .. .. 18.8 sec.

ACCELERATION TIMES Through Gears from Rolling Start
10-30 m.p.h. .. .. 3.1 sec.
20-40 m.p.h. .. .. 3.8 sec.
30-50 m.p.h. .. .. 5.8 sec.
40-60 m.p.h. .. .. 6.6 sec.
50-70 m.p.h. .. .. 9.0 sec.
60-80 m.p.h. .. .. 12.6 sec.
70-90 m.p.h. .. .. 16.1 sec.

WEIGHT
Unladen Kerb weight .. .. 38 cwt.
Front/rear weight distribution 48/52
Weight laden as tested .. 41.5 cwt.

HILL CLIMBING (at steady speeds)
Max. gradient on top gear .. .. .. 1 in 7.7 (Tapley 290 lb./ton)
Max. gradient on 3rd gear .. .. .. 1 in 5.2 (Tapley 420 lb./ton)
Max. gradient on 2nd gear .. .. .. 1 in 3.5 (Tapley 630 lb./ton)

BRAKES at 30 m.p.h.
0.45g retardation .. (= 67 ft. stopping distance) with 25 lb. pedal pressure.
0.72g retardation .. (= 41.6 ft. stopping distance) with 50 lb. pedal pressure.
0.85g retardation .. (= 35.3 ft. stopping distance) with 65 lb. pedal pressure.

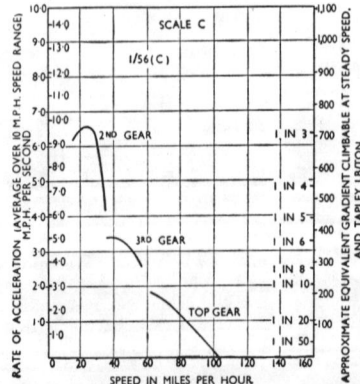

Drag at 10 m.p.h. .. ..59 lb.
Drag at 60 m.p.h. .. 211 lb.
Specific Fuel Consumption when cruising at 80% of maximum speed (i.e. 82.3 m.p.h.) on level road, based on power delivered to rear wheels .. .. 0.72 pints/b.h.p./hr.

Maintenance

Sump: 16 pints, S.A.E. 20. **Gearbox:** 20 pints, Type AQ/ATF. **Rear Axle:** Castrol HI Press S.C. **Steering gear:** S.A.E. 30. **Radiator:** 28 pints (2 drain taps). **Central Chassis Lubrication:** By pedal beneath facia every 200 miles. **Ignition timing:** 2° B.T.D.C. **Spark plug gap:** 0.025 in. **Contact breaker gap:** 0.019–0.021 in. **Valve timing:** I.O. 26° B.T.D.C.; **Tappet clearances:** (Cold) Inlet 0.006 in. Exhaust .012 in. **Front wheel toe-in:** $\frac{1}{16}$–$\frac{3}{32}$ in. **Camber angle:** 0–$\frac{1}{4}$° postive. **Castor angle:** 1° negative. **Tyre pressures:** Front 19 lb. Rear 26 lb. **Brake fluid:** Wakefield Girling Crimson. **Battery:** P.R.55 amp.-hr.

The Motor ROAD TESTS OF 1956 CARS

The ROLLS-ROYCE Silver Cloud

A Superb All-rounder Based on Fifty Years' Experience in High-Quality Manufacture

LINE BREEDING.—The Silver Cloud Rolls-Royce combines swept wing styling with pressed steel bodywork to give greater accommodation than previous models with comparable wind resistance and absence of wind noise. The classic radiator is retained.

WHEN C. S. Rolls & Co. issued the first Rolls-Royce catalogue in 1905 they said it was a car which incorporated "what experience has proved to be the best features in the leading types of cars, and to these have been added notable and most valuable improvements." During the 50 years in which the make has asserted the right to be called "The Best Car in the World," this policy has been followed consistently; that is to say, the Rolls-Royce has never incorporated revolutionary, or even very advanced, design features, but has relied entirely upon what experience has proved to be the best, and has added thereto superlative detail design, great care in the choice of materials, and a fine reputation for finish and workmanship.

The latest Rolls-Royce, the Silver Cloud, was introduced in April 1955 to supplement (but not supersede) the long wheelbase model which is supplied to coachbuilders for the mounting of special bodywork. It has a pressed steel body and the example submitted recently to us for road test proved a worthy successor to a distinguished line of ancestors. As many potential owners will either be in possession of a Rolls-Royce, or will have owned one in relatively recent years, it seems logical to start with the broad statement that this model is roomier, quieter, faster and as economic as the preceding types, and should certainly be their equal in reliability.

Exceptional Comfort

The rear seats, now over 54 in. wide, are capable of carrying three persons and with one or two in the back comfort is maintained by a broad centre armrest. The upholstery is carried up particularly high to give a first-class headrest, and other unusual amenities include smoking and ladies' companions mounted in the rear quarters, with an individual light in each which enables the rear passenger to read without disturbing the driver.

The passengers are accommodated in unusual comfort (although footrests for the rear seat would be a welcome further refinement), and they have behind them a luggage locker with a minimum height of 15¼ inches and a maximum length of 46 inches. This enables really large bags to be carried.

It is thus readily demonstrable, without leaving the showroom, that the driver and passengers in the Silver Cloud are accommodated in greater comfort than in any previous Rolls-Royce and that there is more space for baggage.

On the road it does not take long to prove that this is the fastest car to bear this name. Although a single-seater 7.7-litre Silver Ghost model was driven by the present chairman of the company at over 100 m.p.h. at Brooklands as long ago as 1911, the side valve models to Sir Henry Royce's design did not normally exceed 70 m.p.h. on the flat and had a cruising speed of around 55 m.p.h. In the twenty years 1919/1939 these performances slowly rose so that the 7-litre 12-cylinder Phantom III model of 1939 had a true maximum in excess of 90 m.p.h. and a cruising speed of a little over 70 m.p.h.

The Silver Cloud, by contrast, has a cruising speed of 80 m.p.h. upwards, will reach this speed from rest in a fraction over 26 seconds, and as a corollary has an all-out maximum substantially in excess of 100 m.p.h., being the first closed model Rolls-Royce in which such a claim can be made.

Owners of this class of car are, however, more concerned with the rate of travel that can be sustained in comfort viewed either from the mechanical or human point of view. In this aspect mechanical and wind noises and controllability play an important part. The enjoyment of mechanical quietness is something which every owner of this class of car has a right to expect, and it need scarcely be said that the Silver Cloud meets every requirement. The prevention of wind noise presents far greater difficulties, but on the Silver Cloud, with a qualification which will be later referred to, the overall sound level including wind noise rises almost imperceptibly up to 95 m.p.h.

The Silver Cloud is one of the few

In Brief

Price: £3,385 plus purchase tax £1,693 17s. 0d. equals £5,078 17s. 0d.
Capacity ... ... 4,887 c.c.
Unladen kerb weight ... 38 cwt.
Fuel consumption ... 14.5 m.p.q.
Maximum speed ... 102.9 m.p.h.
Maximum speed on 1 in 20
  gradient ... ... 84 m.p.h.
Maximum top gear gradient 1 in 7.7
Acceleration:
  10-30 m.p.h. in 2nd ... — sec.
    (auto change)
  0-50 m.p.h. through gears 10 sec.
Gearing: 24.8 m.p.h. in top at 1,000 r.p.m.; 82.5 m.p.h. at 2,500 ft. per min. piston speed.

CARRYING POWER.—The large luggage locker of the Silver Cloud is indicated in this three-quarter rear view of the car; the wide rear window has internal electric heating which keeps it clear during the winter months.

highly priced and powerful cars in the world today which denies the driver servo-assisted steering, and the lightness needed for city manoeuvring is obtained by highly skilled design coupled with a steering box ratio which commands 4¾ turns of the wheel to cover the whole of the generous lock. It is normal for a mechanism of this kind to have inadequate response in open road conditions but the overall geometry and rates of the Rolls-Royce suspension have been so contrived that with rising speed there is increasing response to the helm so that at high speeds very small movements of the wheel suffice to displace the car through considerable angles. Hence despite the low gearing the driver has no sense of time lag, and certainly none of lost motion. The latter characteristic is particularly important when driving the car on winding roads with slippery surfaces, for used in conjunction with modest roll the car can be driven in almost sports-car fashion. True, it will not normally be handled in this way, but such driving indicates the inherent safety of the vehicle and an ability to overcome critical circumstances.

High-quality Braking

The good handling is matched by the remarkable quality of the brakes, which have no equal in the world of motoring today. The time-tested Rolls-Royce servo mechanism has been speeded up and although the slight lag before the brakes come on at walking speeds has not been wholly eliminated it has been diminished to 17 inches. In all other respects the brakes stand out as of especial merit, for a mere 25 lb. of pedal pressure suffices for all conditions commonly met, braking in the shortest possible distance is effected "all square" and the car can be driven extremely hard without brake fade or the scarcely less tolerable alternatives of squeak, shudder or stench.

In these factors of controllability in exactness of steering and in steady stopping the Silver Cloud is quite simply in a different class from all other large cars in the world today. If it has one special ability it is the combination in unique degree of content in normal, and confidence in critical, motoring.

In some respects the full merit of the car cannot be appreciated in English roads. In dense traffic it seems somewhat long and wide, and the extraordinary quality of quietness at over 80 m.p.h. can be enjoyed but briefly. By contrast on the less thickly populated Continental road the speed can be maintained at up to 90 m.p.h. for miles

on end, and without apparent effort average speeds nearer to 60 than 50 m.p.h. are realized even with large towns en route.

At these rates of speed the car is consuming fuel at a rate in excess of five gallons per hour and must therefore be stopped for refuelling at least every three hours. It may be thought that this indicates the need for a larger fuel tank.

This brings to mind the small capacity of the battery, for should the dynamo fail (and one did during our tests) three hours of travel by night almost certainly result in unstartability the following day, and there is no starting handle. The case has also to be considered of a chauffeur-driven car used intermittently by the owner between, say, 7 p.m. and 2 a.m., but with the driver making steady demands on the battery for side, tail and interior lamps and, part-time, for radio and heater.

On the second-class surfaces prevalent in many parts of Europe the rear seat occupants are aware of the unsprung mass of the rear axle beneath them and the driver finds noticeable steering reaction to small changes in road level caused by the sections of concrete roads, by strong road camber, and by cross winds. The physical absorption of substantial road irregularities is well handled and we experienced no deterioration in damper performance but cannot say that the car is unaffected by the road upon which it runs. A two-position switch instantly selects firm or soft shock absorber settings, and should be made use of if the best results are to be obtained at both town and country speeds.

A similar combination of good and bad is to be found in the somewhat elaborate heating and ventilating system. Wholly good is the rear window de-mister, fine wires being embodied in the glass itself, but the placing of the ignition-linked switch behind the rear seats, argues the intention that the heating element should be used invariably in winter motoring.

There are separate ducts and heat exchangers from one of which air emerges under the scuttle and beneath the front seat or from the other behind the windscreen. These systems are separately controlled with the temperature options of hot or cold, and the three volume choices

ALL THE AMENITIES.—The backs of the front seats are adjustable for rake and behind them folding picnic trays are placed for the benefit of the rear seat passengers.

of ram, low-speed and high-speed fans.

They can also be played off against each other, that is to say maximum hot air delivery from beneath the seat can be blended with partial or maximum cold air delivery to the windscreen and thence across the faces of the front seat passengers. With both systems supplying the maximum volume of hot air the system fulfils the prime target of keeping the interior of the car warm, but the arrangement is inflexible and the controls must be changed fairly frequently and with some skill if all the interior occupants are to agree upon the habitability at any given time. This is all the more to be regretted since the Silver Cloud is dependent upon the ventilating system, as any opening of any window at speeds above 50 m.p.h. completely destroys the quality of the car by the resultant wind noise. Indeed, an open window at, say, 70 m.p.h., results in physically painful buffeting around the ears of the passengers. This penalty of opening windows when running at high speeds in the summer is a disadvantage the more serious because there is now no sliding roof but the fact that the car must be run with all windows shut mitigates the surprise felt that power-operated windows are not available, even as an optional extra. Similarly, the front seat adjustment depends on human muscle and needs some force to move with two persons sitting on it. Otherwise the physical demands made upon the driver are slight; the lightness of the brakes has already been

mentioned, and so has the comparatively small effort required on the steering wheel although in parking there is no denying that servo assistance would be welcome.

No physical effort is needed for the clutch which has been replaced by a fluid coupling which is part of the four-speed automatic transmission.

This arrangement changes gear without regard to the driver's skill or intelligence, but some of the earlier examples (indeed some existing vehicles where the transmission and engine characteristics are not well suited) exemplify, indeed reinforce, the 60-year-old remark of Levassor, "C'est brutal mais ça marche."

This arises from the inability of the driver to prevent downward changes of gear on moderate throttle openings and even on the Silver Cloud a clumsy driver can catch an echo of this complaint if he pushes the accelerator right down at, say, 55 m.p.h. This will cause a redundant shift to third speed, top gear being engaged automatically almost immediately afterwards when the speed rises to 63 m.p.h.

However the balance between engine torque and gear selection has been so well contrived on this Royce that a sympathetic driver can maintain it in top gear constantly at over 30 m.p.h. with no appreciable loss of performance. The intelligent driver can also make good use of the manually engaged third gear to slow the car before reaching a corner or to increase control over a sinuous section of road where the speed will not exceed 60 m.p.h.

To sum up, although one looks through the specification and the road test data of this car in vain to see any feature that cannot be found on vehicles costing far less, we know of none that matches the Rolls-Royce in quality of road behaviour and in the impression given to driver and passengers that here is a car that offers the owner a lifetime of service, and provides daily motoring in cities, on the main highways, and in winding lanes in a fashion that simultaneously satisfies the senses and justifies the first cost.

POWERFUL SIX.—The 4.9-litre, 6-cylinder engine is the largest of its type in current world production and one of the most powerful.

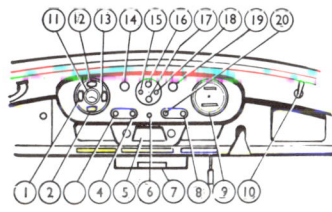

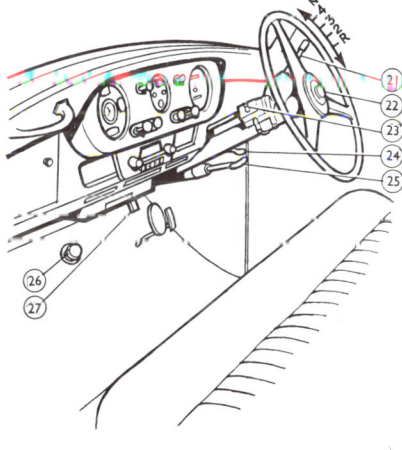

1, Water thermometer; 2, Ammeter; 3, Demister control (pull and twist); 4, Windscreen wiper switch (two-speed, twist) and windscreen washer (push); 5, Radio; 6, Oil level indicator (push for reading on fuel gauge); 7, Ashtray; 8, Heater (pull and twist); 9, Speedometer and distance recorder; 10, Direction indicator switch; 11, Clock; 12, Fuel gauge; 13, Oil pressure gauge; 14, Cigar lighter; 15, Dynamo charge warning light; 16, Lights switch; 17, Ignition key; 18, Fuel reserve light; 19, Petrol filler door release switch; 20, Panel light switch; 21, Gear lever; 22, Horn; 23, Ride control switch; 24, Bonnet release (one each side); 25, Hand-brake; 26, Dip switch; 27, Chassis lubricator pedal.

Mechanical Specification

Engine
Cylinders	6
Bore	95.25 mm.
Stroke	114.5 mm.
Cubic capacity	4,887 c.c.
Piston area	66.25 sq. in.
Valves	I. or E.
Compression ratio	6.6/1
Max. power	— b.h.p.
at	— r.p.m.
Piston speed at max. b.h.p.	— ft. per min.
Carburetter	Twin S.U.
Ignition	Lucas coil
Sparking plugs	Champion H.8BR or Lodge HNLP
Fuel pump	Twin S.U.
Oil filter	Full flow

Transmission
Clutch	Nil
Top gear	3.42
3rd gear	4.96
2nd gear	9.0
1st gear	13.06
Propeller shaft	Open divided
Final drive	Hypoid bevel
Top gear m.p.h. at 1,000 r.p.m.	24.8
Top gear m.p.h. at 1,000 ft./min. piston speed	33

Chassis
Brakes	Hydro-mechanical
Brake drum diameter	11 in.
Friction lining area	240 sq. in.
Suspension:	
Front	Wishbone and open coil
Rear	Semi-elliptic
Shock absorbers:	
Front	Rolls-Royce
Rear	Rolls-Royce
Tyres	8.20—15

Steering
Steering gear	Cam and roller
Turning circle between kerbs:	
Left	38 feet
Right	37½ feet
Turns of steering wheel, lock to lock	4¼

Performance factors (at laden weight as tested):
Piston area, sq. in. per ton	32
Brake lining area, sq. in. per ton	116
Specific displacement, litres per ton mile	2,950

Fully described in *The Motor*, April 27, 1955

Coachwork and Equipment

Starting handle ... No
Battery mounting R.h. side below rear locker
Jack ... Screw type
Jacking points ... 2 below centre pillar
Standard tool kit: Jack, inspection lamp, socket spanner for wheel nuts, tommybar, tyre pump, screwdriver, adjustable spanner, pliers, tyre pressure gauge, tappet adjusting spanner combined with 2, 3 and 5 B.A. open jaw spanner, exhaust tappet adjusting lock, feeler gauges, distributor adjustment spanner, plug box spanner, 1 sump, gearbox, torus cover and rear axle drain plug spanner, 1 tail and stop lamp bulb, 1 headlamp bulb, 2 fog lamp/flasher bulbs and 1 side lamp bulb.
Exterior lights ... 2 head, 2 side, 2 fog, 2 tail, 2 stop
Direction indicators ... Flashing stop and fog lamps
Windscreen wipers ... Two-speed, self-parking electrical
Sun vizors ... Two
Instruments ... Speedometer, clock, ammeter, fuel level, oil level.
Warning lights: Direction flashers, dynamo charge, low fuel (3 galls. or less).
Locks:
 With ignition key ... Doors
 With other keys Glove box and rear locker
Glove lockers: One on left-hand side with door, one on right-hand side, open.
Map pockets ... One in each front door
Picnic shelves One central below facia, two fitted behind front seats
Ashtrays ... One front, two rear
Cigar lighters One front, one right-hand back
Interior lights: Facia, map, roof, two companion in rear quarters, luggage locker.
Interior heater Two—one for screen channel, one for interior
Car radio ... Optional extra
Upholstery material Leather, seven colour choices
Floor covering ... Pile carpet
Exterior colours standardized: Six single colours; four choices dual colours.
Alternative body styles ... None by chassis manufacturer

There are no external changes on the Rolls-Royce Silver Cloud four-door saloon. Air conditioning and power-assisted steering are available as optional extras in export models

| NEW CARS DESCRIBED | ## REFINEMENTS |

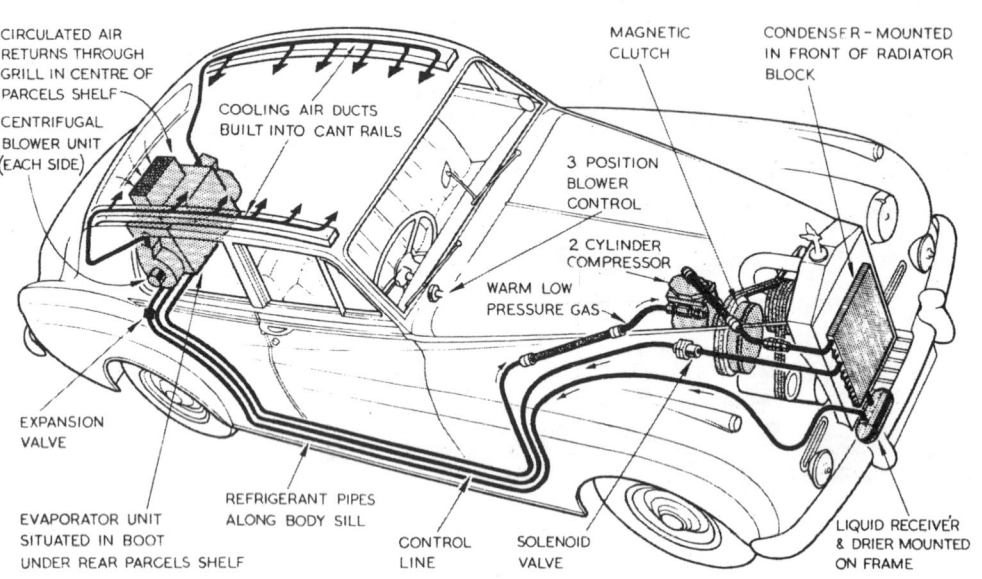

The cooling air of the conditioning unit is circulated through ducts built into the roof cant rails. Under maximum operating conditions the engine-driven compressor of the refrigeration system requires 5 h.p. to drive it. With the air-conditioning system inoperative, the compressor is disconnected by means of the magnetic clutch

AIR CONDITIONING AND POWER-ASSISTED STEERING FOR EXPORT ROLLS-ROYCE AND BENTLEY : INCREASED POWER FOR CONTINENTAL

WITH a large proportion of the Rolls-Royce and Bentley production sold in overseas markets, most of them crossing the Atlantic, it is not surprising that air conditioning and power assistance for the steering are to be made available. The air conditioning system is available as an optional extra on the Rolls-Royce Silver Cloud and Silver Wraith, and on the S series Bentley. Similarly, power-assisted steering is available as an export optional extra on all these models, but also can be supplied on any Silver Wraith chassis sold in the United Kingdom. The complete refrigeration plant for the air conditioning unit is imported from the U.S.A. On the Bentley Continental, for which increased engine power is now provided, only the power-assisted steering is available as an optional export extra.

The air conditioning or refrigerator unit works on the vapour cycle system, which uses a straightforward series layout of compressor, condenser and evaporator. The gaseous cooling medium (Freon 12) is quite harmless and practically non-toxic, though it requires a few simple precautions to guard against a higher-leak tendency than those of most other refrigerants.

It is not necessary for the refrigeration plant to provide a large drop of air temperature. In countries where the system is needed, the occupants of the car usually dress according to the climate—the ladies wear light filmsy dresses and the men cool linen suits, often without jackets. Investigation has shown that the inside air temperature of a car tends to run about 10 deg F higher than ambient. Thus, if the outside air temperature is between 78 and 82 deg F the comfortable inside temperature is around 68 deg F. The worst conditions are when a car is left standing in the sun with all doors and windows closed. Even under these circumstances, if the engine is running at fast idle, the inside temperature can be reduced to the normal operating range within three minutes with the blower control set in its high position. The air is changed at a rate of 300 cu ft per minute at this setting, which gives a complete change of the inside air every 1½ minutes.

On switching on the refrigeration unit, the engine is automatically set to fast idle, to prevent the engine from stalling, due to the power required to drive the compressor. Maximum refrigeration rate is reached at a road speed of 30 m.p.h.

The cycle of operations in the refrigeration begins with the compressor inlet valve, through which the vapour enters at a relatively low temperature and pressure. On the down-stroke of the compressor piston the automatic suction valve is opened by the pressure differential, and vapour enters the cylinder. On the up-stroke the suction valve closes and the vapour is compressed until it opens the discharge valve, so that the high pressure refrigerant is forced into the condenser. Here the heat of compression and the heat absorbed by the refrigerant in the cooling coil is imparted to the air flowing over the tubes, and the refrigerant liquifies.

The liquid refrigerant next flows into the receiver, which acts as a storage tank and contains a strainer-drier to remove dirt and moisture. Still in a liquid state, the refrigerant enters the expansion valve, the function of which is to meter the flow to the evaporator by throttling it, so that some of the refrigerant flashes into vapour. The remaining liquid is cooled down to saturation temperature at the expansion valve outlet. This valve is also the dividing point between the high and low pressure and temperature sides of the system. No adjustment of the valve is provided, and it will maintain superheat from −30 deg F to +50 deg F.

Among the outstanding features of the Silver Cloud chassis is the box section frame, cross braced for stiffness. There are two completely independent braking systems, each having a separate reservoir of transparent plastic for immediate visual check of fluid level

FROM CREWE

Steering power assistance is obtained by a slight initial axial movement of the worm, which circulates the oil under pressure from the pump to a chassis-mounted actuating cylinder

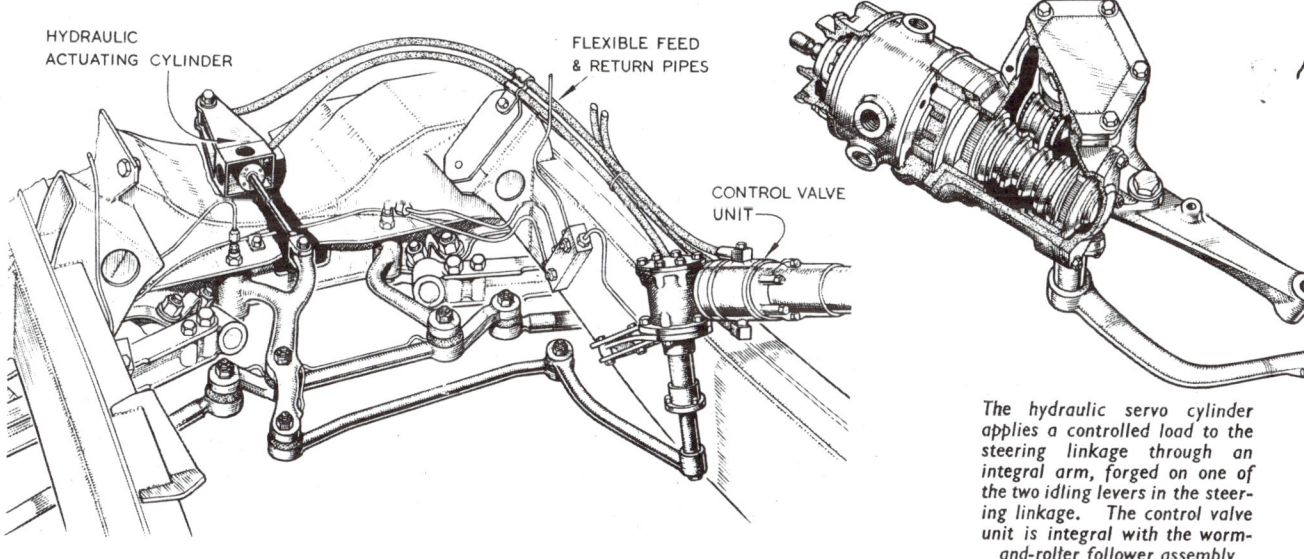

The hydraulic servo cylinder applies a controlled load to the steering linkage through an integral arm, forged on one of the two idling levers in the steering linkage. The control valve unit is integral with the worm-and-roller follower assembly

The liquid and vapour next enter the cooling coil of the evaporator unit, where the remaining liquid evaporates, absorbing heat from the air blown through it from the inside of the car. This is the refrigeration stage, in which the air is cooled by the latent heat of vaporization of the refrigerant in the coil. The completely vaporized refrigerant is finally drawn back to the compressor.

Triple wedge-type vee-belts from the crankshaft pulley drive the dynamo and fan pulley on the water pump shaft. The fan pulley has three additional grooves, two for the compressor drive and one for the power steering pump when fitted. A two-cylinder reciprocating type compressor with its own lubrication system is used. The suction and exhaust valves are of reed type, discharging through a common valve plate located between the block and head.

A magnetic clutch is provided in the compressor drive pulley, with a spring-

The two-cylinder compressor of the refrigeration system is belt-driven from the fan pulley. Great attention has been paid to the installation and fittings of the refrigerant pipes to eliminate possible leaks. The hydraulic circuit of the power-assisted steering has its own self-contained oil and filtration system

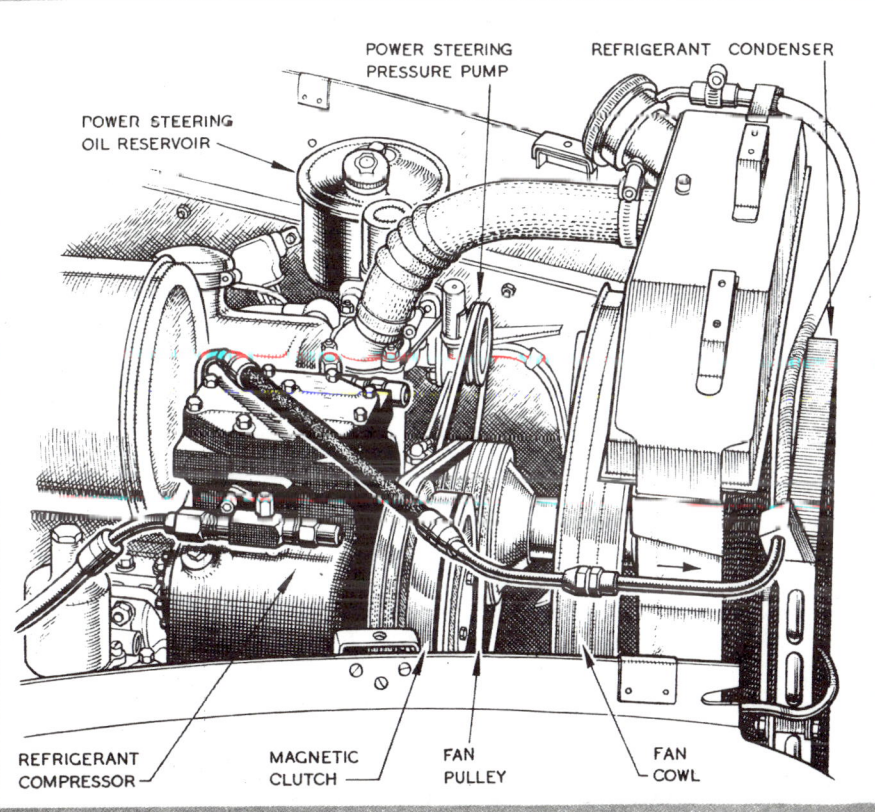

An example of the specialist coachwork fitted to the Bentley Continental chassis is this elegant two-door sports saloon by Park Ward. Engine power has been increased to enhance the already outstanding performance

Refinements from Crewe...

loaded drive plate attached to the compressor crankshaft. When the electromagnetic coil (mounted in the pulley) is not energized, the pulley assembly freewheels on a double roller ball bearing. With the car blower control switched on and the electromagnet simultaneously energized, the drive plate is attracted to the pulley and the compressor brought into action.

The condenser unit, made of finned tubes for rapid heat dissipation, is located in front of the engine radiator block to take maximum advantage of ram effect. For low road speed conditions, the engine fan has been increased in capacity and provided with a cowl. Housed under the parcel shelf in the luggage boot, the evaporator unit contains the twin air blowers and has a substantial filter.

The solenoid valve is fitted in a by-pass line between the pressure side of the compressor and the inlet of the evaporator. It controls the flow of gaseous refrigerant in the by-pass, and mixes it with liquid refrigerant from the expansion valve; the proportion of mixing regulates the degree of cooling imparted to the air in the car. The blower speed and temperature control switch is an integral unit, mounted on the dash; switching on energizes the magnetic clutch and the solenoid valve.

Power-assisted Steering

True hydraulic power assistance, as distinct from power operation, is achieved in the Rolls-Royce designed steering mechanism. It does not become operative until the driver applies one pound of turning effort to the rim of the steering wheel; thereafter equal assistance is provided, i.e., for each pound of driver effort there is one pound of assistance. This ratio obtains until the driver exerts about 8 lb, after which an increased rate of assistance is provided. Under parking conditions with little or no motion of the road wheels, it might be necessary to exert a maximum rim effort of 12 lb which would not tax the frailest of drivers. The important feature of the Rolls-Royce system is that the driver retains a desirable degree of feel, even when motoring on ice or mud.

With power assistance, higher-geared steering is used with a steering box ratio of 18.7 to 1—the unaided manual ratio is 20.6 to 1.

Hydraulic pressure is supplied by a Hoborn-Eaton pump driven by a single wedge belt from the fan pulley, and it has its own built-in filter. There is a separate oil tank for the system, and when the engine is running 1¼ gallons of oil circulate continuously. With no steering wheel movement, this flow is by-passed through the control valve system.

The control valve assembly is mounted in tandem with a Marles hour-glass worm-and-roller follower. When the steering wheel is turned, resistance is met by the worm which is displaced axially (the movement is .008 to .010in), permitting the oil to flow through the distributor to the actuating ram which is mounted on the chassis and connected to one of the centre track rod idling levers. The hydraulic ram is mounted on a single pivot point to accommodate the small arcuate movement of its geometry.

The valves in the control unit are deliberately set to operate at high loads, so that they are not sensitive to any dirt which may escape the pump filters. A piston attached to the worm shaft presses on a valve plate through a series of springs and plungers. These springs control the steering wheel rim load at which assistance becomes operative. Oil pressure on the valve from the piston thrust load tends to centre the valve plate and maintains feel in the steering.

Should the hydraulic assistance system fail, mechanical connection is still retained, with a slight increase in back lash at the steering wheel. This results from the 0.008 to 0.010in axial movement built into the worm to operate the hydraulic control valve. A safety valve in the circuit automatically cuts out the servo should the wheel be turned against a kerb or similar obstacle.

Continental Power

Power output has been increased on the Bentley Continental engine for this year. It is not the policy of Rolls-Royce to publish engine output figures, but the modifications have provided an increase of 13 per cent over previous power. The compression ratio has been increased to 8 to 1 (previously 7.25 to 1) and the inlet valve heads are larger. The twin S.U. diaphragm-type carburettors now have 2in diameter throats, instead of the 1¾in diameter on the earlier engines.

There are no changes in the Rolls-Royce Silver Cloud this year, but the six-cylinder engine of the Silver Wraith is now equipped with twin S.U. carburettors to improve engine breathing.

The bodies for the Silver Cloud and "S" Series Bentley are assembled at the Crewe factory of the Rolls-Royce Company from individual pressings and sub-assemblies supplied by the Pressed Steel Company of Oxford. In each range, specialist coachwork is available from H. J. Mulliner, Park Ward, James Young, Hooper, and Freestone and Webb.

ROLLS-ROYCE

New Long-wheelbase Silver Cloud Limousine Added to Range

September 25, 1957

ADDED length makes the Silver Cloud limousine a more imposing car even than the all-steel saloon model from which it derives.

ADDED to the Rolls-Royce range of cars for the 1958 season is a new Limousine, on a long-wheelbase version of the Silver Cloud chassis. No changes are being announced in other models of Rolls-Royce cars.

Basis of this new model, which costs £4,595 plus £2,298 17s. purchase tax, is the all-steel body shell of the Silver Cloud saloon. Conversion of this to suit a longer wheelbase is undertaken by Park Ward and Co., Ltd., the added length of body being inserted behind the central pillars to provide wider rear doors and increased legroom for rear seat passengers. The extended body is mounted on a chassis of wheelbase 4 inches longer than that used for other Silver Cloud models, 127 inches as compared with 123 inches for the Silver Cloud or 133 inches for the Silver Wraith chassis as supplied to coachbuilders. Extra length and limousine fittings add approximately 1½ cwt. to the weight of the Silver Cloud.

Primarily the new model is a limousine with a division between front and rear compartments, giving the owner privacy when he is being driven by a chauffeur. The dividing glass panel may be lowered out of sight, however, to allow use of the car as an owner-driven saloon at week-ends or during holidays if required. Furnished in the quietly luxurious manner expected by Rolls-Royce buyers, this new model has all the contemporary technical features of the make, including as normal equipment a fully automatic two-pedal transmission with over-riding manual control, servo brakes of high fade resistance, a two-position ride-control switch, and centralized chassis lubrication. Optionally, it may be fitted with power-assisted steering and with air conditioning equipment.

All Rolls-Royce models are powered by a six-cylinder engine of 4,887 c.c. size, which with a compression ratio of 8 : 1 and two S.U. carburetters allows 100 m.p.h. to be exceeded in quiet comfort. Based on an original American design but built in England by Rolls-Royce, Ltd., the automatic transmission combines a fluid coupling with a four-speed epicyclic gearbox, the driver being able to call for use of 3rd or 2nd gears if he desires or to leave ratio selection entirely to the automatic control system.

Conventional in layout but showing exceptionally thorough detail work in the interests of a long trouble-free and noise-free life, the chassis is supported at the front on independent coil springs, the rear suspension being semi-elliptic. Optional power-assisted steering provides ultra-light control right down to the lowest speeds, despite the lightest model weighing almost two tons, to match the easy retardation available from a braking system which incorporates a gearbox-driven servo motor.

LEGROOM in the limousine rear compartment benefits from the extra length, as does ease of access through rear doors which are 4 inches wider than on the saloon.

CAR LIFE CONSUMER ANALYSIS

"The Best Car in the World"

ROLLS-ROYCE

BY JIM WHIPPLE

I SUPPOSE that when anyone first drives a Rolls he feels as I did—that somehow the legendary vehicle will be unlike any other automobile ever made. I've always thought that a Rolls-Royce was sort of a Magic Carpet, Coronation Coach and Cleopatra's Barge all rolled into one

After a lifetime of respectful admiration from the curbside—having been kept at arm's length by the haughty stares of silver-haired chauffeurs and the bored indifference of mink-coated dowagers — it was almost an anticlimax to slip behind the wheel of a $12,700 1957 Silver Cloud saloon put at our disposal by J. S. Inskip, Rolls-Royce distributor for the eastern U.S.

But, after the initial shock of finding that behind the generations of glamour and swank and beneath the thousands of dollars of craftsmanship, the Rolls was still after all an automobile whose wheels must touch the streets as any other's and whose ignition is operated by a key rather than a magic wand, I began to enjoy my session with the big "Cloud".

Like any auto enthusiast with fond memories of the mighty Packards, Pierce Arrows, Lincolns and Duesenbergs of the early 1930's, I approached the Rolls with a small patriotic chip on my shoulder and a "show me" attitude towards Rolls' slogan, "The Best Car In The World."

After a few hours behind the wheel of the Silver Cloud I realized that Rolls has greater justification for that slogan than any car currently produced, although the Mercedes Benz comes close to matching it.

The first thing that impressed me, as it would most people, is the quality of workmanship found throughout the car from fog lights to tailpipe.

Rolls perfection begins in the body design. The car is sleek without any sacrifice of function, has an atmosphere of traditional elegance without a trace of stodginess.

The only lines that seem out of harmony with the graceful curve of the fenders and sweeping roof and rear deck lines are those of the angular vertical radiator grille — which have remained unchanged for fifty years!

That slablike radiator cover has come to symbolize a half-century of almost uncompromising devotion to perfection, and anyone lucky enough to possess a Rolls is quite willing to overlook the fact that it makes no more concession to aerodynamics than a GI footlocker.

Driving the Silver Cloud was almost pure pleasure. The fabled quiet operation did not disappoint me and I found that I was relying on the generator warning light to tell me if the engine was running. So well-designed and well-balanced is the Rolls overhead valve six that I defy anyone to guess blindfolded that it is anything less than an eight.

The transmission is an automatic of four-forward-speeds-and-reverse con-

Price, "Silver Cloud" model $12,700 (port of entry)

ROLLS-ROYCE

is the car

for you

if... You can afford to invest $12,700 in personal transportation of unequalled reliability.

if... You want the world's number one prestige car — one that will not "go out of style" in a year or two.

if... You really appreciate quality and attention to small but important details, and value craftsmanship more than gadgets and gimmicks.

if... You are looking for the last word in driver-ease and passenger-comfort and are willing to do without flashy performance.

ROLLS-ROYCE "SILVER CLOUD" SPECIFICATIONS

ENGINE	6-cylinder
Bore and stroke	3.75 in. x 4.50 in.
Displacement	298.2 cu. in.
Compression ratio	6.6:1
DIMENSIONS	
Wheelbase	123 in.
Overall length	212 in.
Overall width	74.5 in.
Overall height	64.5 in.
TRANSMISSIONS	Rolls-Royce automatic

From any angle, Rolls is the ultimate in conservative design. Twin traditions of style and craftsmanship have endured for over 50 years.

Jim Whipple admires mechanical splendors of Rolls "Silver Cloud." Under old-fashioned bonnet is superbly smooth 6-cylinder engine.

trolled by a steering column lever and quadrant.

If I ever want to walk off with anything more than I did that Rolls front seat they will have to lock me up. It is the most perfect device ever created for holding the human body in a seated position at the controls of a passenger vehicle. (We will except the sports car bucket seat in the case of competition driving.)

Upholstery on the Silver Cloud is fawn leather as soft as your favorite pair of gloves. What those English coach builders stuff seat cushions with defies my imagination. It may be foam rubber, or even whipped cream. The only thing that could possibly be softer is a cherub's backside.

And yet despite the surface softness the designers seem to know more about the business of supporting the human anatomy than any designers on this side of the Atlantic, because underneath the soft padding of the Rolls cushions there is firm support where you need it .

No matter what I did in the course of normal driving in traffic or on the open road I couldn't make the transmission lurch or misbehave. Even on full-throttle acceleration the automatic gear change was a gentle surge that never proved annoying.

The steering wheel isn't adjustable but we found its position and angle just right. There's an old-fashioned horn button instead of a ring, but somehow we found it worked out fine. This could be the beginning of a great new trend — rim and spokes for steering, center hub for horn blowing.

Careful design in making extremely thin but strong windshield corner posts, door pillars and window frames gave me perfect vision from the driver's seat, and eliminated the need for distorted and expensive wraparound glass, with its frame designed to bruise the knees of long-legged occupants.

In spite of their massive elegance, I found that the Silver Cloud's grille and hood did not block my vision.

The only item in the entire control and instrument setup that I could find fault with was the steeply slanted and unsupported accelerator pedal, which became uncomfortable on long stretches. This defect seemed all the more regrettable because everything else in the driving compartment had been so carefully thought out and was so completely "right".

Rolls' power steering is something I'll never tire of praising. It's as efficient and unobtrusive as a good English butler, and must have brought tears of joy to battalions of elderly chauffeurs as their owners switched to late model Rolls's.

In any type of driving the power assist never intereferred with my feel of the road. Rolls steering is best described as "natural"; it feels merely as if you were steering a very much lighter car with manual steering —

Rolls-Royce cars have had power-assisted brakes for almost as long as we've had four-wheel brakes in this country.

The Rolls brakes are easy to apply smoothly, requiring slight effort, and are entirely predictable in action.

As far as the ride is concerned — no automobile produced today is perfect and free from vibration or bounce, and Rolls is no exception. The Silver Cloud is a heavy car and — in order to provide comfortable exit and entry and plenty of head room — a high one.

As a result, the suspension engineers were forced to make somewhat of a compromise. In order to provide a ride that was free from swaying, bouncing or pitching, they had to design their springing on the firm side.

True, the rear shock absorbers have a normal and firm adjustment, controlled by a switch on the instrument panel — spring action is generally stiff in action on small bumps and surface irregularities.

The result is not discomfort, because the soft tires take the sting out of cobblestones and washboard roads, and of the wheels and heavy unsprung rear axle is transmitted to the body.

ROLLS-ROYCE CHECK LIST
FIVE CHECKS MEAN TOP RATING IN ITS PRICE CLASS

Ample trunk shows meticulous detail-work. Compartment is lined like jewel box.

Rolls interior blends luxuries like tray with severity of "no-nonsense" dash.

PERFORMANCE	Entirely adequate for a large touring car but a good deal less powerful than current U.S. medium and high-priced cars selling at one-third Rolls price.	✓✓ ✓✓
STYLING	A very tasteful combination of traditionally elegant lines and an up to date streamlining that adds up to a car that will look "in style" years after the fin fad passes.	✓✓✓ ✓✓
RIDING COMFORT	Rolls-Royce is very smooth and steady going on all types of roads. Only compromise in the direction of firm springing which permits wheel bounce on bad surfaces keeps ride from ideal.	✓✓ ✓✓
INTERIOR DESIGN	In all respects, vision, seating comfort, driving position, ease of entrance and exit, ventilation, storage space and placement of controls for driver and all passengers Rolls is far beyond other cars. Silver Cloud is more suited to five than six passenger comfort.	✓✓✓ ✓✓
ROADABILITY	Rolls has all the roadability required for fast comfortable touring, but the car's higher than U.S. center of gravity makes it more susceptible to lean on curves than low slung late model U.S. cars.	✓✓ ✓✓
EASE OF CONTROL	An extremely smooth flexible engine coupled to one of the world's most precisely engineered and well behaved automatic transmissions, perfect power assisted steering and power brakes put the Rolls way out front.	✓✓✓ ✓✓
ECONOMY	Buyers of Rolls Royce will not be looking for budget car gasoline economy from Rolls but conservatively driven the Silver Cloud will deliver 14-16 mpg.	✓✓✓ ✓✓
SERVICEABILITY	Surprisingly for a six cylinder car Rolls engine compartment is crowded and rather inaccessible. However the foot operated chassis lubrication system makes for a long run between service stops.	✓✓✓ ✓✓
WORKMANSHIP	Standards of craftsmanship are uniformly high in every detail from upholstery to finish of engine parts. Nothing like its quality has been seen in a production car in the U.S. since World War II.	✓✓✓ ✓✓
VALUE PER DOLLAR	At $12,700 the Silver Cloud is not for the budget conscious. However from a percentage standpoint depreciation is lower than that of U.S. cars. For long term ownership (10 years) a Rolls will cost no more than a high-priced U.S. car traded every two years.	✓✓✓ ✓✓

ROLLS-ROYCE OVERALL RATING...4.7 CHECKS

On average U.S. highways with all types of surface, the Rolls is one of the absolutely top comfort cars in the world. On long trips I think that its silence and the comfort of upholstery give it a positive edge.

What does happen is that on small, sharp jolts like heat-rippled macadam or badly laid Belgian blocks, some of the vibration from the rapid bouncing what they miss is absorbed by the remarkable seat cushioning. Meanwhile the firm springing has done a really superb job of controlling sway and bounce. The car doesn't roll or wallow on winding high-crowned roads.

SUMMING UP: The Rolls-Royce Silver Cloud is probably the most carefully made and most comfortable car in the world. A good many other cars have more powerful performance, but none are as restful and quiet. A few cars have riding qualities that are more nearly perfect under all road conditions, while some others, particularly certain late-model American cars, have better roadability under severe conditions. But, no car offers so much in the way of luxurious comfort and driving ease, combined with ultra high quality of manufacture and unequalled mechanical reliability. For the needs of the average motorist the Rolls is "The Best Car In The World."

NOVEMBER, 1957

December 18, 1957

Your Motoring Made Easy

No. 3. Rolls-Royce Automatic Gearbox

Standard equipment on: All Rolls-Royce and Bentley models.

Similar transmission available on: British Motor Corporation Princess, Armstrong Siddeley Sapphire, Cadillac, Oldsmobile, Pontiac, Nash, Hudson, Rambler.

A WIDELY believed fallacy is that fully automatic transmission leaves the driver with no control over his destiny apart from the accelerator pedal for "go" and the brake for "stop." It is quite untrue that any currently available transmission reduces driving compulsorily to such simple terms. Although the degree of control provided varies a great deal between the different systems, a skilful driver can apply the best of his art (or alternatively none at all) in operating the automatic gearbox developed by Rolls-Royce* from the Hydra-Matic transmission of General Motors.

The remarks which follow apply directly to the standard Rolls-Royce Silver Cloud and Bentley "S" series saloons, but they are true except in minor details of the other British models listed above and, with further reservations, of the American cars when fitted with Dual-Range Hydra-Matic.

By way of technical description, it is sufficient to say that this transmission comprises a fluid flywheel and a four-speed epicyclic gearbox, with a mechanical "brain" which can (but need not) decide automatically when to make a gear change. The brain works by balancing the requirements of speed and throttle opening, so that the harder the accelerator is pressed, the higher will be the road speed at which each upward gear change is made. Built-in refinements allow for a rapid kick-down to a lower gear, for holding second or third gear above the normal changing points, and for parking with the transmission locked.

The driver's means of control are the accelerator pedal and a steering column lever moving in a quadrant, with safeguards to prevent its being accidentally moved past any desired position. The positions marked, from top to bottom, are N-4-3-2-R; corresponding to Neutral—Fully automatic in four gears—Automatic up to third gear only—Fixed second gear—Reverse. Their functions follow in greater detail.

Neutral.—This is the only position at which the self-starter will operate, and the lever can be moved into or out of it only by pressing a button at the end. It need be engaged only for very long halts (traffic jams, for example).

Four.—It is perfectly practical, and no doubt for many owners quite normal, to make a long journey without ever changing out of the four-gear, automatic drive position. The full performance of the car is available entirely under the control of the accelerator, and provided traffic is not too dense this

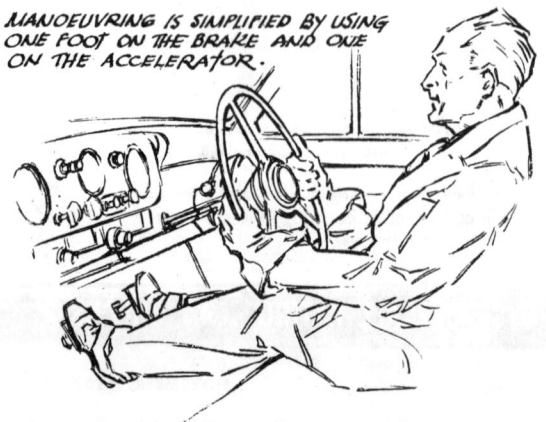

The selector lever on the Rolls-Royce, which functions as described in the text. To select neutral or reverse the button in the end of the lever must be pressed.

drive range can be used for fast, as well as leisurely, cross-country trips. When the lever is moved to Four with the car at rest there should be no perceptible tendency to creep forward unless the engine is cold and the automatic choke is causing it to idle slightly faster than usual. In this case, or when the throttle is just touched lightly, there is a slow rate of creep which can be easily checked with the left foot on the brake, or by the efficient handbrake, and which forms the simplest method of manoeuvring in a confined space. The creep is not sufficient to stop the car running backwards, if halted on a slight up-grade, but it is on the other hand so readily controlled by the throttle that it can easily be used for restarting under these conditions.

Moving Off

Bottom gear is automatically selected when the car is stationary in this range. Pressure on the accelerator will cause the fluid flywheel to take up gradually, and on a level road with the gentlest possible treatment changes into second, third and top gears occur at about 6 m.p.h., 11 m.p.h. and 20 m.p.h. respectively. The speeds at which these changes are made depend directly upon the position of the accelerator pedal (irrespective of whether the car is going up or downhill, except in so far as this influences the driver's use of the throttle), and at the other extreme the speeds can be raised to about 18 m.p.h., 31 m.p.h. and 65 m.p.h.

The automatic control will change down, of course, as well as up, but the mechanism is adjusted so that only at quite low speed or on steep hills will it change out of top gear without the use of the "kick-down." If the car is brought to rest with the throttle closed, the change from top to third is just perceptible at about 14 m.p.h., and the other two at almost a walking pace. However, in addition to the automatic brain, this range provides a quick change under the direct control of the driver, through a kick-down switch which operates when the accelerator is given a "second pressure" beyond the normal full-throttle position.

The point of arranging control of the gearbox thus is that in ordinary main-road motoring a succession of gear changes, each with its almost inevitable slight jerk, is avoided. Extra

Your Motoring Made Easy

Second gear can be manually selected, to assist the brakes in exceptional conditions.

acceleration is always on tap if it is needed suddenly, when a momentary interruption of the smooth progress will be accepted. The kick-down will select third gear at any speed up to the 65 m.p.h. possible in that ratio, or with third engaged it will drop to second if the speed is below 18 m.p.h. At very low speed a kick-down into bottom gear is possible.

Three.—Every driver of experience is familiar with the situation when he pulls out to overtake a slower car, but then has to pause for a few seconds before the road is completely clear and he can accelerate once more. The "Three" range is intended for this and similar conditions. If the lever were left at Four and the kick-down used for overtaking, the pause would allow the gearbox to revert to top. By selecting Three, third gear is engaged and held right up to the 65 m.p.h. limit, when the transmission changes into top to spare the engine. Nevertheless, in this range the gearbox is fully automatic over the lower three gears, so that it may be used for smooth town driving. If the lever is moved into this position with the car travelling fast, a safety device prevents third being actually engaged above the limiting speed.

Two.—This is not really a driving *range* at all, for although bottom gear can be engaged by using the kick-down the car will otherwise remain solely in second gear. The objects are two: to provide engine braking on extremely steep hills, and to make possible smooth and very slow progress through the thickest traffic, where frequent stops and starts would entail constant changes into and out of bottom gear in either of the other ranges. To reach Two, the selector lever must be moved forwards into a different plane from that of Three and Four, avoiding any possibility of its accidental engagement. In any case, second gear will only engage when the speed falls to about 30 m.p.h.

Reverse.—In the same plane as "Two," but guarded like neutral by the spring-loaded button on the lever, reverse can be selected while the car is still moving slowly forward, when the fluid coupling will absorb the shock and stop the car before starting it in the other direction. A special usefulness of this is for rocking the car out of soft ground, by keeping the throttle partly open and moving the lever alternately between Reverse and Two. An additional function of the Reverse position is to provide a transmission brake for parking. With the engine switched off, the brakes of the epicyclic gears operate in such a way that the transmission is locked—which it will be seen is even more positive than the normal practice of leaving a car in gear.

Special Occasions

So much for driving. Transmissions of Hydra-Matic type have been developed more thoroughly than perhaps any other, and consequently require less learning than more primitive types in which the driver's skill may still compensate for imperfections. It is, nevertheless, possible to learn a few tricks and one or two points of caution. In cramped quarters, for example, it may be necessary to select reverse immediately after starting the engine, with no room for manoeuvring ahead. There is no difficulty about this, but because the lever will have to pass through the forward range positions on the way, automatically selecting bottom gear, it is most essential to keep the handbrake applied until the transmission is properly in reverse. Again, the slight creep produced by bottom gear if the car is halted in traffic on a slightly downhill road must be checked with care, for the Rolls-Royce servo brakes allow a few inches of movement along the road before they will take up.

At the other extreme, the use of the third gear range is encouraged for fast driving, and with it an indetectable change from top to third can be made by adjusting the throttle accordingly. The change can be pre-selected on the approach to a bend, engaging only when the speed drops to 65 m.p.h. For economy the Four range should always be selected, and pressure on the accelerator kept light to ensure that upward changes occur as soon as possible.

Finally, there is the question of towing in a major emergency, and tow- or push-starting in a minor one. The first is extremely unlikely except following an accident, but provided the gearbox itself has suffered no damage it is possible, with one precaution. Because there is an inadequate supply of lubricating oil when the engine is not running, it is necessary to slacken the adjusting screw of the rear brake band in the gearbox, and then to tow the car with the lever at Neutral, at a speed between 15 and 25 m.p.h. By the same token, coasting is definitely forbidden practice with this transmission.

For the short distance necessary to start the engine by towing, pushing or coasting downhill the lubrication is, however, sufficient. (The two last-named methods are preferred, because of the sudden acceleration which may follow the starting of an engine with automatic choke.) The procedure is to slip the lever from Neutral to Four at about 15 m.p.h., bearing in mind that until the engine starts there is no power assistance for the steering on cars so fitted.

Instructions on matters of this sort, affecting the welfare of the transmission, may vary somewhat from make to make, and in any case the owner is very well advised to study any special notes supplied with his car. For reference, the adjoining table shows the equivalents used on other types for the N-4-3-2-R of Rolls-Royce. It should be noted that they may not always occur in the same order.

Rolls-Royce Bentley Princess	N	4	3	2	R
Armstrong Siddeley	Neutral	Normal	Fast	Fixed 2nd	Reverse
Cadillac* Pontiac*	N	Dr 1	Dr 2	Lo	R
Oldsmobile*	N	Dr	S	Lo	R
Hudson* Nash* Rambler*	N	D 4	D 3	L	R

* On these American models a separate position marked P engages a parking pawl in the transmission.

By selecting "Two" and "Reverse" in quick succession the car can be "rocked" to release it from mud or snow.

STILL

OUT OF TRADITION?—Current Rolls-Royce Silver Cloud.

It isn't just a car, it's a legend. As the advertisements say, it's "the best car in the world."

Well, it was — but motor-car engineers are beginning to say that the Rolls-Royce today would make the man who designed it, Sir Henry Royce, turn in his grave.

Today the very cheapest Rolls-Royce, the standard Silver Cloud, costs £5693stg. The dearest, the Park Ward or Hooper limousine, costs £8708.

People who pay this sort of money have a right to expect the best in the world.

What do they get?

Off-the-Hook Items

To begin with, the dynamo is no longer gear-driven from the engine. It is a standard job, from the same factory that supplies B.M.C. (Morris and Austin), Standard, and many other car firms.

It is belt-driven, and if the belt breaks "The Best Car in the World" is stranded.

"But surely they fit enormous batteries to a Rolls?" you say.

They used to do so. Exide had a huge battery made specially to Rolls-Royce specification.

Today these enormous cars have a 57-amp-hour battery at the back, no bigger than that in an English Ford!

When a British motoring magazine, **The Motor,** tested a Rolls-Royce Silver Cloud, the dynamo did happen to fail, and three hours of night driving meant no start in the morning.

When Charles Gretton's article first appeared in an English Sunday paper, "The People," it drew the accompanying reply from Rolls-Royce Ltd. But did Rolls really answer Gretton?

"But you could always use the handle?" you ask.

No. They didn't supply a starting handle as standard equipment.

Phony Mascot, Hubs

Back in 1911 Royce wanted a suitable mascot. Charles Sykes, R.A., designed a winged figure, the famous "Spirit of Ecstasy."

It wasn't just a radiator mascot. It crowned a cap which was a filler and also a steam-valve.

The "flying floosie" — as Rolls chauffeurs call it — still adorns Rolls radiators today. But it's a dummy.

It is just a solid casting, useless except for decoration. To fill the radiator you have to open the bonnet and turn an ordinary pressed-steel cap, just as you do on a mass-production car.

Sir Henry invented a wonderful wheel hub. No Rolls was ever stranded with wheel-nut trouble. You just used a special spanner on this self-locking hub nut.

They're so proud of that wheel hub at Rolls-Royce today that they have retained it. But it's only a dummy.

Undo this flimsy pressed-steel copy of old Sir Henry's hub. Off comes the wheel disc—and there are five ordinary wheel nuts inside!

Steering, Suspension

Rolls-Royce steering used to be proverbial. A worm-and-nut steering-box was used, hand-assembled, the nut lined with white metal and ground in so that all steering parts fitted with silky smoothness.

Today the car has the ordinary Marles cam-and-roller steering, as on dozens of other makes.

Every Royce-designed car from 1929 (except the 20-h.p. cars) had hydraulic "ride control."

A little lever on the steering column could be flicked from hard to soft, overriding an automatic governor controlling oil pressure on each of the shock-absorbers. As the car went faster, the gear pump automatically made the springs harder.

In "The Best Car in the World" today there is still this lever on the steering column. But it is only an electric switch, either on or off, which is supposed to put a little extra pressure on the rear dampers only.

It is not made by Rolls-Royce. An independent test by **The Autocar** reported: "Disappointing . . . virtually no difference in the ride could be felt when this was operated."

Gearbox, Wheels

Rolls-Royce is fitted with an automatic gearbox as standard.

The pity is that the Rolls-Royce today uses an American-style automatic

MODERN MOTOR — April 1958

THE BEST?

Once R-R meant perfection, because every part was tailor-made for it. Now many of them come "off the hook"—and experts are shaking their heads, claims Charles Gretton

box, made under licence from General Motors.

Rolls engineers have made some alterations, but there are mixed views as to whether these are improvements.

Sir Henry believed in large wheels for his cars, so that tyre rims ran slowly, kept cool, and treads lasted for upwards of 40,000 miles.

Tyre-makers produced special "Rolls-Royce quality" tyres in the special 19-inch size.

Today they fit button-size 15-inch wheels, the same size as on many Morris and Austin cars.

Other Changes

Old Rolls-Royces were built to last. It was nothing to go more than a quarter-of-a-million miles between rebores. They never used any material like rubber, which could perish.

Yet today's Silver Cloud has a rubber diaphragm holding such a vital component as the carburettor jet block, subject to heat and to fuel corrosion.

Rubber bushes are used throughout the front suspension, and there is rubber interlining between the four top leaves of the rear springs.

All this makes a quiet and cosy car. But will it last?

Sooner or later every car comes to a halt with a puncture. The Rolls Wraith, the last of the series made before the war, had inbuilt hydraulic jacks. You merely moved a small lever near the front seat and the car rose up.

Today you have to lift two tons of motor-car with an ordinary mechanical jack.

Britain needs the R-R reputation. But if we have to yield pride of place in car quality, couldn't we at least get the benefit of lower prices?

If not, I have a suggestion to make to Rolls-Royce Ltd. In future, call your product "the dearest car in the world."

BETTER THAN EVER, SAY ROLLS

AND this is what Rolls-Royce Ltd. had to say on the points raised by Charles Gretton:

THE BELT DRIVE.—In the days of gear drive, that was the only satisfactory method. Today, with the improvement in the belt system and materials, and better engine suspension, belt drive is considered best.

THE BATTERY.—The electrical equipment is designed to use the 57-amp-hour battery, leaving a margin of safety considered adequate by Rolls-Royce engineers. The car that failed on The Motor's test was found to have a faulty dynamo.

THE MASCOT. — Rolls-Royce owners complained that the mascots were often stolen. Owners almost always removed the mascot when leaving their cars.

Today the mascot is fixed and an efficient radiator cap is incorporated under the bonnet.

THE HUB NUT.—The self-locking hub nut was designed for use with wire wheels. With today's independent suspension tremendous shearing forces would be exerted on that point

Continued on page 48

R-R MASCOT — once it topped a functional radiator cap, but now serves purely as a dummy ornament.

IN TRADITION: Two of the older "hand-made" Rolls, the 1951 Silver Wraith and famous 1911 Alpine Eagle.

The cavalcade of 22 Rolls-Royces and Bentleys, led by the 1905 two-cylinder 10 h.p. Rolls, on the new Fort Worth to Dallas turnpike. Special permission had to be obtained to travel below the normal minimum speed of 40 m.p.h.

PIPELINES to PROSPERITY

ROLLS-ROYCE

SALES of Rolls-Royce motorcars have in the past two years made tremendous strides in the U.S.A., but the success has been almost entirely limited to the Eastern States centred round New York and the West Coast in the Los Angeles and San Francisco areas. The large mid and south-western territories have barely been explored—particularly the territory that lies west of the Mississippi and east of the Rocky Mountains. The leading centre of this area is Texas—vast in size, with tremendous wealth from oil and cattle, and a standard of living which is almost as fabulous as its remarkable people claim it to be.

Texas—the Lone Star State—is by repute the most rugged and independent member of the Union. Geographically and historically it has no affinity to Europe and the Old World, so it provides a separate challenge to the manufacturer who wants to import and sell his goods into the south-western area.

Rolls-Royce took up this challenge last year, when a new distributor was appointed for the states of Texas, Oklahoma, New Mexico, Arkansas and Louisiana. The problem was to establish the distributor in such a way that he would be given every chance of breaking into this difficult market.

Research had indicated that even in the Eastern States the Americans do not really appreciate what Rolls-Royce stands for today. Too many people, if they knew the name at all, merely associated it with *old* motorcars and there was little indication that the considerable aero-engine achievements of the company were known at all.

It was decided therefore that a sales promotion campaign would be organized in Texas to do two things— to bring American thinking about the name "Rolls-Royce" up to date and to promote sales of the Silver Cloud.

The suitability of the Rolls-Royce Silver Cloud to American conditions had already been established and confirmed by the tremendous upsurge of sales in the U.S.A. since it replaced the Silver Dawn in 1955. It is the appearance of the car—its restrained, classically modern lines—more than anything else that seems to appeal to the Americans, many of whom are tending to react from a surfeit of chrome, glass and finnery. As soon as it was realized that this modern Rolls also had the performance

Forming part of the special display in the lobby of the Statler Hilton Hotel, Dallas, the Bentley Park Ward Continental saloon draws attention from guests.

in TEXAS

By

Dr. F. LLEWELLYN SMITH

Managing Director of the Motor Car Division, Rolls-Royce Ltd.

and all necessary addenda such as power steering, refrigeration, etc., then curiosity turned into genuine interest.

The problem of getting an entrée into the top-level Texas social and business circles was largely solved when it was agreed to join forces with Neiman-Marcus, the famous fashion store of Dallas, and to promote jointly their Fall Fashion Exposition of 1957. This proved to be a particularly happy liaison, with the fortuitous coincidence that both Rolls-Royce and Neiman-Marcus were celebrating their fiftieth anniversary of trading in the U.S.A.—the Hon. C. S. Rolls himself having sold the first Rolls-Royce in America in 1907. It was found that Neiman-Marcus set themselves the same high standards in their own sphere that Rolls-Royce set in engineering practice, and right from the start sales and publicity executives worked in complete harmony and understanding in developing the plans for the joint promotion.

First of all, the distributor—Overseas Motors Corporation of Fort Worth, Texas, appointed more than twenty dealers throughout the five-state territory. Then Rolls-Royce engineers were sent out to Texas to hold courses of instruction on servicing for the selected mechanics from each dealer.

In July, a shipment of 23 Rolls-Royce and Bentley motorcars valued at $300,000 went aboard the S.S. *James Lykes* and sailed out of Liverpool bound for Houston, on the Gulf of Mexico. When these cars were off-loaded in Texas, banner headlines announced "Rolls-Royce arrives —Texas style"—as never before had there been such a

ROLLS-ROYCE

in TEXAS—*Contd.*

Centre of attraction in the Neiman-Marcus store was the Rolls-Royce Silver Cloud, fitted with power steering, ride control, automatic transmission and power brakes.

large single shipment in the history of the company.

The opportunity was taken to stir up more interest in Rolls-Royce by means of the Press, radio and television. The need for Rolls-Royce to use such mass media as radio and television was self-evident when it was realized how much ground had to be covered. The great interest shown by the American journalists led to a very considerable amount of editorial publicity—all of which it was hoped would soften up the rugged Texans and persuade them to buy a $12,900 car.

By the end of the Fashion Exposition Week there must have been few Americans in the South West who had not gathered the fact that Rolls-Royce were very much to the fore, that they produced fine aero engines and motorcars, and that they were one of the world's great engineering concerns. But because the Texan has no inbred respect for the imported product, it was necessary to convince him of Rolls-Royce quality and reliability. And even having convinced him, he did not lightly part with his dollars.

There were various highlights in the sales promotion campaign and perhaps the one which caught the imagination most was the cavalcade of 22 Rolls-Royce and Bentley cars led by the 1905 2-cylinder 10 h.p. Rolls. This long line of motorcars wound its way through the city of Fort Worth out on to the brand new turnpike road joining Fort Worth to Dallas. The two cities are only some 30 miles apart, but the bitter rivalry that exists between the two has to be experienced to be believed. Each city is spending millions of dollars on its own airport simply because they will not get together and share one between them.

But the "Cavalcade" went undaunted through a dust storm, followed by a thunderstorm, led by the 2-cylinder Rolls at 23 m.p.h. (Special police permission had to be granted to travel at this speed on the turnpike where the normal minimum speed allowed is 40 m.p.h.).

The occasion of the Fashion Exposition was a fabulous experience, and the hard-bitten Texans were able to have a critical look at the Silver Cloud and Bentley Continental Park Ward Convertible on the ground floor department of the Neiman-Marcus store. Their comments were forthright and appreciative, and they were quite unawed by the name or the price.

There was a sectioned Dart propeller-turbine aero engine in the Men's Department on the second floor. This engine in the Vickers Viscount has brought a new level of passenger comfort to U.S.A. air travellers. In the nearby Statler Hilton Hotel another Silver Cloud and a Bentley saloon were driven into the lobby, where they stood on display for all the world to see.

V.I.P's were transported around Dallas in other Rolls-Royces from one function to another—in the shimmering heat the porcelain white cars with their refrigeration units on full fan provided cool comfort in the fierce heat of the Texas summer.

All the activities were reported in newspapers well beyond the State border, so that few people were left in doubt that Rolls-Royce had arrived and intended to stay. From the number of cars sold during the Exposition Week the distributor is now able to build up sales in the territory, taking advantage of the host of excellent contacts which were made with the splendid co-operation of Neiman-Marcus. There is much work still to be done, and considerable effort is being made this year to develop the market.

The Autocar,
16 May 1958

The lines of the Silver Cloud are well balanced; the bumpers wrap round at the corners. The radio aerial can be lowered from inside. There are sills protected by the doors. These visitors to the Brussels International Exhibition clearly are interested in the car

Autocar ROAD TESTS 1683

Rolls-Royce Silver Cloud

THIS is the first occasion on which a full road test has been carried out on the latest model of the Silver Cloud, although on 7 October, 1955, we published a test report on the Series S Bentley, which differed from its contemporary Silver Cloud only in frontal appearance. Since the Bentley test neither model has changed markedly, but developments and improvements, of which there are many minor ones, include a higher compression ratio (8.0 instead of 6.6 to 1) and to the actual car now under review is fitted Rolls-Royce power-assisted steering. The differences resulting from the higher power output and assisted steering proved to be considerable. Brake horsepower is not quoted by Rolls-Royce, but the improvement in the already high performance suggests that it has been stepped up appreciably.

Belgium and Holland were used as the testing grounds for high-speed driving, with French roads to recapture appreciation of the model's "grand touring" character. Although sheer speed is not a major consideration in a car of this type, it is no less pleasing to find that the peak is now 106 m.p.h. (101 was the best achieved with the Series S).

Like the other car, the Silver Cloud reached exactly the same maximum in each direction, and its consistency was further confirmed while making the series of each-way runs necessary for the compilation of the performance data as a whole. This consistency is an impressive feature of the car.

In the more important field of acceleration the improvement makes 90 m.p.h. possible from a standing start in 34.1 seconds (39.4 previously) and the standing quarter-mile takes 18.8 instead of 19.7sec. The power continues to be delivered smoothly from the moment of getting away from rest. At traffic speeds of 16 to 20 m.p.h., when automatic high ratio is still retained, there is sometimes a slight engine harshness at pick-up; again at maximum engine speeds in all ratios the presence of the engine can be felt. However, in the wide cruising range it is particularly sweet.

The automatic transmission, standard on these cars, has maxima on the three lower gears of 22, 34 and 63 m.p.h. respectively. No cruising speed can be quoted with any precision, for there is no fuss even at the maximum. Certainly, 90 m.p.h. can be maintained on suitable roads from one tankful to the next—barely 200 miles, allowing a reason-

The back window is sufficiently large to give adequate rearward visibility, while appropriate privacy is achieved in the rear compartment for occasions on which the car is chauffeur-driven

THE AUTOCAR, 16 MAY 1958

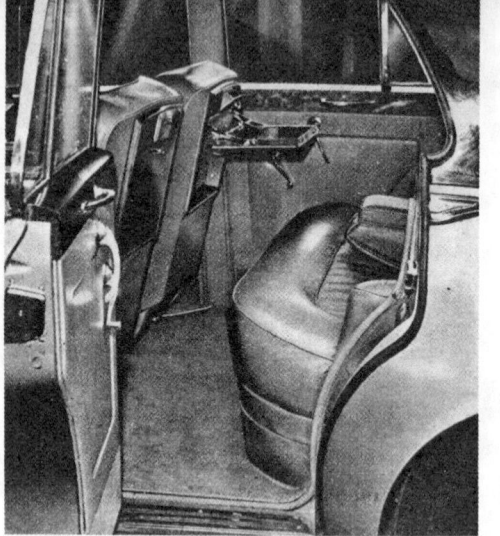

Rolls-Royce

There are separate, central armrests on the divided backrests of the front seats, and a small lever is provided for each occupant for adjustment of backrest rake. The folding tables in the rear compartment have ashtrays above them, and the sides of the rear seat sweep round to support occupants' shoulders and then to form armrests. Facing picture: The traditional Rolls-Royce radiator and mascot are continued. Fog lamps play a dual role as winking indicators

able margin. The tank capacity could, with advantage, be increased. On two occasions this speed was maintained almost without pause between Brussels and Ostend, the car's occupants enjoying what must surely be a unique degree of quietness. As long as the windows are shut, conversation may be carried on in drawing-room tones or the radio heard clearly at low volume.

There is a small price to pay for the extra power—the tickover on this car did not have quite the silkiness of old. This subtle change must be related carefully to the car; it amounted to no more than that the driver can detect the engine tick-over by feel rather than sound—with a little less difficulty than before.

The transmission control has five positions and, in concert with throttle pedal kick-down, a variety of techniques may be used for gear selection. At speeds of less than 60 m.p.h. it is possible to change to third by pressing the pedal fully down, or into second if the speed is very low. As the lower ratio then takes up coincident with the application of full engine power, the result is a little abrupt. This changing technique is normally reserved for emergency use.

With the quadrant set at position 4, all changes are automatic, and should a change-down be wanted while top is in engagement then instead of kicking down one may simply move the lever one notch to position 3. The quadrant lever never requires to be used in this way, but it is useful if, say, the driver wants to prepare for quick overtaking; while held up and, therefore, using a light throttle opening, he can make the car change down by moving the lever. With 3 selected the box remains automatic on all gears but now engages third gear at once, as and when the speed falls below 60 m.p.h. Similarly, in position 2, second is held, but this is an emergency position for very steep, long descents.

The other positions on the quadrant, neutral and reverse, are safeguarded by a button on the tip of the lever; the button must be pressed to get out of neutral or into reverse. Another customary safety factor is that the engine cannot be started unless neutral is selected. The particular car tested did have a tendency to creep sometimes from a standstill with any one of the driving positions selected. There was no difficulty when starting from cold; the automatic choke, interconnected to the throttle, set the engine speed slow enough even first thing in the morning, so that manœuvring out of a garage, for example, presented no problem.

The power-operated steering is a joy; just enough effort is required of the driver for some feel to be retained, but all the work is taken out of parking manœuvres. The wheel gives little kick-back from road surfaces whatever their condition. The difference which the assistance makes to town driving is well worth having, and no snags were encountered to offset the advantages. The steering is more sensitive than previously, and one cannot detect the operation of the servo at any time. In the event of failure of the power mechanism, normal steering remains. Power-assisted steering should not be judged on less than a day's driving. The tyres' increased propensity to squeal probably results from cornering faster, unconsciously, with the aid of the power-steering, and the low pressure recommended for the front tyres further to reduce road-shock transmission to the steering wheel.

With a weight of some two tons, and with the high speed capability, something very special is required in the matter of brakes and suspension if safety is to be achieved. Of Rolls-Royce mechanical servo drum brakes there is little which has not already been said; little which is not known, if only by repute. For a maximum retardation on dry concrete of 94 per cent a pedal pressure of only 80lb is required. This pressure is not much in terms of driver effort, while 94 per cent is the sort of braking power that will fling any oddments on the seats to the floor. More than 60 per cent efficiency—the equivalent of the hardest braking power ever likely to be used in normal driving—requires only 50lb pressure. The newcomer to the model may startle himself by the result of using too high a pressure on the pedal, but it requires no skill, only an hour or two at the controls, to master the art of the soft touch, to enjoy the benefits of smoothly progressive four-square, fade-free stopping power

The instruments, surrounded by polished walnut, are neatly lettered in white on black. Below the radio controls is a pull-out table. The ride control switch is on the steering column. The luggage locker is deceptively long, and the spare wheel has a separate compartment. Illumination is now provided automatically when the lid is open

Silver Cloud . . .

which Rolls-Royce provide in return for so little effort. The hand brake is under the facia, reasonably easy to reach (if not entirely to hand), light, and fully effective in action.

The suspension of the Silver Cloud proved a little softer than that of the Bentley. The comfort of the ride remains of an extremely high standard, for combined with the soft, almost spongy feel experienced at slow and medium speeds is sufficient stability to enable remarkably high averages to be maintained even on winding roads of indifferent quality.

If a corner is taken fast, or in emergency it is necessary to make a quick change in direction at high speed, the initial feeling is that the suspension may prove too soft and imprecise—but this is not, in fact, confirmed by the car's behaviour; it will rise to the sudden difficulty well. There is more roll than was found with the Series S, though its extent is far short of giving any suggestion of wallow. There is a Normal and Hard switch on the steering column which varies the setting of the rear dampers to a slight degree. In practice the difference between the settings is difficult to detect; many owners would prefer a rather harder optional setting for use when driving fast.

The standard Silver Cloud, though without a fixed division, is equally well suited to chauffeur- or owner-driving. Front and rear compartments have similar choice leather upholstery and deep pile carpets; the facia and window surrounds are of superbly matched and polished walnut. The extraordinarily comfortable seats wrap round the shoulders at the rear. At the front there is a single seat with separate backrests, each instantly adjustable for rake. There is a wide central armrest at the rear, while a pair are used at the front so that, if desired, the front passenger can have maximum comfort while the driver enjoys all the elbow room he requires.

The steering column is not adjustable, but is well sited for most drivers—and a little too long for some. Particularly when power assistance is fitted, some drivers might also prefer a smaller wheel. The positioning of the two pedals is comfortable for a driver of any height. To the left of them, in addition to the dip switch, is the miniature pedal for the automatic chassis lubrication. The speedometer can be seen easily through the wheel. It is very nearly accurate, reading at high speed only one per cent slower than the car's true speed. The needle moves smoothly past the clear, white-on-black numerals.

Minor instruments, grouped in a dial on the left, while not quite so easy to read are, at least, comprehensive. The fuel gauge has a warning light which operates when petrol runs low, and a line on it for oil level; when the driver presses a control button the oil level reading is given. (During the test, including all the high-speed Continental motoring, the sump required no topping up.) Other gauges include water temperature, oil pressure, and the ammeter.

There are some differences in the operation of the switches, chiefly for ignition and lights, which have their own central, circular panel. Removal of the key now acts as the master switch; without it even the side lights or radio cannot be put on. However, it is possible to switch on whatever may be required and then remove the key; with the key out, nothing can be turned off. Therefore, in addition to the benefit of a complete master control, no one (including children left in the car) can tamper with settings when the key is removed. The lights include fog lamps with yellow bulbs which also act as powerful winkers, under-bonnet and luggage locker lights and, in addition to the usual interior arrangements, lights for vanity mirrors in the rear corners.

Extra touches of luxury are provided by a pull-out table mounted centrally under the facia (incorporating a wide ashtray) and hinged tables let into the rear of the front seat backrests. All the tables are of polished walnut matching the facia. Visibility is very good for a car of this character. As many of these cars are frequently chauffeur-driven, the privacy afforded the rear passengers seems appropriate. It is not achieved at the cost of any serious blind spots, and the slim windscreen pillars of the curved screen scarcely interrupt forward vision.

The heating and ventilating system fitted as standard operates silently, and is capable of precise adjustment. The rear compartment is now equipped with vents, and the temperature and distribution of air can be adjusted to meet any requirement at the screen, and in the front or rear.

A centrally hinged bonnet enables one side or the other to be opened. The layout beneath is one of the car's most impressive aspects to the enthusiast; not only is the density of components in the large space productive of awe, but the cleanliness gives the air rather of a laboratory than of machinery. One wonders how room was found for the quite

On each side of the rear seat is a vanity mirror with its own light, and with space for storing cosmetics. The engine compartment is well filled and the cleanliness of its contents is unrivalled. Although most owners would be unlikely to attempt servicing, components requiring regular attention remain easy to reach

Rolls-Royce Silver Cloud ...

massive needs of the steering power assistance—and marvels that units for high-capacity refrigeration equipment can be added here.

The carpeted luggage locker is deceptively large, particularly in fore and aft length. The shape permits ordinary suitcases to be carried, but best advantage of the space would be taken with fitted cases. The spare wheel has its own compartment under the floor, with the jack. It can be reached without disturbing luggage. On the car tested the lock was sometimes difficult to operate.

The Rolls-Royce is a car of tradition, and few of its owners would welcome radical changes or unconventional design features. At a time when, for a number of cumulative reasons, quality of finish and attention to details of trim and fit cannot receive as much attention on production cars as they did in the past, this make continues to attain the highest standards—little short even of the best seen in any of its predecessors.

A difficulty in appraising the Rolls-Royce lies in the exceptionally wide range of attributes which its designers have set out to achieve. Size and weight have not been allowed to reduce cornering safety; high performance is combined with silence. In these and other matters compromises may have been made, but few people once familiar with the car would not agree that the decisions have been taken with wisdom. There is no doubt that the car fully justifies the admiration, even glamour, which attaches to its name throughout the world in countries where cars of quality are understood and appreciated.

ROLLS-ROYCE SILVER CLOUD

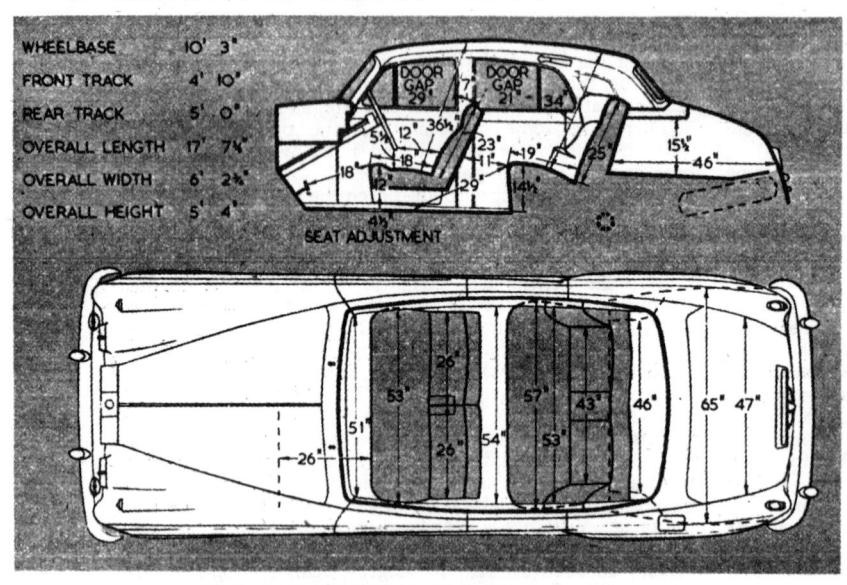

Measurements in these ¼in to 1ft scale body diagrams are taken with the driving seat in the central position of fore and aft adjustment and with the seat cushions uncompressed

PERFORMANCE

ACCELERATION: from constant speeds.
Speed Range, *Gear Ratios and Time in sec.

M.P.H.	3.42 to 1	4.96 to 1	9.00 to 1	13.06 to 1
10—30	—	—	3.6	—
20—40	—	5.9	—	—
30—50	8.9	6.0	—	—
40—60	9.4	6.7	—	—
50—70	11.1	—	—	—
60—80	13.0	—	—	—

*Figures obtained avoiding kicking-down.

From rest through gears to:

M.P.H.	sec.
30	4.1
50	9.4
60	13.0
70	18.4
80	25.0
90	34.1
100	50.6

Standing quarter mile, 18.8 sec.

SPEEDS ON GEARS:

Gear		M.P.H. (max.)	K.P.H. (normal and max.)
Top	(mean)	106	170.6
	(best)	106	170.6
3rd		63	101.4
2nd		34	54.7
1st		22	35.4

SPEEDOMETER CORRECTION: M.P.H.

Car speedometer:	10	20	30	40	50	60	70	80	90	100
True Speed:	10	21	31	42	52	62	71	81	91	101

TRACTIVE RESISTANCE: 20 lb per ton at 10 M.P.H.

TRACTIVE EFFORT:

	Pull (lb per ton)	Equivalent Gradient
Top	245	1 in 9.1
Third	415	1 in 5.3
Second	540	1 in 3.9

BRAKES (at 30 m.p.h. in neutral):

Efficiency	Pedal Pressure (lb)
23 per cent	25
62 per cent	50
90 per cent	60
94 per cent	80

FUEL CONSUMPTION:
12 m.p.g. overall for 960 miles. (23.5 litres per 100 km).
Approximate normal range 10–15 m.p.g. (28–19 litres per 100 km).
Fuel, Premium grade.

WEATHER: Sunny, no wind.
Air Temperature 56 deg F.
Acceleration figures are the means of several runs in opposite directions.
Tractive effort and resistance obtained by Tapley meter.
Model described in *The Autocar* of 29 April 1955.

DATA

PRICE (basic), with standard saloon body, £3,795.
British purchase tax, £1,898 17s.
Total (in Great Britain), £5,693 17s.
Extras: Power-assisted steering, £110 plus £55 purchase tax. Refrigeration, £385 plus £192 10s purchase tax.

ENGINE: Capacity: 4,887 c.c. (298.2 cu in).
Number of cylinders: 6.
Bore and stroke: 95.2 × 114.3 mm (3.7 × 4.5in).
Valve gear: o.h.v. inlet, s.v. exhaust.
Compression ratio: 8 to 1.
M.P.H. per 1,000 r.p.m. on top gear, 25.

WEIGHT: (with 5 gals. fuel, 37 cwt (4,144lb).
Weight distribution (per cent): F, 52; R, 48.
Laden as tested: 40 cwt (4,480lb).
Lb per c.c. (laden): 0.92.

BRAKES: Type: Rolls-Royce/Girling.
Method of operation: hydro-mechanical, servo assisted.
Drum dimensions: F, 11¼in diameter; 3in wide. R, 11¼in diameter; 3in wide.
Lining area: F, 120 sq in. R, 120 sq in. (120 sq in per ton laden).

TYRES: 8.20—15.00in.
Pressures (lb sq in): F, 19; R, 26 (normal).

TANK CAPACITY: 18 Imperial gallons.
Oil sump, 16 pints.
Cooling system, 28 pints.

TURNING CIRCLE: 41ft 8in (L and R).
Steering wheel turns (lock to lock): 4¼.

DIMENSIONS: Wheelbase: 10ft 3in.
Track: F, 4ft 10in; R, 5ft.
Length (overall): 17ft 7½in.
Height: 5ft 4in.
Width: 6ft 2½in.
Ground clearance: 7in.
Frontal area: 26¼ sq ft (approximately).

ELECTRICAL SYSTEM: 12-volt; 57 ampère-hour battery.
Head lights: Double dip; 60-36 watt bulbs.

SUSPENSION: Front, Independent, coil springs and wishbones. Rear, Semi-elliptic. Anti-roll bars front and rear.

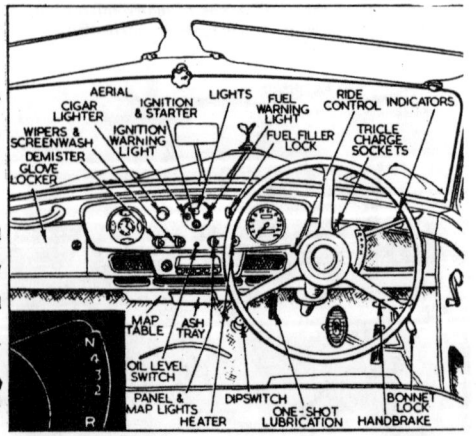

WHEELS, December, 1958

For the first time in many years, it's now possible to buy a Rolls Royce drop head coupe. This one was shown at the London Motor show, has a body by H. J. Mulliner.

Again, a Rolls-Royce Drop Head

And still they come. More new cars from England and France and our correspondent Gordon Wilkins has just sent us further details of some of the models which caused great interest at the London Motor Show.

As well as the standard Rolls Royce Silver Cloud and Bentley "S" cars which were seen at the London Motor Show, there were several interesting coachwork exhibits, and for the first time for many years, there was a Rolls Royce with drophead coupe body.

H. J. Mulliner showed a Silver Cloud with drophead coupe body in Whitehall grey over rose beige with red hide upholstery. There was a Rolls Royce with the 127 in. long wheelbase with division fitted by Park Ward and a Silver Wraith on 133 in. wheelbase, seven passenger limousine, both with coachwork by Park Ward.

Park Ward were exhibiting two Continental Bentleys, one saloon and one drophead and there was the Continental by H. J. Mulliner, called the Flying-Spur four door saloon.

Stable mate of the Rolls Royce is the Bentley. Park Ward and Co. Ltd. made this glorious two door saloon in a 120 m.p.h. Continental chassis.

Modern Motor Road Test

yes —

Rolls-Royce seldom release cars for road-testing, but they gave one to Jerry Ames, to prove or disprove their famous slogan

THREE months ago we ran a story called "Still the Best?" questioning some features of the current standard-model Rolls-Royce as compared with its predecessors. The Rolls people promptly gave us a chance to find out for ourselves, by offering our regular correspondent, Jerry Ames, the rare opportunity of road-testing a Rolls-Royce Silver Cloud. Here's what Jerry wrote about it:

THE old order changeth; stately homes of England are fighting for survival, and many are now in business trying to attract visitors at half-a-crown a time. Fewer dowagers can afford to sit behind a chauffeur and footman in their Rolls-Royces when paying social calls.

The private chauffeur is a rare sight nowadays—and even among those well-heeled persons who can still employ one, there are many who occasionally prefer to do their own driving.

These changes were foreseen by Rolls-Royce, and the latest Silver Cloud is built with that in mind—yet my road test convinced me that the knockers are wrong: the Rolls is still the finest car in a changing world, make no mistake about that.

Several modern touches have been subtly introduced to the Silver Cloud of today to keep it in the forefront of motor engineering, but not at the expense of its best-loved features, so well remembered by a large and wealthy clientele.

Gone is the old 7-litre engine, its place being taken by a more highly-developed modern unit of 4.9 litres. Transmission is now fully automatic, and I will say without fear or favor that it is the best in the world. Power steering is available for those who want it, for an extra £165stg. (basic)—say £A350 in Australia.

If I were placed blindfolded in the latest Silver Cloud, I would know I was in a Rolls-Royce. With the engine running, there's no mistaking that subdued, even purr with complete mechanical silence, or the uncanny smoothness with which the car glides away; these are features which no other make has managed to equal.

The traditional radiator grille is still topped by that charming mascot, famous throughout the world. Even policemen, I found, will hold up other traffic with a smile in order not to hamper the progress of the Silver Cloud. No other car seems to get quite the same courtesy; there IS magic in Rolls-Royce motoring.

Through City Traffic

I collected my test car from the London service depot at Fulham, which was the original workshop of the Hon. Charlie Rolls. Until he tried the early masterpiece of Henry Royce he wouldn't have anything to do with British cars, but from that moment the two formed a partnership, and thus began a tradition that has remained legendary for more than half a century.

The business of taking 17½ft. of Rolls-Royce through heavy traffic and along some of London's narrowest streets was one reason why almost every owner employed a chauffeur in the old days. But my car, equipped with power-steering and automatic transmission, proved as easy to handle as a modern family saloon.

Even apart from these two features, which take all the effort out of driving in congested streets, the driver is never conscious of the fact that this is a very large car, because it is so beautifully proportioned. When cornering, the steering wheel spins lightly and easily. Very soon I found

MODERN MOTOR — July 1958

myself threading the car through unbelievably tight spaces with the utmost confidence.

Well clear of London, I was able to make full use of the power of the Silver Cloud.

Sting in Its Tail

A Rolls-Royce never appears to hurry; nevertheless it has a tremendous performance. Maximum speed I found to be a genuine 105 m.p.h., and the cruising pace an effortless 80. A check on the speedometer showed that it was spot-on throughout its range.

Acceleration is outstanding for such a heavy luxury car; 50 m.p.h. could be reached from a standstill in 9.8 seconds, while 90 m.p.h. would come up quite often on comparatively short straights. Indeed, it pays to keep a wary eye on the speedometer, for speed can be more than usually deceptive on this car.

Steering, of course, is very light at all speeds, and some care is advisable when handling the car at more than 65 m.p.h., as there is a degree of oversteer. When negotiating fast curves at high cruising speeds with passengers in the rear seats, a perceptible amount of roll oversteer is noticeable.

Although the steering power-assistance is reduced appreciably in the upper ranges, I would have preferred a means of switching it off altogether at speeds in excess of 60 m.p.h. Incidentally, I found that the longer-wheelbase Silver Cloud I drove recently at Goodwood with power-steering had less oversteer at speed than the normal model lent to me for this full road test.

Transmission, Brakes

The fully automatic transmission, originally imported from America, has

STILL THE BEST

FULL Rolls-Royce luxury means walnut trim, folding tables in adjustable seat-backs, lush upholstery. BELOW: Body is a masterpiece, blending dashing modernity with classic tradition.

MAIN SPECIFICATIONS

ENGINE: 6-cylinder; overhead inlet, side exhaust valves; bore 95.25mm., stroke 114.3mm., capacity 4887 c.c.; compression ratio 6.6:1; twin S.U. carburettors.
TRANSMISSION: Fully automatic, with override control; ratios 3.42, 4.96, 9.0 and 13.06 to 1: hypoid bevel final drive.
SUSPENSION: Front independent, by coil springs, wishbones and anti-roll bar; semi-elliptics with Rolls-Royce double-acting adjustable piston-type dampers at rear.
STEERING: Cam-and-roller (power-assistance optional).
BRAKES: Hydraulic, servo-assisted.
DIMENSIONS: Wheelbase 10ft. 3in.; track front 4ft. 10in., rear 5ft.; length 17ft. 8½in., width 6ft. 2¾in., height 5ft. 4in.; kerb weight, 38cwt.
FUEL TANK: 18 gallons.

PERFORMANCE ON TEST

MAXIMUM SPEED: 105.2 m.p.h.
MAXIMUM in indirect gears: Second 39 m.p.h.; third, 64.
ACCELERATION: 0-30, 4.9s.; 0-40, 6.4s.; 0-50, 9.8s.; 0-60, 12.5s.; 0-70, 18.2s.; 0-80, 24.9s.; 0-90, 36.4s.
PETROL CONSUMPTION: Normal driving, 16-18 m.p.g.; hard driving, 14 m.p.g.

been considerably improved by Rolls-Royce engineers—I haven't seen another to equal it.

A fluid flywheel coupling is linked with a four-speed epicyclic gearbox, but the driver is provided with an overriding control which enables him to select second and third gears at will.

On most cars with automatic transmission, one can kick down and engage a lower gear by jabbing the accelerator hard; but as soon as this pedal is released, top gear is automatically re-engaged, usually at the wrong moment. Not so on the Rolls-Royce, if one uses the overriding control.

On approaching a really tight main road corner, the gear lever can be moved down a notch on the quadrant and third gear is automatically selected and held. One therefore has the advantage of assistance from the engine. Of course, most of the driving is done with the lever in fourth gear—in fact, it is usual for the lever to be left in this position all day for town or country driving.

A great deal of power is required to stop a heavy car moving rapidly, but the Rolls-Royce's hydraulic brakes, with servo assistance, cope admirably. There is no suggestion of judder or fade, the brakes being wonderfully smooth and even, whether they are used at speeds around 100 m.p.h. or from normal touring gait. You can also make an emergency stop on wet roads with complete confidence.

At speeds below five m.p.h. the servo motor is not effective and it is better to use the handbrake—one of the twist-and-pull variety, placed under the right-hand side of the dash.

Comfort Plus

The ride is quite firm and free from pitch. A small adjustment can be made to the rear shock-absorbers by an electrically-operated control on the handsome walnut dash, so that all the occupants can enjoy a most comfortable ride even over bad surfaces; the luxury travel of the occupants, sunk deep in the soft Rolls-Royce leather seating, is indeed no fable.

Many changes have been made to the luxurious interior of the latest Silver Cloud, and in every respect this is very much a car for the owner-driver, as well as for the professional chauffeur.

Two armrests are provided in the centre of the divided front seat, while those on the door are adjustable for position. Rake of the front seats can be instantly altered by means of a lever on either side, which helps also to provide even more leg-room for the driver; this is, of course, in addition to a generous fore-and-aft adjustment of the seat.

The driving position is unusually good and most comfortable, all controls being within easy reach. The seating is high enough to give the driver excellent visibility—both front wings can be seen, and the door pillars cause no awkward blind spots.

A great deal of care and thought has obviously gone into the arrangement of the driving position. The nicely raked, three-spoked steering wheel closely approaches the ideal; in the centre is a horn button, and just above it the lever controlling the automatic transmission. One minor criticism here: this lever was somewhat in the way of the dash control operating the winking direction indicators.

There is, of course, no clutch pedal; brake and accelerator pedals are nicely spaced, and there is ample room to rest your left foot.

Like all the interior woodwork, the dash is in beautiful walnut veneer,

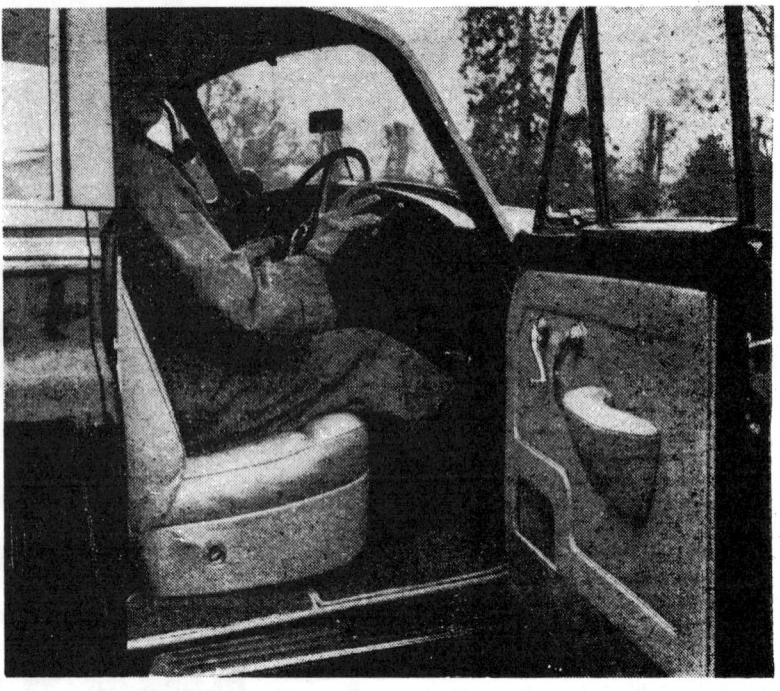

JERRY AMES behind wheel of Silver Cloud—a spot he hated to leave.

DESPITE old-fashoined split bonnet, engine-room accessibility is good.

STILL THE BEST

with sensible black instruments. The very accurate speedometer reads up to 110 m.p.h. Surprisingly, there is no rev-counter, and the oil-pressure gauge is merely marked High and Low, and the water-temperature gauge Hot and Cold. By pressing a button, the oil level can be read off on the fuel gauge. The petrol filler flap is unlocked by a switch on the dash.

Typical of Rolls-Royce thoroughness is the provision of a master key that will turn all the locks, and a second key, which operates only the doors and the ignition. Thus an owner can securely lock up any valuables or confidential papers in the glovebox and leave the door-and-ignition key with a garage, safe in the knowledge that only he holds the key to the locker.

The rear seats are every bit as luxurious as those in the front, and allow full leg-room and head-room. Folding picnic tables are provided, and ashtrays and cigar lighters are within easy reach of all the occupants.

Mention must be made of the demisting arrangements for the rear window—an important driving convenience overlooked by most manufacturers. The device consists of a number of wires so fine as to be almost invisible to the naked eye, and is brought into use by the demister control close by the efficient heater below the dash.

A boot of truly gigantic proportions for luggage is provided. The spare wheel, tools and battery are accommodated under the boot floor.

Finally, what about those recent criticisms which have suggested that the modern Rolls-Royce represents little progress when compared with its predecessors?

To check on these, after testing the Silver Cloud, I took a ride in a superbly-kept 12-cylinder Rolls-Royce P III, built in 1937.

Although a small fortune had been spent to keep this car in tip-top condition, I found that it was well behind the Silver Cloud in every respect—comfort, performance, ride and handling. From the viewpoint of the owner-driver especially, the Silver Cloud represents a big and genuine advance.

● ● ●

ROLLS-ROYCE *Continued from Page 15*

lands in 1911 and clocked 101 mph. These tests were done more than forty years ago, yet bearing in mind the weight of the chassis, these performance figures are extremely good even when compared to modern automobiles.

By 1914, the name of the company was used, all over the world, as a synonym for excellence and its cars were held in such wide esteem that it was obvious the company would be asked to take a major part in producing transportation for the armed forces, when the First World War began that summer.

Henry Royce began work on the design of an airplane motor and the Eagle engine he designed saw service on every front. Similarly, streams of armored cars, staff cars and motors left the Derby plant for every theater of war and it is in the classic tradition of the firm that there were no instances of any of its products ever failing.

In the period between the wars, Royce designed the airplane motor with which the Schneider Trophy was won and the world's air speed record smashed. His engines also powered racing vehicles with which successful attacks were made on the land and water speed records. Eyston's 'Thunderbolt', for example, hurtled across the Bonneville Salt Flats to record 394.1 mph.—and, in one run, he clocked over 400 mph.—propelled by the thrust from two Rolls-Royce motors.

In 1931, Rolls-Royce acquired the Bentley company and produced a sport, or semi-sport car known as the 'Rolls-Bentley' which combined speed and performance with elegance and engineering of the very highest standard. And, of course, cars bearing the 'R.R.' insignia continued to be sold all over the globe.

As the art of body-styling matured, so the Rolls-Royce became more and more distinguished in appearance, though there can be little doubt that the retention of the well-known radiator shape must have posed some tough problems for the designers since that shape is not particularly suitable for modern body contours.

The story of Rolls-Royce in World War II is largely one of airplane motor production; the 'Merlin' engine, which powered the Spitfires in the Battle of Britain, was almost continuously being improved so that, by the end of hostilities, over 150,000 of them had been produced.

A large number of these were manufactured under license in the United States by Packard and it is worthy to note that the association between Rolls-Royce and Packard was always one of great mutual respect.

When the blueprints of the 'Merlin' were received by Packard, for example, the top-level engineers of that firm did not believe that the tolerances Rolls stipulated could ever be achieved in a production motor. Then when the 'Merlin' arrived from England, for Packard to use as a yardstick, they stripped it and found that every tolerance was as shown on the blueprint—and that motor had come off the Rolls-Royce production line in Derby.

Probably almost everyone who is at all interested in automobiles has asked, at one time or another, just what is the secret of Rolls-Royce? Apparently this question always amuses the company and they claim that there is no secret at all—other than hard work.

This is probably true but only if one relates hard work in its absolute sense to the thought, design and manufacture of every component—however trivial. In that perspective it is genius for, by definition, genius is an infinite capacity for taking pains.

And this pride in craftsmanship, these exhaustive tests, this perfectionist spirit are all embodied in the finished article—the car whose every line and every movement is stamped with the unmistakable signs of elegance and breeding.

Although Henry Royce died in 1933, his influence is still as strong today as when he was alive. It was he who laid down the high and exacting standards to which they work; it was he whose life exemplified his favorite maxim — "Whatever is rightly done, however humble, is noble." ☆☆

STILL THE BEST?

Continued from page 35

of the axle where the hub-nut keyway was cut—making it the weakest point.

The hub nut does not go with modern wheels. The modern bolting system would have been used in Royce's day had it been known. We consider the modern system better. We retain the hub nut for appearance sake.

THE STEERING. — The Rolls-Royce cam-and-roller steering is by far the best. It makes possible the more flexible lock required for modern parking. It may be heavier at parking speeds, but is the best possible over the whole range.

THE RIDE CONTROL. — The electrical system has been installed because it is instant and does not need the build-up time of the hydraulic. It has been designed to give the smoothest ride, taking into consideration all suspension factors.

THE GEARBOX.—When the demand for auto-change came, we looked around for a box technically up to our standards. We found the General Motors box and adapted it for our own use, and our own standards.

THE SMALLER WHEELS.—We did not reduce wheel size purely for the sake of appearance, though the smaller wheel does enable us to lower the car and so keep abreast of modern trends.

Smaller wheels also mean that the car can have a lower centre of gravity and better weight distribution.

Improvements in tyres make the tyre-life factor less important.

THE RUBBER PARTS.—Because maintenance facilities throughout the world were not what they are today, we had to make things to last the life of the car, even though they might be less efficient.

Today we can count on regular maintenance, so parts are made in the best available material, even though the parts may need replacing at some time in the future.

Rubber is so vastly improved as compared with pre-war that we can now use it.

THE JACK.—The hydraulic jack was discontinued because there were too many complaints of it failing. Our engineers considered it was better to have a 100-percent reliable system than one which could fail.

Besides, punctures are very rare today, and incorporating the hydra-jack would add great weight to the unsprung parts of the car.

THE LAST WORD.—It is not true that we are failing to keep the lead in the motor-car world. We now have a higher proportion of development staff than at any other time in our history.

We do not bring out a new car every year. When we do, it has to be so far in advance of anything else that it will not suffer by comparison before the next model

WHO WINS, THEN?

It isn't easy to say who is the winner in this argument. While Rolls-Royce have answered some of Gretton's criticisms satisfactorily, they seem to have sidestepped the real issue on others.

For instance, it is hard to see why belt drive for the dynamo should be "considered the best." It's certainly much cheaper than gear drive—but the latter is more positive and virtually everlasting. As for the dynamo failure, so glibly dismissed by Rolls-Royce, surely it was to guard against just such eventualities that Rolls used to fit their special big batteries.

Those phony hubs aren't properly explained, either. Quick-action hubs are still considered safe enough for modern independently-sprung Grand Prix racing machines, on which they are subjected to tremendous stresses. If Royce's original design wasn't suitable for independent suspension, why couldn't it have been modified?

But perhaps the least satisfactory answer is the one dealing with ride control. The quicker action of the electric system can't have been the real reason for its adoption in place of the far more versatile and efficient hydraulic set-up. Here, cheapness and simplicity of installation must have been the main considerations. For proof, re-read what **The Autocar's** tester had to say about the electric control.

None of this proves Charles Gretton's contention that the Rolls is no longer "The Best Car in the World" —we, at least, aren't prepared to say so until we know of a better one.

But—let's face it—the standard-model Silver Cloud isn't the old "hand-made" article, either. The old ceremonial of special appointments, inspection, and check-ups on the buyer's respectability is gone, too. Nowadays, in England, you can order and buy your Rolls over the phone, like any other car.

It's still a great car, of course, and a thing of luxury. But the luxury is mass-produced, and must perforce make use of some of the short-cuts and limitations imposed by mass-production.

• • •

By ROYCE to ROME

A Personal Account of
3,360 Miles in
The Best Car In The World

By

LAURENCE POMEROY,

F.R.S.A., M.S.A.E.

"... at 60 m.p.h. the noisiest thing
on the car is the clock..."
LAURENCE POMEROY, *The Motor*, February 19, 1958

CONTINUING TRADITION.—The façade of St. Peter's, completed in 1612, continued the principles embodied in the Parthenon built 2000 years previously. Both in shape of radiator, and in quality of construction, the Rolls-Royce Silver Cloud continues the same traditions 350 years later.

THE conductor raps the rostrum with his baton, the violins strike their tremulous tones; and (on the fourth bar?) the woodwinds softly join them. The Prelude to Act I of *Tristan und Isolde* is launched, but we listen not sitting in a box at Bayreuth but driving at somewhere between 60 and 80 m.p.h. on the Dover road *en route* by Royce to Rome. In addition to the music we enjoy power steering and the benefit of some re-thinking at Crewe which has immensely improved the four-speed, automatic-shift transmission that has always had a hold-in third position which retains this ratio down to about 15 m.p.h. and up to 70 m.p.h. irrespective of throttle opening.

But earlier models with the lever in fourth could, if the throttle were opened at, say, 50 m.p.h. produce a disagreeable change-down which was neither desired by the driver nor warranted by ambient circumstance. An adjustment now ensures that full throttle can be used at all speeds over 30 m.p.h. without a change-down, although the driver can at his own volition obtain third speed below 70 m.p.h. simply by moving the gear lever, to obtain acceleration or retardation as requisite.

The pleasure afforded by this sensible improvement was, I will admit, soon erased by the steward telling me I should be roused at 2.30 a.m. to make a 3-a.m. exit from the Dover-Dunkirk night ferry, and that at this point the local clocks would read 4 a.m. This was not a good time to leave one's bed, an even worse one at which to wait for a further hour while the Customs formalities were attended to, but it made possible breakfast in Liège, lunch at the ex-Empress Victoria's palace at Kronberg and arrival in Stuttgart at 5.30 p.m. some 12 hours after leaving Dunkirk. I had been behind the wheel for some 9½ hours in covering 463 miles or, say, 49 m.p.h. average speed.

I was wholly free from physical fatigue despite having had barely three hours' sleep the previous night, and this is as big a tribute as one can pay to a superlatively good motorcar. The silence in which it travels does not merely improve appreciation of classical concerts coming through the radio set, but more importantly seals one from assault by the outer world, so enabling high cruising speeds to be maintained in a mental calm which is equalled only by a high degree of physical comfort and an extraordinarily low factor of muscular effort.

To obtain the best ride over really rough roads independent suspension to all four wheels may be needed, but over modern European highways the suspension characteristics of the Royce are hard to fault, particularly if intelligent use is made of the electric switch which offers an alternative of soft or hard damper settings. The seats are exceptionally well designed and give real support to the shoulders, although with power steering one drives more with the wrist than with the biceps. Similarly the brake pedal calls for no more than an inclination of the ankle, as a ½-g. stop, which is the highest needed except in dire emergency, demands but 30-lb. pedal pressure and all normal braking can be done with less than 20 lb.

With this much merit I was familiar from past experience; the power steering came to me almost as a novelty and the best testimonial to it is that I was never for one moment conscious that it existed! It is simply normal steering with the effort about halved, and although a graph shows that the driver must try harder in the centre of the range than he need on full lock there is no realization of a sudden break through into the power sector such as one feels on some American systems. Nor on the other hand must one learn to drive by position alone, which is a characteristic of the 100% power assistance given on yet other American cars.

My enthusiastic approval of the Crewe contribution in this matter was endorsed by Uhlenhaut of whom I can say that no equal engineer can drive as fast or as well, and no driver of comparable skill has anything approaching his

A. and M.—These pictures show the Rolls-Royce adjacent to the Colosseum which was used for circuses, mock naval battles, and the martyrdom of early Christians in the first hundred years A.D. and also a sample of the new buildings being erected south-west of Rome to form a 20th century annexe to the Eternal City.

By ROYCE to ROME

knowledge of engineering. He assessed the power steering as the best he had experienced and observed, on a tea-time jaunt to Freudenstadt, which we had on Sunday, November 2, that the road reaction felt on the steering wheel had been eliminated for the first time in over 50 years of R-R production. He thought that on some rough roads the car needed driver correction to keep it on a straight line and this sophisticated observation may follow from some rear-steering effect, although I must admit that any such was too slight for my notice.

However, some rear-traction effects were brought very vividly to my notice on the following day (Monday), when we lunched at Zurich, and, reaching the St. Gotthard at about 3 p.m., resolutely went past the notices which said, "St. Gotthard mit Ketten." As we reached the top of the Pass we found a few slippery bits but all was going well until we were within about 300 yards of the summit, where we encountered hard-packed snow and icy surface combined with a left-hand corner. Although passed by a 600 Fiat, a VW, and a 1900 Alfa, the Royce just would not take this at all; in fact, it was quite hard to hold it from sliding backwards when it had come to rest. The prospect of reversing down in the dark was not very appetizing, but mercifully three road men got some grit from a heap about 100 yards up and 10 Swiss francs from my wallet, and with a bit of a push we were round this critical corner, up to the summit and on the way down.

And so at the end of the second day of real motoring we reached the Esplanade Hotel at Locarno with a deep appreciation of the steering, the brakes, and the silence, in a slightly critical frame of mind about the rear-wheel adhesion, and with even more sombre reflections about the standard ventilating system, as the car needs both a driver and a *chauffeur* if the interior is to be kept in a condition acceptable alike to driver and passengers.

The contrast between the quietness of travel with windows shut, and the instantly noticeable noise when they are opened by as much as a fraction, provides a strong incentive to keep them shut at all times and rely wholly upon the ventilating system as an insurance against asphyxiation.

The venting on the Silver Cloud is divided between the de-misting stream coming from the base of the windscreen and a secondary supply rising from the floor, both having their own two-speed fans and both a hot or cold position. In the depths of winter when all the air is heated as much as possible all is well; in summer with no heat added the volume of air is somewhat inadequate. In the middle range of temperature one is supposed to blend the streams but in fact the cold air inlet cannot be used all the time, in part because ram effect is such that at high speeds the car soon becomes too cold and turning up the hot air coming from the floor is something like sitting huddled around the fire in a cold and draughty house. With no cold, one is soon too hot and although by adding a mere 10% to the British purchase tax one may enjoy all the advantages of full air conditioning, I still feel that the standard system can be improved with advantage to all.

On a brighter note a check at Locarno showed that we had done a very creditable 15.8 m.p.g. and an easy run to Turin brought us to the excitements of a Motor Show in which the efficient organization by Dr. Giovannetti and his small staff is matched by a splendid exhibition hall, beneath the single-span roof of which the most beautiful cars in the world are congregated.

I have never been one to swallow the foolish doctrine that streamlining is effective only in excess of 60 m.p.h., but our run south to Rome from Turin was certainly a remarkable example of the limits placed upon the modern car by the circumstances in which it is forced to operate.

On the first stage a combination of dense traffic, darkness, fog, and a stop for coffee did not prevent us from covering the 199 miles to Modena in exactly five hours, but from Modena to Rome was another affair again. We started by motoring slowly over the Raticosa and Futa passes and thus took three hours to cover 90 miles before taking a Florentine luncheon.

Leaving at 3 o'clock, 184 miles lay before us, a distance covered (in the reverse sense) by Moss with twice our gross engine power; and four times our b.h.p. per laden ton, in 2 hr. 24 min. 15 sec., at an average speed of

76.8 m.p.h. This compares significantly with the 118 m.p.h. that he averaged for the 188 miles between Brescia and Ravenna driving in the same 1955 event, in that his average speed fell by about one-third over the seemingly interminable Radicofani pass, which we took in failing light amidst the most desolate country. We were averaging barely 30 m.p.h., or only about half what we might expect to do in favourable circumstances, and after we had taken 6 hr. 15 min. for the journey I recalled Thomas Coddrington who, in 1903, wrote in his book "Roman Roads in Britain": "The straightness of Roman roads... has been, perhaps, too much insisted upon but in a broken country, or along the valleys, a winding course to suit the ground is usually followed, and in a hilly country straightness is sometimes not a characteristic at all."

That "all roads lead to Rome" is a matter in which I have a personal interest, for I was born in Dunstable and if the straight line of the Roman road from Stony Stratford to Dunstable were projected it would pass over New Romney, a little to the west of Rheims, rather to the west of Besançon, straight over Novarra and Lucca and thence over the dome of St. Peter's! Whether this is by accident or intention I cannot say but I can record my extreme regret that 49 years elapsed between my departure from Dunstable and my arrival in Rome.

Although we were comfortably housed there, low cloud and driving rain made miserable the exploration of the Colosseum, where lions were catered for by the Christians, and the Circus Maximus, where two straights joined by hairpins at each end were used for seven-lap 14-km. races by two-wheeled vehicles with propulsion units developing normally 2 or 4 h.p., but sometimes 3 or even 8 h.p. according to the formula of the day. These races were run anti-clockwise, as in the first Grand Prix races, at Brooklands during the whole of its life, and at Indianapolis today, thus showing once more how traditionally minded, with their retention of the Queen Anne gallon and Shakespearian spelling, the Americans really are.

To see this and the profusion of other classical objects in such wretched conditions was a misfortune; to see St. Peter's at all so splendid an experience as to make all else of no consequence. Neither Forrest nor I could find words for the Baroque magnificence which unfolded before us, and as we looked up to the vast dome and saw (in Latin) the words "Thou art Peter and upon this rock I will build My Church and I will give unto thee the keys of the Kingdom of Heaven," we could not but feel that we were in a building made by man which was truly worthy as the centre of an Eternal City.

These solemn reflections could not blind us to the fact that Tuesday morning was as cold and wet as Monday evening and we were therefore confirmed in our "half warmed fish" to take the opportunity to see Naples although, we hoped, not immediately to die.

With this in mind we departed from the Flora Hotel at 2.30 leaving some laundry behind, and after passing through the truly monumental buildings in ferro-concrete which are making a new Rome south-west of the old City, we encountered one of the most violent storms

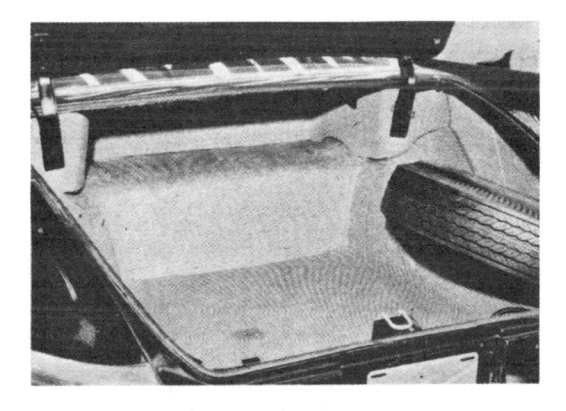

SCALE EFFECT.—The relative sizes of a custom-built Fiat 600, a standard Fiat 500 and the Rolls-Royce Silver Cloud are interestingly indicated by this picture taken adjacent to the Turin Motor Show.

TAILS FROM TURIN.—The Buick (*above*) shows how the modern American car exchanges length for depth in the luggage boot, it no longer being possible to stand the spare wheel upright within it. Left is seen the Farina exemplification of the fast back on the Ferrari, and below the latest trend in tails with a full-width flat window as displayed by Ghia on a Fiat chassis.

By ROYCE to ROME

BIRDS OF A FEATHER.—All over the world the quality of the Rolls-Royce is acknowledged; this shows a Milan-registered car waiting outside the Savoy Hotel at Rapallo in company with the car referred to in this article.

which has been known in this area in recent times, hail falling so thickly as to bring visibility down to a level where 20 m.p.h. was hazardous, rain falling in such volume that quite steep gradients took on the appearance of a waterfall and the wind blowing with such strength as to uproot trees and cast them across the road, this last phenomenon, fortunately, never occurring when we were going too fast to stop. In the course of the hour or so which this lasted the Royce did not leak a single drop of water, but although Naples is separated from Rome by a mere 140 miles, it took us 4 hours for the journey.

On Wednesday morning the warmth of the atmosphere, some 60° F, was matched by low clouds and continuing rain which pursued us through a brief interlude at Pompeii just as throughout our stay in these parts we were pursued by touts of every kind wishing to sell us everything from guaranteed gold watches at 50s. each through religious charms to by no means doubtful pictures.

Nevertheless we had one piece of good fortune, for as I cast around near the railway station on the way to the Pompeii autostrada I was accosted by a young man who said that owing to a bus strike he was unable to reach his factory on the motor road and could I help him on the way. This I did and he took us into the showrooms of a company called Donaddio which was full of cameos and other jewellery which one could see being made in representative workshops nearby. The products were obviously of high quality, and we were given the opportunity, which we took, of buying some of them at far below shop prices—a point which, being something of a sceptic in these affairs, I have since firmly established.

We resolved to start early, i.e. at 8 a.m. on Thursday and to call firstly at Rome and to take a picture of the car in front of St. Peter's and to get back my washing, passing then through Siena en route to Florence.

All this was accomplished, starting with an extremely fast run to Rome which we accomplished in three hours, making an average speed of just under 50 m.p.h. for the journey. In the course of this trip there is a straight starting in the seaside town of Formice which runs level for 20 miles beside a canal with only one curve and one village to prevent one averaging 100 or even 120 m.p.h. for the full distance. After an hour's interval in Rome we set off for Florence, this time taking the coast road to Grossetto, after which we were back in mountainous country across to Siena, which is a town of completely medieval construction and layout, so that with a car of R-R size the front wings have but little clearance between the houses, except in the vast central square where twice yearly there are ceremonial horse races with the riders decked in 16th-century dress. From Siena to Florence the road continues to twist and turn and as it was now dark our average speed slowly fell, and after a total time of 12 hours, and running time of 9¼ hours we had completed 356 miles at an average of only 38 m.p.h.

I commend the Savoy Hotel in Florence and even more strongly the way in which Mr. Beppino runs his restaurant called the Campidoglio in the street of the same name. We were also lucky in being given the name of a small leather manufactory called Ricci at 8 Via Romana, where they are prepared not only to show you some beautiful work being hand-made, but also to sell it at very far below shop prices, so that the finest handbags can be had for £8 and wallets and purses for £1 apiece, or even less.

Our departure from Florence brought the novelty of the trip to an end so far as I was concerned, for after taking the autostrada to Pisa I was back on the road over which I drove last summer to Rapallo. Here a dynamo bearing seizure enforced a day's delay at another Savoy Hotel, and this was far from unwelcome, for the sun blazed down from a cloudless sky on to a waveless sea and the November warmth exceeded anything experienced in England this year. As a further example of how Rolls-Royce cater for individual taste, we were delayed yet another day in Nice by an extraordinary failure of the rubber boot which seals the splined end of the universal joint. This opened up at over 30 m.p.h. to give the impression that the entire transmission system would at any moment disgorge itself on to the road, so we again sat in the sun, on the terrace of the Ruhl et Anglais. On Monday morning the company's service station quickly discovered the trouble and we had no difficulty in covering the 295 miles to Lyon in 7 hours 50 minutes despite dense patches of fog in the dark of the last two hours.

From Lyon to Paris we had an incentive to avoid darkness by reason of what was now a total dynamo failure and with N.6 under our wheels we did 46, 59, and 61 miles in the first three hours, followed by 21 miles in the next 20 minutes before stopping for lunch at Auxerre. In the first hour after luncheon we again did 61 miles and a last hour with 43 miles in it brought us right into Paris. The next day we had merely the routine run to Boulogne and a crossing by the new ferry-boat *Compiegne*, a well-planned vessel, although one does reflect that this planning has had in mind those people for whom the holiday begins when they take off their ties. On English roads the smooth silence of the Royce was even further reinforced, and retrospectively knowing that it averaged 14.8 m.p.g. for over 3,300 miles, it is obvious that increased running costs are no bar to ownership. In fact an inability to raise £6,000 is really the only excuse that any man can offer for not enjoying the highest standard of terrestrial transport so far attained in the recorded history of the world.

"The loudest noise in *this* Rolls-Royce doesn't come from the electric clock."—Drawing by Stevenson. Copyright 1958, The New Yorker Magazine Inc.

1959 ROAD TEST
ROLLS-ROYCE Silver Cloud

For the first time Motor Life exposes one of the automotive world's most sacred vehicles to the hard criterion of an unbiased American road test

IF THERE are any legends as such in the present-day world of automobiles, one must be recognized as foremost: the complete public identification of the Rolls-Royce as the ultimate in luxurious motoring. To most of us the sight of its perpetual vertical grille is the symbol of undeniable wealth and elegance.

But, how does the legend stand up when this British monarch is subjected to the impudent hands of American road testers? The MOTOR LIFE verdict: luxury it has, but it is of a quieter variety than that displayed by some of the Rolls' flashier competitors.

Because of its very success the car suffers from a disadvantage at first meeting. Passengers enter with a "you've-gotta-show-me" attitude and are bound to be disappointed. For real evaluation the Rolls must be measured in terms of the whole and not in bits and pieces. The car is not a super bomb by American standards, yet it is surprisingly lively and agile. It is a large car by the current European yardsticks but a compact one in comparison to American limousines.

Its 123-inch wheelbase is seven inches shorter than the Cadillac, yet it has more leg and shoulder room and has a general feeling of more interior spaciousness. The interior appointments are luxurious, but gaudy glitter has not been allowed to substitute for functionalism.

The high-seated driving position is quite comfortable, giving an excellent view of the traffic about you. The front seat cushion is a bench type, however the seatbacks are separated and adjustable for rake. Twin fold-down armrests divide the seat to give a bucket effect if desired. The armrests on the doors are both adjustable for height or can be detached entirely. The fore and aft seat adjustment is a simple mechanical one and there appears to be little necessity for a six-way power seat to jockey the driver into position.

Driving controls and switches fall readily to hand. The instruments are almost austere in design by American standards and are grouped centrally on the polished walnut dash. While legible enough, their location could be improved by positioning them American-fashion in front of the driver. A novel feature of the instrument panel is a line on the fuel gauge which indicates the crankcase oil level when a button is pushed. Another lever releases the hinged gas tank flap for re-fueling purposes.

The automatic transmission on the Silver Cloud has a five-position quadrant which selects the four forward speed sequence. With the quadrant in position 4, all gear changes are automatic. At speeds of less than 60 mph a complete depression of the gas pedal will kick the transmission into third, or into second if the speed is very low. The same affect can be obtained by moving the selector to position 3, without jumping on the gas pedal. In position 3, the transmission also remains automatic in all gears

STEERING LINKAGE presents a complex display of arms and bellcranks. Pencil points to link from power steering's hydraulic cylinder that projects through front frame member.

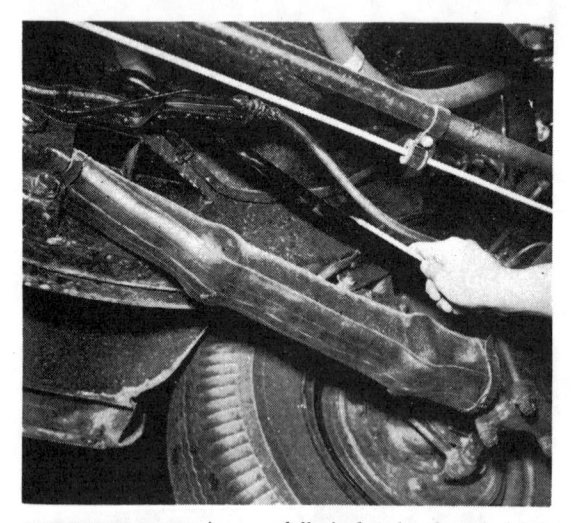

SEMI-ELLIPTIC rear springs are fully jacketed to keep in lubricant and shut out dirt. Black bar pointed to is a radius rod that prevents spring windup induced by rear axle torque.

ALUMINUM VALVE COVER displaying Rolls-Royce name is finished in black porcelain as are the intake manifold, carburetors, exhaust manifold. Round object is intake silencer.

but an immediate shift-down is made from fourth to third when the speed falls below 60 mph. In position 2, the same action takes place and second gear is held. This position proved to be an excellent means of utilizing engine braking on steep downgrades. Although the driver can feel the shifting within the automatic transmission, which is standard on all Rolls-Royce cars now, there is no disconcerting lurching or other roughness to accompany the shifts.

A Rolls-Royce looks a bit incongruous on a drag strip, but acceleration runs proved it no slouch for an imported car. From a standing start, 45 mph was reached in 8.5 seconds and 60 mph in 13.5 seconds. Watching the speedometer during these runs and especially when higher speeds were reached was an almost unbelievable experience. With the windows closed, the astounding quietness with which this car runs almost completely cancelled out any sensation of speed. Although top speed runs were not attempted, the Rolls glides along at 60, 70, even 80 miles per hour with the greatest of ease, making a reported top speed of over 100 mph highly believable.

The test car was equipped with extra-cost power steering, a relatively new accessory for Rolls-Royce. There are widely different opinions on how much "feel" should be present in power steering; in the case of the Rolls, the driver is conscious of some road feel. The steering characteristics give a definite feeling of sensitivity, yet inspire the driver with a complete sense of mastery over the car at all times. The 4½ turns lock-to-lock is the one feature which seems to call for some unwarranted action on the wheel, which, in a minor degree, detracted from the excellent steering qualities of this luxury car.

Rolls-Royce cars have long used a mechanical servo assist on their braking system and it has proven to be excellent. Drive for the brake servo is taken from the transmission and the degree of assist is proportional to the speed of the car. Hard stops from high speeds were made in an almost casual manner. The pedal action was light, progressive and controllable. Stops were made without the slightest hint of grab or swerving, conditions which continue to hound American-built cars in the low and high-priced fields. Stops in traffic from slower speeds were equally smooth and the general impression gained by the driver is that the required effort to stop is governed automatically.

A round trip in a Rolls is an experience in sheer comfort with a feeling of safety and security. The chair-height seats are marvels of comfort and in the rear, downright luxurious because of the backs which wrap around the shoulders of the occupants. A first impression of the ride, which was gained on city streets, is that it is not as soft as most American luxury cars. On the highway the car is very stable. Hard corners may be taken at high speeds; there's no feeling of mushiness and very little body roll or lean. One reason for the greater feel of solidity is that the Rolls does not have rubber-ball joints on the front suspension system. On the highway, the Rolls goes exactly where it's pointed; you can take your hands off the steering wheel and the car will travel a straight line if the road is flat almost indefinitely without correction. There is an absence of wallowing, bottoming and oscillation on poor pavement surfaces and no dangerous nose-dive when the brakes are applied. A radius rod like a traction-master bar on the rear keeps the rear end from racking.

A two-position switch on the steering column controls the resistance of the rear shock absorbers. Under ordinary conditions, little difference in the ride was noticed when a switch-over was made from "normal" to "hard." On very tight corners at high speeds, the "hard" position seemed to reduce the very slight tendency toward body lean.

The Rolls-Royce may not be entirely in keeping with the American concept of a luxury class car. This would seem to be due to the purpose of its builders. Their goal, which quite obviously has been fulfilled, is to produce an automobile that will remain a vehicle of undated prestige for years to come. To do this certain compromises must be made between current untried advances and known long-life techniques.

MOTOR LIFE

CLASSIC LINES of the Silver Cloud are a symbol of authority wherever they appear. The perpetual vertical grille and flying lady set it off from a sister car of identical features, the Bentley.

INSTRUMENT PANEL has a full complement of instruments resting in a wonderland of walnut woodcraftsmanship. Drawer-type board pulls out from under panel for a table or writing surface.

REAR PASSENGER area contains the utmost in comfort. Center arm rest comes complete with a writing pad, tobacco supply and milady's compact. Picnic tables retract quickly into seat backs.

TEST DATA

Test Car: Silver Cloud
Body Type: four-door sedan
Basic Price: $13,750 P.O.E.
Engine: six-cylinder ohv
Carburetion: twin Solex downdraft
Displacement: 298 cubic inches
Bore & Stroke: 3.75 x 4.50
Compression Ratio: 8-to-1
Horsepower: not available; only factory statement is that it is substantial. Estimate: somewhat over 200 hp.
Test Weight: 4400 lbs. without driver
Transmission: four-speed automatic
Rear Axle Ratio: 3.42-to-1
Steering: 4½ turns lock-to-lock
Dimensions: overall length 212 inches, width 75, height 64, wheelbase 123, tread 58 front, 60 rear
Springs: independent coil, front; semi-elliptic rear
Tires: 8.20 x 15
Gas Mileage: 11.7 mpg
Speedometer Error: indicated 30, 45 and 60 mph are actual 31, 46 and 61 mph, respectively
Acceleration: 0-30 mph in 4.4 seconds, 0-45 mph in 8.5 and 0-60 mph in 13.5 seconds

1960 CARS

6¼-LITRE V-8 ROLLS-ROYCE and BENTLEY SALOONS

Improved Bentley S2 and Rolls-Royce Silver Cloud II Models Powered with a Big New Aluminium Engine

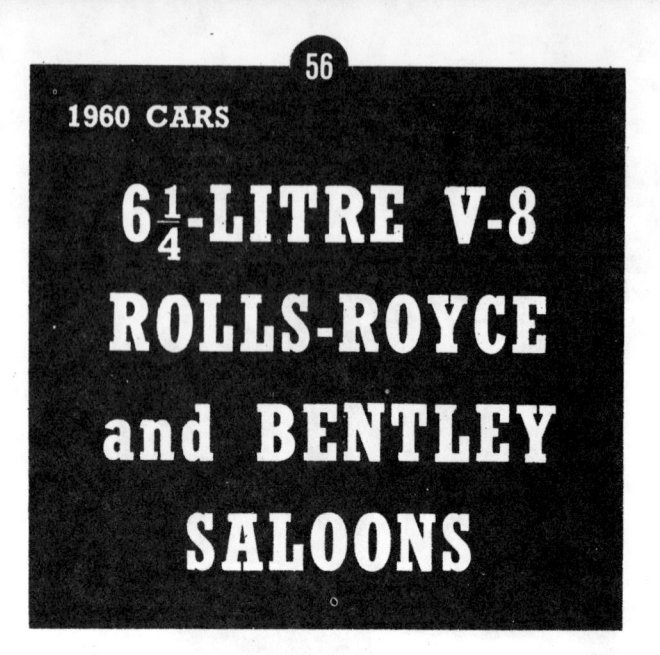

STILL happily able to pay more regard to quality of detail and less attention to cost than can any other large-scale car manufacturers in the world, Rolls-Royce Ltd. are continuously active in improving their models but introduce completely new designs only at very long intervals. Today's announcement of a completely new 6¼-litre V-8 engine, replacing an in-line "six" which has been in use (with progressive enlargements and improvements) since 1946, is therefore a rare and important occasion. Stepping up the engine displacement of both Bentley and Rolls-Royce models by 27½%, and increasing the even more significant piston area dimension by 59½%, the primary objective sought and achieved has been an increase in refinement with some slight increase in acceleration and top speed as an incidental. It is not the policy of the company to quote power output figures for their engines, but whilst this new engine is obviously not yet being asked to deliver anything approaching its true potential output, few engineers studying its details would be willing to give credence to any torque figure below 325 lb. ft., and even with two rather small carburetters the maximum power output must comfortably exceed a genuine 200 b.h.p. In the car, it is geared to give approximately 110 m.p.h. at 4,000 r.p.m. in top gear.

When the now-superseded "six" engine was introduced in 1946 it was a 4,256 c.c. unit with 88.9 mm. by 114.3 mm. cylinders and a compression ratio of 6.4 to 1. As successive Rolls-Royce and Bentley cars became roomier, more accessory-laden yet faster, it was necessary to enlarge this engine by stages to 4,887 c.c., by increasing the cylinder bore first to 92.0 mm. and later to 95.25 mm., and to raise the compression ratio to 8 to 1. As long as five years ago, the eventual need for a new engine was foreseen, so that further power increases would be possible without departure from the factory's high standards of car smoothness and silence, and the straight-eight engine fitted to a few long-chassis cars was considered to be too long and too heavy for use in other models.

Compact Unit

Designed so that it fits into existing well-proven cars without demanding major chassis alterations, and doubtless also readily applicable to other chassis and body designs which may appear in a near or distant future, the new and potentially much more powerful engine is nevertheless compact in both length and width. Thanks to design features and to the wide use of aluminium alloys in its construction, it is as light as the smaller "six" which it replaces, although some transmission parts have had to be made stronger and heavier so that extra engine torque can be applied to the road wheels. Some chassis and body changes also introduced for the 1960 season will be mentioned after description of the new engine.

At the heart of this new engine is a forged steel four-throw crankshaft, carried in five steel-backed copper-lead-indium main bearings of 2·5 in. diameter and with six counterbalance weights to ensure smooth running. A torsional vibration damper on the nose of this short crankshaft has comparatively little work to do. The main engine block in which the crankshaft runs is a Birmal high-silicon aluminium alloy casting, which extends downwards well below the main bearings and upwards to form the two cylinder blocks which are inclined at 90° one to the other. Not all aluminium engines have been quiet, but this one promises to be notably rigid structurally. The two "halves" of the V engine are for example linked not merely at crankcase level but by cross webs at the ends of the engine and where two intermediate bearings support the camshaft.

Above the V of the engine, where the two banks of cylinders form an oil bath, there is one camshaft of Monikrom cast iron, the tips of 16 cams just dipping in the oil. Drive to this camshaft is by helical gears at the front of the engine, the "wavy web" construction of the light-alloy camshaft gear recalling some recent racing car road wheels; the mating gear on the crankshaft nose is of steel. Hydraulic self-adjusting tappets, with spherical bases of 35-in. radius rotated by slightly conical cams, operate the engine's overhead valves through pushrods and rockers, the valve timing diagram showing very little overlap indeed. Inside the self-adjusting tappets, flat-seating valves are more resistant to dirt on their seatings than are ball-valves, and oil from the tappets passes up hollow pushrods to the o.h.v. rockers.

In the light alloy cylinder blocks there are individual wet cylinder liners of cast iron completely surrounded by thin water jackets. Held in position by the cylinder head and its corrugated metal gasket, each liner has a rubber coolant sealing ring around its top which keeps any coolant-borne dirt away from the liner locating flange; two similar rings around the liner's base have an external drain-hole between them, to disclose any passage of oil or water whilst rendering the leakage harmless.

Firing Area

Two light-alloy cylinder heads with austenitic steel valve seat inserts bolt on to the main engine casting, and carry in-line overhead valves which are closer to the vertical than are the cylinders. Fully machined, the combustion chambers taper laterally (a squish area faces each sparking plug) and also fore-and-aft (the inlet valves are larger in diameter than the exhaust valves). Both cylinder heads are identical one with the other, inlet and exhaust valves alternating along the length of each. Long individual inlet ports of rectangular shape face the inside of the V, and four short exhaust ports on the outer side of each cylinder head transfer a minimum of waste heat to the cooling water. A refinement of oil sealing in the overhead valvegear is the bonding of a tiny synthetic rubber seal on to half of each split valve-cotter, only single valve springs being used.

Above the centre of the engine, an aluminium casting forms the double induction system. Two inward-facing S.U. horizontal carburetters of 1¾-in. size (smaller in bore than those used on the 4.9-litre "six") feed through 90° elbows downwards to separate induction pipes which are in the same casting but are connected only by a tiny drain-hole. With the variable-choke carburetters, obtaining mixture enrichment during snap acceleration by dashpot-delayed choke enlargement instead of by an accelerating pump, the risk of hot-starting difficulties such as have afflicted some V-8 engines seems slight. Each carburetter feeds cylinders 1 and 4 of one bank and cylinders 2 and 3 of the opposite bank, so that suction impulses are evenly spaced—the firing order is 1R, 1L, 4R, 4L, 2L, 3R, 3L, 2R, etc. During the warming-up period when the coolant thermostat is closed, water from the cylinder heads is directed through jackets below the carburetter elbows to check fuel condensation, and as hitherto there is an automatic choke of progressive pattern. Both carburetters draw air from a common intake cleaner/silencer installation, through a flexible duct which has enabled the impregnated-paper cleaner and the silencer to lift with the bonnet as an aid to engine accessibility.

To ensure that it will run without protest on French and other petrols of modest anti-knock quality, this new engine is being built with a compression ratio of only 8 to 1. As an insurance against pre-ignition, Brightray treatment is applied to the heads of stellite-faced KE965 exhaust valves, the inlet valves being of S65 steel. Heplex pistons have slight recesses in their crowns, and are of semi-slipper form with flanks cut away to clear crankshaft balance weights. A solid skirt has two heat-barrier slots separating it from the piston crown, and each piston carries three

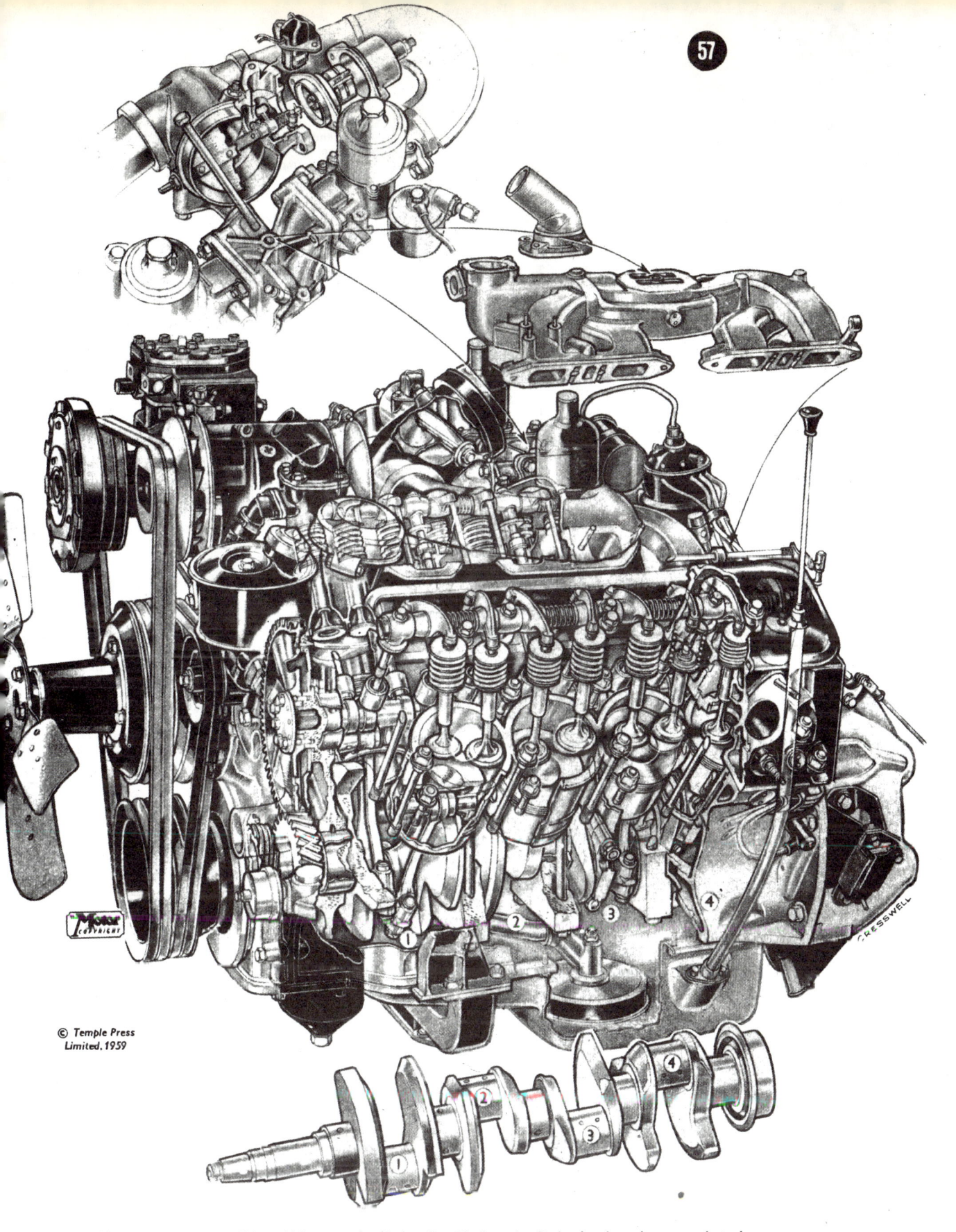

The new oversquare 6¼-litre V-8 engine has light-alloy block and cylinder heads and a pressed steel sump. The five-bearing crankshaft carries two side-by-side big-ends on each crankpin. Hydraulic self-adjusting tappets are used in the valvegear. On the front of the far bank of the block can be seen the freon compressor, with magnetic clutch, for the air conditioning system and on the nearside bank is the hydraulic pump for the power steering. The drawing inset above the engine shows the automatic choke set in the common intake to the two carburetters: a bimetal strip "tastes" exhaust gas temperature and a solenoid operates the strangler. A single 6-in. bolt attaches the whole carburation system to the water-heated cross-over manifold seen at top right.

The 1960 cars have a new facia, with adjustable fresh-air outlet louvres at each end of the top facia rail. Two control knobs, to the right of the steering wheel on this l.h.d. model, separately control fresh and recirculated air temperatures and fan speeds.

6¼-LITRE V-8 ROLLS-ROYCE AND BENTLEY SALOONS

narrow but radially deep compression rings (the top one chrome plated) and a composite oil scraper ring above the circlip-located gudgeon pin. Big-end bolts are integral with the connecting rods, each side-by-side big-end bearing being of 2.25 in. diameter and 0.8 in. length.

Engine lubrication is by a helical gear pump, driven from the nose of the crankshaft so that any pump noise is as far as possible from passengers in the car, the 12½-pint oil sump being a steel pressing and a full-flow filter looking after oil cleanliness. Ignition for this eight-cylinder engine is by a single coil, but the Delco distributor which is driven from the back of the camshaft has two contact-breakers in parallel: centrifugal timing control is accompanied by an "octane selector" vernier but it has not been found advantageous to use the now-fashionable load-sensitive vacuum timing control. At the front of the engine there are four belts, two driving the hydraulic pump for a power steering system, two a dynamo which starts to charge at hardly more than tick-over r.p.m., and all four helping to drive the centrifugal cooling water pump and a fan which has its five blades unevenly spaced in the interests of silence. Water from the pump is forced into the cylinder blocks, upwards through jets to the sparking plug bosses and exhaust valve seatings, and out of the engine through passages in the induction manifold casting. Placing of the sparking plugs beneath the exhaust manifolds, where they are tilted over beyond the horizontal, means that access panels below the front wings have had to be provided, but a very long plug life without cleaning or gap adjustment is claimed. The engine is started by a unit of which the pinion is pre-engaged magnetically before the motor starts to rotate, thus eliminating a source of noise and ensuring that the pinion does not disengage prematurely.

Inclined at an angle of only 2° when in the chassis (and positioned so that the gearbox is just over an inch further forward than with the six-cylinder engine) the new V-8 has a three-point mounting. Below No. 2 main bearing there is one central pad of rubber/metal sandwich construction, interposed between a stirrup under the engine and a U-section removable chassis cross-member. Alongside the fluid coupling of the R.-R. automatic four-speed transmission there are two similar flexible mountings, inclined to take engine torque in shear, and with a fore-and-aft rubber buffer to steady the power unit during emergency braking. With an eight-cylinder engine built to Rolls-Royce standards of static and dynamic balance, extreme flexibility of engine mountings is not necessary.

Each bank of cylinders has its own exhaust manifold, but from the right-hand manifold the tail-pipe runs across to the left side of the car and all cylinders then exhaust into a common silencing system. Two silencers in series break down low-frequency exhaust noise, and a third absorbs high-frequency noise.

Power steering which was formerly an optional extra has now become standard equipment on all Rolls-Royce and Bentley cars, this permitting a 1° increase in castor angle to ensure satisfactory "feel". Extra width of the V-8 engine as compared with an in-line "six" has required the steering gearbox to be moved further sideways, so to retain the desired steering wheel position and angle a pair of precision gears has been inserted in the steering column base. Rubber mountings spaced 17 in. apart insulate the steering from any vibration of the chassis without destroying accuracy of control.

Refinement of the I.F.S. system has included the use of sturdier wishbones and a stiffer anti-roll torsion bar, to suit increased car performance. Rubber bushings are not used in the front suspension except for the torsion bar, and it is at first surprising to find that a familiar one-shot centralized chassis lubrication system has been discontinued. The reason for this is however that by improved sealing of bearings, provision of extra lubricant capacity and use of a Ragosine grease containing 20% of molybdenum disulphide it has been possible to suggest a 10,000-mile interval between chassis greasings in the knowledge that test cars have successfully been run for much greater periods without attention.

Damper Control

Rear suspension continues to be by half-elliptic leaf springs, rubber interleaving and gaiters to keep out dirt ensuring their consistent behaviour. The Z-bar which has recently provided roll stiffness as well as halving rear axle wind-up under torque reaction is now rubber-bushed and serves only the latter purpose. Spring stiffness corresponds to static deflections of approximately 10 in. at the front and 9 in. at the rear, a two-position switch on the steering column giving alternative bleed settings to the Rolls-Royce hydraulic rear shock absorbers: production tolerances on damping characteristics of the latter are stated to be less than half those of proprietary shock absorbers.

With metal-to-metal bearings used in the suspension in the interests of precise controllability, rubber body mountings are depended on to insulate passengers from road noise. The chassis is an X-braced structure, with box-section steel side members 7 in. deep and 3½ in. wide, and the steel saloon body is supported at 15 points by rubber-in-shear flexible mountings of which the flexibility varies according to the risk of noise transmission. Individual mounting points are adjustable, and these adjustments are only locked-up finally after the load distribution between them has been pre-set pneumatically on the factory assembly line. Thus silence is obtained yet the body shell contributes extra torsional stiffness to the complete car.

In external appearance, the V-8 engined cars which are to be known as the Rolls-Royce Silver Cloud II and the Bentley S2 are scarcely distinguishable from their immediate predecessors. Inside the steel saloon bodywork some rearrangements are immediately apparent however, associated with a new ventilation and heating system designed to include air conditioning (i.e., refrigeration and/or drying of uncomfortably hot or humid atmospheres) as an optional but fully integrated part of the installation.

The new integrated air conditioning system has two air circuits. The recirculation system draws air from the rear interior of the car through an intake fan, whence it passes forward through ducts to cooler and heater and is expelled through the downward-facing outlets. Fresh air is drawn (by an intake fan at low car speeds) through a front grille, then diverted by shutter through heater or cooler, or both, and then passed to de-misters and adjustable louvres on the facia. The air circuits are fitted into all the new cars, but refrigeration is an optional extra.

REFRIGERATION HONEYCOMB — BLOWER — RECIRCULATING AIR INTAKE — BLOWER — HEATER — REFRIGERATION HONEYCOMB — HEATER CONTROL FLAP — FRESH AIR INTAKE

September 23, 1959

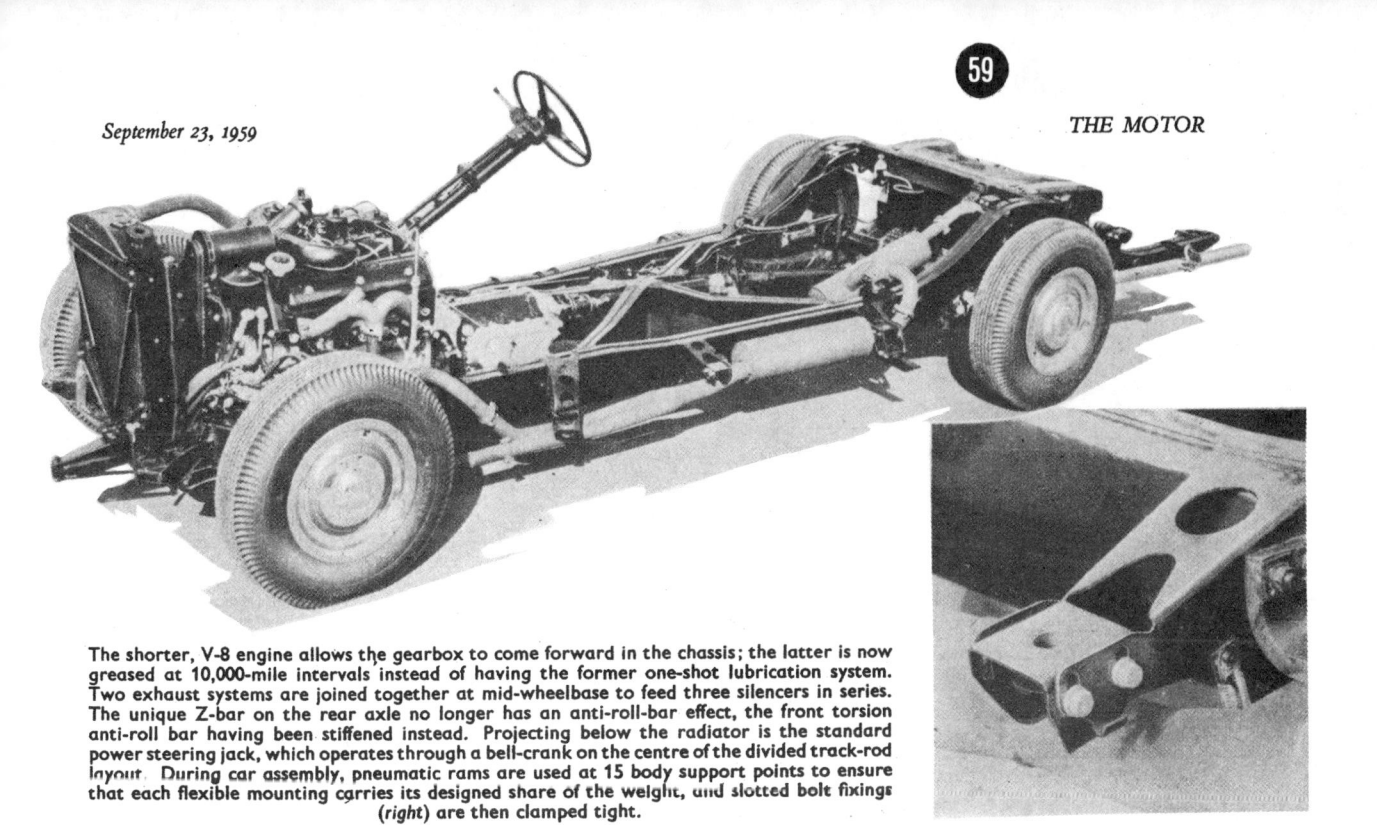

THE MOTOR

The shorter, V-8 engine allows the gearbox to come forward in the chassis; the latter is now greased at 10,000-mile intervals instead of having the former one-shot lubrication system. Two exhaust systems are joined together at mid-wheelbase to feed three silencers in series. The unique Z-bar on the rear axle no longer has an anti-roll-bar effect, the front torsion anti-roll bar having been stiffened instead. Projecting below the radiator is the standard power steering jack, which operates through a bell-crank on the centre of the divided track-rod layout. During car assembly, pneumatic rams are used at 15 body support points to ensure that each flexible mounting carries its designed share of the weight, and slotted bolt fixings (right) are then clamped tight.

Physically, the new air conditioning equipment is centred in the right front wing where the air ducts converge and the air heating and cooling elements are grouped—if refrigeration is specified, as it surely will be by most buyers even for temperate lands such as Britain, there are also a freon compressor in front of the engine's right-hand bank of cylinders, and a condenser unit in the main engine cooling air stream.

To provide genuine comfort under widely varied conditions, the air conditioning system is divided into two parts, one of which takes in fresh air from a grille below one headlamp and the other of which re-circulates air that is already inside the car. Each half of the system is simply and effectively controlled, as to both temperature and volume of air delivered, by a delicately sensitive control which twists left for cooling or right for heating —the fresh-air system's control also has a push-pull action to bring a booster fan into action.

The low-level heater has a variable-speed electric fan to take in air from the rear compartment, pass it first through a cooler element and then through part of the heater element (either or both of which may be operative) and deliver the warmed, cooled, or dried air to several outlets serving the front and rear seat passengers. Re-circulating through a heater element air which is already inside the car, it is possible to provide comfortably adequate heating even in arctic cold—it is also possible to dry the atmosphere in the car by using the refrigeration to condense moisture out of the air, also using heating so that the temperature is not lowered excessively, in very humid weather.

The upper level heater draws in fresh air, which is passed over either heating or cooling elements (or split between them) before being delivered to four de-misting slots along the base of the windscreen and to air outlets which can be set to blow on to the faces of the driver and front passenger or to de-mist the side windows of the body. Thus, one can have plenty of fresh air to breath, cooled or heated as desired, and can have pleasantly warm feet without being obliged to breathe heated air. The Rolls-Royce of today has power-operated windows, but with this air conditioning system, and air permitted to escape gently around the lower edges of the opening windows so that fresh air can replace it, the need for open windows and consequent wind noise during fast driving should be virtually a thing of the past.

Despite inclusion of power steering, the new 6¼-litre V-8 engine and the much improved heating and ventilating system in the standard specification, Bentley and Rolls-Royce show an increase of only £300 (approximately 8%) in their basic prices.

BENTLEY S2 AND ROLLS-ROYCE SILVER CLOUD II.

ENGINE
- Cylinders ... V-8 with wet-lined cylinders at 90° angle in alloy engine block. 5-bearing crankshaft.
- Bore and stroke ... 104.14 mm. x 91.44 mm. (4.1 in. x 3.6 in.).
- Cubic capacity ... 6,230 c.c. (380.2 cu. in.).
- Piston area ... 105.5 sq. in.
- Compression ratio 8.0/1.
- Valvegear ... In-line o.h.v. operated by hydraulic tappets, push-rods and rockers from gear-driven camshaft.
- Carburation ... Two S.U. type HD.6 horizontal diaphragm-type 1¾-in. carburetters, with automatic choke, fed by rear-mounted duplex S.U. electrical pump from 18-gallon tank.
- Ignition ... Delco double-contact breaker with single coil: centrifugal timing control with vernier adjustment 14 mm. Champion RN8 or Lodge CLNP sparking plugs.
- Lubrication ... Helical gear driven from crankshaft nose, and "British Filter" full-flow filter: and 12½-pint sump plus 1½ pints in filter.
- Cooling ... Water cooling with centrifugal pump and five-blade fan driven by twin V-belts. Coolant capacity 21 pints, including heater.
- Electrical system 12-volt 67 amp. hr. battery charged by belt-driven 35 amp generator.
- Power and torque No figures disclosed. (130 lb./sq. in. b.m.e.p. would give 328 lb. ft. torque).

TRANSMISSION
- Clutch ... Fluid coupling.
- Gearbox ... Rolls-Royce automatic 4-speed epicyclic gearbox, with direct top gear.
- Overall ratios ... 3.08, 4.46, 8.10 and 11.75. Reverse 13.25.
- Propeller shaft ... Rolls-Royce-Hardy Spicer divided open, with flexibly mounted central bearing.
- Final drive ... Hypoid bevel gearing in semifloating axle with aluminium alloy centre section.

CHASSIS
- Brakes ... Rolls-Royce-Girling hydro-mechanical with two trailing shoes in front drums, assisted by gearbox-driven mechanical servo. Duplicated operating mechanism.
- Brake dimensions Drums 11¼ in. dia. x 3 in. wide
- Brake areas ... 240 sq. in. of lining area working on 424 sq. in. rubbed area of drums.
- Front suspension Independent by coil springs and unequal length wishbones, with anti-roll torsion bar. Rolls-Royce hydraulic piston-type dampers in upper wishbone pivots. Anti-roll torsion bar.
- Rear suspension... Rigid axle, gaitered semi-elliptic rear springs with single radius arm on right, Rolls-Royce piston-type hydraulic dampers with two-rate remote control.
- Wheels and tyres 8.20–15 tubeless tyres (Dunlop, Avon or Firestone) on 5-stud steel disc wheels.
- Steering ... Rolls-Royce cam and roller with hydraulic power assistance.

DIMENSIONS
- Length ... Overall, 17 ft. 7¾ in.; wheelbase 10 ft. 3 in.
- Width ... Overall 6 ft. 2¾ in.; track 4 ft. 10½ in. front and 5 ft. 0 in. rear.
- Height ... Overall 5 ft. 2 in.; ground clearance 7 in.
- Turning circle ... 41¾ ft.
- Kerb weight ... 39½ cwt. unladen (without fuel but with oil, water, tools, spare wheel, etc.).

EFFECTIVE GEARING
- Top gear ratio ... 27.3 m.p.h. at 1,000 r.p.m., and 45.5 m.p.h. at 1,000 ft./min. piston speed.
- Maximum torque —
- Maximum power —
- Probable top gear pulling power... 300 lb./ton approx. (Computed by The Motor from manufacturer's figures for gear ratio and kerb weight, 130 lb./sq. in. probable b.m.e.p., and with allowances for 3¼ cwt. load, 10% losses and 60 lb./ton drag.)

AUTOSPORT, SEPTEMBER 25, 1959

EVEN MORE MAGNIFICENT than before, the Bentley S2 saloon is powered by the new 6¼-litre V-8 engine.

A New Rolls-Royce Engine
John Bolster Describes the New V-8

THE announcement of a new Rolls-Royce engine must interest every student of automobile engineering. Any other firm must consider cost to some extent, but in this case we see a design unfettered by any such mundane consideration. The task was simply to produce the best power unit for fast luxury cars, and now we are presented with the result of five years' work by a brilliant team.

For years the Rolls-Royce has had a six-cylinder engine, which has gradually been enlarged and developed. It has now reached its apogee, and more cylinders are needed as power requirements increase. The company has a straight-eight engine, but this is too long for the modern car. Space considerations are against it, but above all the loss of rigidity occasioned by extending the front of the chassis cannot be tolerated. A 1959 version of the 12-cylinder Phantom III would be superb, but even tycoons must consider fuel consumption at today's prices. Thus, a V-8 embodying all the latest Rolls-Royce know-how, but of Phantom III inspiration, is the correct and logical answer.

The new engine is based on a light alloy cylinder block and heads. It adds two cylinders and well over a litre of capacity, yet it is a few pounds lighter than the "six", which is a true measure of engineering progress.

The over-square cylinders are inclined at an included angle of 90 deg., and have a bore and stroke of 104.14 mm. x 91.44 mm. (6,230 c.c.). The compression ratio is 8.0 to 1, but with a well-designed light alloy head one may use any medium-grade fuel, and there is no pinking on the Continental brands—an important point.

A return has been made to earlier Rolls-Royce practice in placing all the valves in the head, for the I.O.E. arrangement is not compact enough for a high-compression over-square unit. Valve operation is by pushrods and rockers from a single camshaft in the centre of the vee, and modern oil filtration methods have rendered a return to hydraulic tappets practicable. Incidentally, there are no low-pressure oil circuits in this engine.

Twin S.U. carburetters of the diaphragm type have an automatic choke for cold starting, and receive their fuel from two S.U. electric pumps. The elaborate tuned air silencer contains an impregnated paper filtration unit, and is coupled to the carburetters by a long flexible hose. The engine cannot be tuned with the silencer disconnected because the whole system is matched together. Accordingly, the silencer lifts with the bonnet, but remains connected to the engine, in the interest of accessibility. Three acoustic-type silencers, each tuned to a different range of frequencies, operate in series on the exhaust, and the gases reach the atmosphere through a single tail pipe.

The engine is in unit with the automatic gearbox, and is mounted on rubber at three points. As on all recent Rolls-Royce and Bentley cars, the transmission contains a fluid flywheel and a four-speed epicyclic box, with "kick-down" and overriding manual control. The new engine, having greater torque in the middle ranges, requires less gear-changing than the "six" which it replaces.

This very powerful and advanced unit propels the Rolls-Royce Silver Cloud II and the Bentley S2, which are similar to the preceding models. However, the rear suspension now has a plain torque

OUTSIDE VIEW of the new power-unit. Although it has added two cylinders and well over a litre of capacity, it is yet a few pounds lighter than the "six".

THE ROLLS-ROYCE SILVER CLOUD II saloon, also powered by the new engine.

THE PHANTOM V seven-passenger limousine with Park Ward body, which also uses the new V-8 power unit.

resisting member without roll stiffness, in the interest of more understeer. The front hydraulic dampers have increased piston travel, and the front brakes are of the four-shoe type. Power-assisted steering is standard, and so the diameter of the wheel has been reduced.

In addition, an entirely new Rolls-Royce limousine-type chassis called the Phantom V has been introduced. This has a wheelbase of 12 ft., and the complete car is 19 ft. 10 ins. long and 6 ft. 7 ins. wide. The axle ratio is 3.89 to 1, compared with the 3.08 to 1 of the Silver Cloud II. All the above cars, and the new Continentals, will be seen at the Paris Show and at Earls Court. Prices range from £5,660 14s. 2d. for the Bentley saloon to £9,393 12s. 6d. for the Phantom V limousine.

Road Impressions of the Car

ROAD tests of secret new models are sometimes a problem, because public interest may be embarrassing and awkward questions may be asked. In the case of the Bentley S2, however, there was no danger of letting the cat out of the bag unless the bonnet were opened. The external appearance was unchanged, but 6¼ litres of aluminium V-8 nestled coyly behind the familiar winged B.

The Bentley is a big car, but it is so well proportioned that its size is not apparent. My large suitcase virtually disappeared in the boot, but as I drove away from Conduit Street I was not especially conscious that my two-ton magic carpet was 17 ft. 7¾ ins. long and 6 ft. 2¾ ins. wide. I am lucky enough to have driven every model of Rolls-Royce and Bentley that has been produced in the last 50 years, and as always I was delighted to find that a marked family resemblance had been retained.

At traffic speeds I would say that the "eight" is quieter and smoother than the "six" as far as the occupants of the car are concerned, though the exhaust is a little more audible from outside. The response to the accelerator is completely progressive and smooth, but immense power may be unleashed if required. The power-assisted steering is extremely light, and one is never conscious that the tyres are of 8 ins. section. The brake pedal, too, requires just a gentle touch from the toe, and the famous gearbox-driven servo does the rest. These brakes are not only powerful, but they never make a sound.

On leaving town I was at once conscious of the increased engine power. The greatest enjoyment of the car comes from driving it smoothly and gently, with plenty of power in reserve. By this I mean that if one cruises at 70 to 90 m.p.h., a touch of the accelerator will produce real acceleration for overtaking, and appreciable hills may be surmounted with only a fractional depression of the pedal. The machine is so lively and so utterly silent in this speed range that it would be possible completely to misjudge one's pace; a new owner would be well advised to keep half an eye on the speedometer.

Yet, even though I was driving such a luxurious carriage I could not resist pressing the pedal to the floor, on a clear but by no means smooth stretch of road. In a matter of seconds I was past the 100 m.p.h. mark, and I held 110 m.p.h. for quite a distance before road conditions dictated a return to my 90 m.p.h. cruising gait. There was no increase of sound and normal conversation was maintained by my companions. The verdict, then, is that the "eight" is appreciably faster than the "six", has much greater acceleration in the upper ranges, and is so well mannered that it never joins in the conversation. The automatic gearbox suits it admirably.

Of course, the seats are gloriously comfortable, and there is every sort of accessory for the passenger's convenience. The windows of the test car were raised and lowered electrically, and whereas each door had its individual control button, the driver's door had four buttons to control any one of the windows. It is normal to travel with the windows closed, for this almost eliminates wind noise and one is less conscious of the grinding and clattering of other cars that one overtakes. The very elaborate ventilation installation then comes into play, and in fact it is a double system, for both fresh air and re-circulation—it is best to cut off the fresh air when diesel fumes pollute the atmosphere, and to use only the re-circulating ventilation and heating system. A powerful refrigerating unit may be added to cool the air, but this extra was not on the test car.

To judge a large and heavy car by the standards of smaller machines is to demand an agility that is difficult to achieve. Nevertheless, the S2 may be swung through fast curves at a considerable velocity, and the absence of tyre noise is praiseworthy indeed. There is an electrical two-position control for hardening the rear suspension, and this may be used with advantage on roads where there are sharp descents and sudden hump backs. In general, though, the normal setting gives the best ride.

The Bentley S2, and the Rolls-Royce Silver Cloud II which it resembles mechanically, are a sheer delight to the connoisseur of motoring. For a basic price in the region of £4,000, one expects something out of the ordinary, and I can state with certainty that there is nothing cheaper which is "just as good". As always, there is the magic of a name, but above all these cars represent an investment in prestige, superb British engineering, and motoring enjoyment that has no peer.

ANOTHER VIEW of the massive, 6¼-litre, overhead-valve engine.

wheels SPECIAL TEST

ROLLS-ROYCE—

ROLLS-ROYCE directors, studying the results of a recent market survey in U.S.A. were chagrined to find that many people when asked "What do you associate with the name of Rolls-Royce?" replied "Old automobiles". Not a word about the Merlin engines that powered the Spitfires, Hurricane, Mustang, Mosquito and many other aircraft of World War II. Not a word about their current jet and turbo-prop engines, or their pioneer work in nuclear power. Perhaps they brought it on themselves by building cars that never seem to wear out.

But other cars are improving all the time. Does the current Rolls-Royce maintain the margin of superiority which made the name a synonym for perfection? To find out I took a Silver Cloud sedan on a long trip from London to Monte Carlo and then on into Italy.

Its classic lines, devoid of pointless ornament have a timeless quality and will still look elegant twenty years from now. You ascend into the interior without squirming or stooping and sink into seats which provide armchair comfort for four, or enough space for five.

From the fairly high driving position, the front view is comprehensive but the rear end of the car is not visible through the back window, so parking is done mainly by instinct.

The test car had the excellent power-assisted steering which is optional equipment. It has enough road feel for confident cornering, but requires negligible effort when parking. It needs four and a quarter turns from lock to lock for a 41 ft. 8 in. turning circle which will seem normal to American users, but is not strictly necessary. The engine has more than enough power to produce an occasional flick of the tail when accelerating hard in the wet, which could be more quickly checked if the steering were more direct.

The ride is superb and the stability most impressive. Passengers simply do not notice bad roads, which neither shake them nor assail their ears with drumming noises. Side winds do not deflect the car and it can be steered up and down steep cambers with finger and thumb. A switch on the steering column gives a harder setting for the rear dampers, but it is rarely required, except perhaps when travelling very fast on winding roads.

With an engine capacity of 6,230 c.c. the V/8 o.h.v. engine is similar in capacity to some current American units and the initial rush away from the traffic lights is near spectacular and acceleration is still something many sports car owners might envy. Ninety miles per hour is passed in little over half a minute and 100 m.p.h. is reached in well under the minute. And this car has brakes and tyres designed to make 100 m.p.h. a regularly usable speed. There is no ideal cruising speed. With windows closed and the excellent heater-ventilation system regulating the temp-

Gordon Wilkins, his wife and the Silver Cloud pause for breath while crossing the Mont Cenis Pass.

—Still the world's best

... reports GORDON WILKINS after doing a 2,500 mile European test behind the wheel of the famed Silver Cloud.

erature any speed up to 100 m.p.h. is maintained without noise or effort. With the full air conditioning system consuming additional power I gather there was sometimes a feeling that extra cubic inches would be welcome in the old six but this extra was not fitted to the test V/8.

The transmission (developed from the General Motors Hydramatic) changes its four speeds as speed and load demand, but a lever permits the driver some over-riding control. The downward shift from top to third can be abrupt if made by kicking the accelerator, but imperceptible if made by moving the lever. At full throttle, first gear gives 22 m.p.h., second 35 m.p.h. and third 65 m.p.h. Third gear can be held for winding or hilly roads. In town traffic second can be selected. The car then starts in this gear and unnecessary changes between first and second are avoided. In short, you have fully automatic control or part-manual, as you wish.

The brakes alone set the car apart from most other big closed cars built today. A touch on the pedal brings in a mechanical servo from a clutch, driven off the gearbox, which applies front and rear brakes through two separate master cylinders. Rear brakes are also applied by direct mechanical linkage, so there are three separate braking systems, besides the handbrake. Speed is subdued swiftly, silently and without a tremor. The hardest use in the Alps never produced a trace of fade.

You must drive this car in the Alps to savour its astonishing dual character. Leaving Dunkirk at five in the morning I lunched in Geneva, spent the afternoon working in Switzerland and continued southwards to bring the day's mileage to well over 600 without feeling at all tired. This two-ton town carriage can be tossed into corners like a small sports car, holding its course without appreciable roll, and with only the squeal of tyres to indicate the stresses involved.

Certainly there are some things which could be improved in a car of this type. I would like to be able to sound the horns (which are a little too discreet for the performance, anyway) and work the indicators and flash the headlamps without moving a hand from the wheel. It would be pleasant to raise or lower the windows by touching a switch instead of turning a handle. A fully reclining passenger's seat would be welcome on long runs, and a station-seeking radio which eliminates twid-

Although some of the conditions Wilkins encountered on his European trip were severe, the comfort of the Rolls was such that the occupants were immune from outside elements.

ROLLS-ROYCE —STILL THE WORLDS BEST

...continued

dling the knobs. But the finish is impeccable in every detail. The rear window is electrically defrosted by an almost invisible grid of hair-fine wires. The toolkit is finished like a canteen of cutlery. The instrument panel furnishes all the information the keen driver requires and the speedometer was accurate to within two per cent right up to maximum.

The care lavished on the finish of every detail, seen and unseen, the silence, and the precise response of every control — these are terribly costly to achieve, but the price of a Rolls-Royce in terms of real purchasing power has steadfastly fallen over the years. Allowing for the fact that the pound sterling is worth one third of its pre-war value, the Silver Cloud at a basic price of £4095 in England costs 25 per cent less than

The Rolls' stately radiator and lines of the body attracted attention wherever the car went, reports Gordon Wilkins, after driving a Silver Cloud in Europe.

Interior appointments in the Rolls offer luxurious comfort for four people, although five can be accommodated reasonably well.

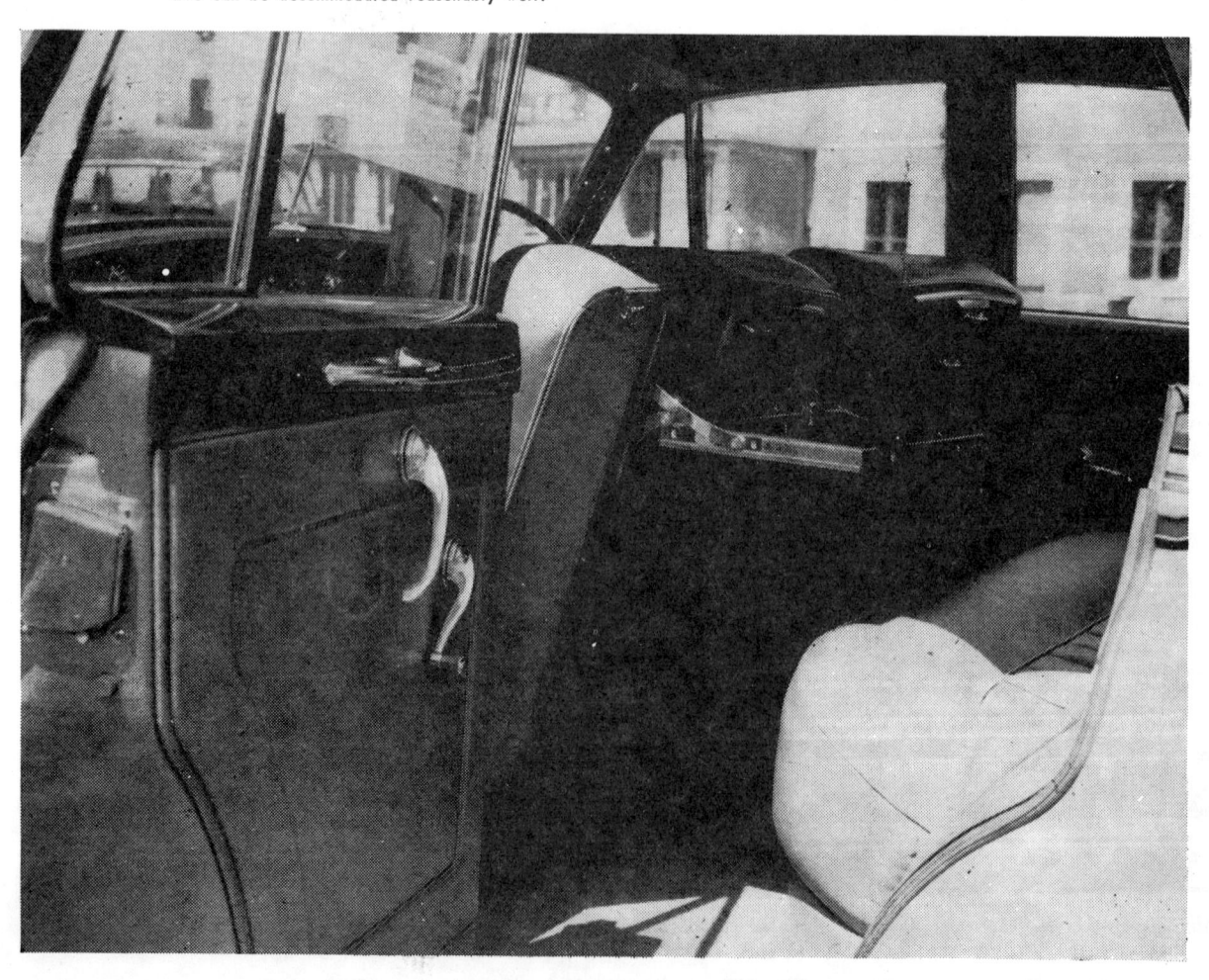

a pre-war Rolls-Royce Wraith and has a performance superior to that of the Phantom III, which cost the equivalent of twice as much.

The luggage trunk is adequate for most needs, but when driving really hard refuelling stops were necessary every 200 miles so a larger tank would be welcome. Neither oil nor water was added in 2,400 miles.

Performance: Acceleration from 0 to 30 m.p.h. was 4 sec.; 0 to 50, 9.5 sec.; 0 to 60, 13.2 sec.; Mean maximum, 106 m.p.h.; Fuel consumption, driven hard, 12.3 m.p.g. Touring speeds, 14 m.p.g. on Premium fuel. Tank capacity is 22 gal. Oil consumption was nil in 2,400 miles. Speedometer error, 30 m.p.h. indicated was a genuine 31. 50 m.p.h. was 51. 100 m.p.h. was 101.

Dimensions: Length, 17 ft. 8 in.; Width, 5 ft. 2½ in.; Height, 5 ft. 4½ in.; Kerb weight, 4,375 lb.

Conclusion: there is no substitute. This is not the biggest nor the fastest of current cars, but for my money it offers the best combination of desirable qualities. It is admired wherever it goes and it remains one of the finest things that money can buy. #

Rolls-Royce and Bentley share the same 6,230 c.c. V/8 engine, which is capable of propelling the big cars at speeds well in excess of 100 m.p.h.

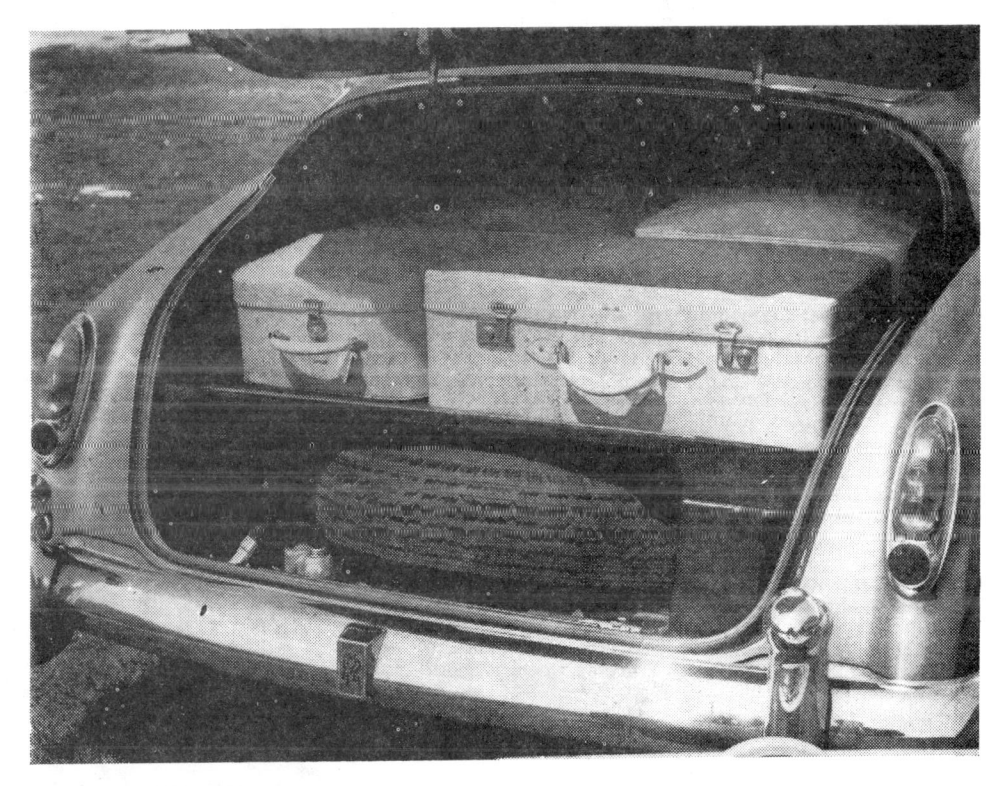

Luggage space in the tail of the Rolls is large enough to pack in a sufficient number of cases for an extended touring holiday. Spare wheel, tools, etc. have a separate compartment.

The Autocar ROAD TESTS
1773
Rolls-Royce Silver Cloud II

Outwardly the eight-cylinder Rolls-Royce Silver Cloud II is almost indistinguishable from its six-cylinder predecessor; but improvements to the chassis complement increase performance, while heating and ventilation of the interior are much more efficient

EVERY year the Best Car in the World gets a little better; sometimes there are only minor modifications in hidden parts of the chassis to carry it forward, while very occasionally a completely new or redesigned type is introduced. The most recent big advance was the introduction of a 6·23-litre vee-8 engine, which now powers the Rolls-Royce Silver Cloud II and Phantom V chassis, as well as the Bentley S2 and Continental. With single camshaft and pushrods to operate the overhead valves via hydraulic tappets, it replaces a 4·9-litre straight six with overhead inlet and side exhaust valves. At the same time the chassis of the Silver Cloud II, subject of this test, was improved in several respects concerning suspension, braking and steering; passengers are kept comfortable by a much more comprehensive heating and ventilation system.

Since a primary interest of this road test for many people will be the increase in performance of the Silver Cloud II relative to its predecessor, of which a report was published on 16 May 1958, a few comparative figures follow, those of the earlier car being bracketed. The latest car proved very considerably faster, achieving a mean maximum of 113·1 m.p.h. (106); it accelerated from zero to 60 in 11·5sec (13·0), to 80 in 21·1sec (25·0) and to 100 in 38·5sec (50·6). Of even greater significance, perhaps, is the very rapid rate of acceleration from quite high speeds in top; the new Silver Cloud required only 8·5sec to reach 80 from 60 m.p.h., a 65 per cent improvement over its predecessor.

Maxima in the indirect gears of the test car were: 21 m.p.h. in first (22), 50 in second (34) and 70 in third (62). For the car to reach its peak in second the automatic change has to be over-ridden by moving the selector lever to position 2 on its quadrant, 50 m.p.h. being the highest speed recommended by the makers' handbook.

It seems remarkable that this significant step-up in performance has been achieved without materially affecting the car's fuel consumption, for the overall figure, covering a mileage exceeding 1,500, was 11·8 m.p.g., whereas the earlier car had returned 12·0 m.p.g. For a 60-mile country run in which 60-65 m.p.h. was not exceeded, the consumption figure was 17·5 m.p.g., whereas with the 1958 car 15 m.p.g. was not bettered.

Before attempting to start the engine from cold one is instructed by the owner's handbook to depress the accelerator pedal fully and release it "to allow the fast idle cams on the carburettor controls to position themselves in relation to engine temperature and thus set the throttle to the correct opening for starting." Then the ignition key is turned to operate the starter. Mild weather during the test gave no opportunity to try the car's quick-starting ability in extreme conditions. When assessing the behaviour of a vehicle in this class, the standard of judgment is naturally raised to a degree where a shortcoming, so small that it could be ignored in a lesser car, cannot pass unnoticed. Thus one must report that, even when the engine had reached normal running temperature, its idling was neither completely smooth nor silent. It tended to run rather hot when the car was held up in dense city traffic, and would then occasionally stall.

As soon as the car moved off from rest, the engine smoothed out and thereafter maintained the standard of refinement expected of a Rolls-Royce, right up to maximum speed. Indeed, if one accelerates on part-throttle and drives generally in the manner of an experienced chauffeur, then the passengers may be almost divorced from awareness of the engine propelling them. Hard driving, using full throttle and maximum speeds in the indirects, brings about a little commotion, most of it apparently from the six-bladed cooling fan. No sound-deadening material lines the undersides of the bonnet tops.

Even from outside the car the single exhaust is quiet, so that no sound of it reaches the passengers. The Silver Cloud II is not constrained on long journeys by any particular cruising speed range, since it remains astonishingly effortless even when held almost to its maximum. It will be seen, in the data tables, that it consumes fuel at the rate of 12·3 m.p.g. at a constant 90 m.p.h., whereas the consumption graph climbs rather steeply thereafter. One might therefore regard 90 m.p.h. as a logical cruising speed (where conditions permit), having regard to fuel consumption and tyre wear. With such a performance potential, combined with first-class stability and superb braking, this Rolls-Royce can record very high journey average speeds. Its 18-gallon fuel tank allows little more than three

From this angle the boot looks much smaller than it really is. An inspection lamp is on the left of the spare wheel

hours driving between refills if an average of, say, 60 m.p.h. is being maintained. A green light on the instrument panel warns the driver when three gallons remain in the tank.

It cannot be said that the Rolls-Royce automatic gearbox has kept pace with current developments across the Atlantic. A four-speed epicyclic unit, mated to a fluid flywheel, it is based on the General Motors' Hydramatic transmission. A limitation of its design lies in the spacing between gear ratios, a wide gap between second and third being apparently unavoidable. Nevertheless, all upward changes are acceptably smooth, the more so when the car is accelerated on part-throttle. Yet the makers recommend in their handbook that deliberate downward changes should be made by flicking the selector lever, rather than by using the throttle pedal kick-down; while this advice seems somewhat contrary to the purpose of a fully automatic transmission, it certainly results in smoother operation.

In particular, a kick-down change from third to second (possible only at 21 m.p.h. or below) resulted in a quite violent jerk—sometimes even rear tyre squeal; by use of the selector lever the operation was still none too smooth with the throttle open to any degree. On the overrun, with throttle closed, its operation was much smoother, although by this method there was a three- to four-second delay between disengagement of third and second taking over. With second selected there is no governor to prevent engine over-speeding, whereas from third there is an automatic change to top at a predetermined limit (70 m.p.h. on this particular car), whether the selector is placed at 3 or 4.

A great asset of the Rolls-Royce transmission is the ability to use the gears to assist in braking this heavy car—especially useful when, for instance, crossing the Alps with a full complement. With reverse selected and the ignition switched off, a positive transmission lock holds the car safely on any gradient—an almost essential device on this car, of which the parking brake would not hold it on a 1-in-4 slope when lightly laden. This lock is released when the lever is moved to neutral, before the engine can be started, when it may be necessary to hold the car with the footbrake.

While most other large British cars, as well as quite a few of medium size, have now adopted disc brakes, Rolls-Royce maintain their allegiance to the drum type. Ever since these manufacturers adopted four-wheel brakes in the 'twenties their cars have been renowned for remarkable braking performance, aided by a mechanical friction-type servo driven off the gearbox. This type of servo is still used, although over the years its speed of rotation has been stepped up to increase efficiency and reduce lag.

Rolls-Royce provide unparalleled security from failure by duplicating their hydraulic lines and by including also a mechanical linkage to the rear wheels, so that there are three independent systems. Probably this installation represents a peak in efficiency and smoothness of drum braking never achieved by any other concern, and the effortless way

A silent, high-speed drawing-room for four: beside the front passenger's seat cushion is a trigger for adjusting rake of the backrest. Back-seat passengers have folding picnic tables; their feet could relax better on a sloping floor

in which this heavy car can be brought to rest from high speeds instils great confidence. Only in endurance against fade are they inferior to discs, but their rate of recovery after overheating is rapid. At very low speed, if one application is followed by another in quick succession, the servo may not have time to come to grips again; consequently a much higher pedal pressure is then required.

Power assistance for steering is now standard Rolls-Royce equipment. An hydraulic pump, belt-driven from the crankshaft pulley, feeds an external ram, of which one end is anchored to the chassis, the other to an idler-lever in the steering linkage. It is controlled by valves incorporated in the cam-and-roller steering box. The steering wheel is now an inch smaller in diameter than those of earlier Silver Clouds, and is mounted slightly nearer the facia to improve the driving position.

Anyone taking the wheel of this car without knowing that the steering was assisted would obviously suspect this because of the low effort required, but would sense no loss of feel. Indeed, this is just how powered steering should be, with assistance in constant proportion to normal effort, and with no lag in reaction or tendency to exaggerate the driver's movements. There is no appreciable kick-back through the wheel. Parking the car in a confined space is made easy by a combination of this assistance, a quite compact turning circle and a clear view of the side lamps on each front wing from the driving seat. At full speed on M1 on a still day, the car ran straight as an arrow with

Left: While almost as sumptuous as it looks, the back seat has very small side armrests. In the quarters are recesses containing mirrors and reading lamps; the one seen here also holds a cigarette lighter. The rear panel of each door window is fixed. Right: With power steering standard, the wheel is of quite small diameter; it has a particularly comfortable rim. The T-handle parking brake is apt to catch a knee when pulled out. A group of four switches enables the driver to control all windows—invaluable when travelling alone. Adjustable air vents at each end of the screen rail are new

Rolls-Royce Silver Cloud II . . .

Truly imposing is the standard coachwork, with lines which should not become démodé for years. Of particular note are the almost flush-fitting rear window (with built-in electric heater element) and flush rain gutters above the doors

practically no call for correction; on a less excellent road surface there was more need for concentration, to control an occasional tendency to wander.

A two-way electric switch on the steering column selects Normal or Hard setting of the rear suspension dampers, the latter for high-speed driving or to compensate for a full load of passengers and baggage. This makes an appreciable difference to cornering stability in particular, without stiffening the ride too much. A very high standard of riding comfort is expected and achieved, that on the rear seat being especially relaxed. So efficient is the damping that even a large bump is felt once only, the car thereafter levelling out without any pitch. It is not prone to wallow, despite low-frequency spring rates, and body sway or roll also is strongly resisted.

While certain forms of road irregularity can be felt more than might have been anticipated, no rear axle hop or other behaviour suggestive of excess unsprung masses was experienced. The large-section Dunlop tubeless tyres were quiet-running, except over the coarser type of road dressing which always induces noice, nor was there much squeal from them when sharp corners were taken fast.

A comprehensive array of black-dialled instruments is carried in a central panel, adaptable for right- or left-hand drive. It includes a push-button whereby the fuel gauge can be made to show oil sump level. The speedometer reading became rather optimistic above 70 m.p.h., showing 110 at a true 102 m.p.h. By pushing in the switch for the two-speed wipers, screen-washing jets are fed by an electric pump. Typically, the switch incorporates a temperature control to break the circuit, should the wipers be overloaded.

Concealed from view above the instrument panel is a small green map-reading lamp, but there is no lamp in the glove locker. Two fog lamps are controlled from the main lighting switch, their double filament bulbs also sharing duty as flashing signals; one cannot have head and fog lamps lit simultaneously. On the test car the head lamp beams were inadequate on both main and dipped filaments for fast

Topping a dummy filler cap on the radiator shell, the famous mascot is still silver-plated. The air-cleaner can be hinged upwards to give access to a pair of S.U. carburettors. To reach the right bank's sparking plugs one must first remove the wheel, then detach a panel behind it

driving; this did not appear altogether attributable to faulty adjustment. There is no provision for signal-flashing the head lamps. Potentially useful red tell-tales above the side lamps passed insufficient light to be seen by the driver.

Heating and ventilation are controlled by two switches on the facia, with such a wide choice of combinations of temperature, boost and direction that a full description is impossible here. There are two adjustable fresh air outlets recessed in the wooden capping above the facia, by which front seat occupants may receive a refreshing cool breeze on their faces in warm weather; there is also a separate recirculating intake from the floor of the rear compartment. The only criticism of an otherwise very efficient and versatile system was that, when very little extra warmth was required, the lowest heat setting gave a bit too much. Having this system allows one to travel comfortably with all windows shut, when the car becomes almost uncannily quiet, except for a little wind whistle behind the front screen pillars.

A much appreciated demisting element, almost invisible in the rear window glass, has a switch beside it—that is, remote from the driver. However, as it is wired through the ignition circuit, the obvious routine is to leave it switched on throughout the cold-weather months.

Separate backrests of the bench front seat have a much finer adjustment for rake than is usual, whereby any driver must find a correct and comfortable angle; they are not of the fully reclining type. It almost goes without saying that the seat is extremely comfortably shaped and upholstered, all trim being in top grain hide. In addition to folding armrests in the squabs, there are rests on each front door, adjustable for height. Three can be squeezed abreast on the front seat, when one appreciates how small is the hump in the floor over the transmission.

Electric window lifts (an optional extra) were fitted to the test car; they are so convenient that one very soon becomes reluctant to return to a car without them. Rather surprisingly, a hearty slam was needed to close the doors, even with a window open. Great travelling comfort for two is afforded by the luxurious rear seat, of which the backrests are just high enough to support the head if one slouches sufficiently. Although there is a wide shelf forward of the back window, no map pockets are included in the door trim; the backs of the front seats are occupied by folding picnic tables. An H.M.V. radio is standard in the Rolls-Royce, with a single rectangular speaker beneath the instrument panel. On this car its quality of reproduction was scarcely up to the standard of the rest of the vehicle.

At first sight the luggage compartment may appear less spacious than it is in fact, because it extends far back behind the lid hinges. Main dimensions are over 5ft long, 3ft 10in. wide and up to 15½in. deep. The spare wheel is in a separate compartment beneath, and secured by a typical Rolls-Royce lever and spring pad mechanism. The boot is lit automatically through a mercury switch when the counter-balanced lid is lifted. Rear lamp, stop lamp, brake-light and signal bulbs are reached through hinged access panels.

While few Rolls-Royce owners are likely to do their own maintenance, the schedule suggested in the handbook is not very demanding. Although centralized chassis lubrication is no longer fitted, main greasing points are fed from individual reservoirs which need topping up only at 10,000-

THE AUTOCAR, 13 MAY 1960

Rolls-Royce Silver Cloud II . . .

mile intervals. These include 10 points on the front suspension, 11 on the steering mechanism. A small set of hand tools, together with spare lamp bulbs and a tyre pressure gauge, are stored in a tray beneath the luggage boot floor. A very efficient hand pump and geared jack are clipped in place beside the spare wheel.

Without purchase tax, the Silver Cloud II costs a little over £4,000. Before the war its mighty forbear, the 7.3-litre vee-12 Phantom III, cost about £2,750 for a typical touring limousine. A two-door Morris Eight was then £128, whereas the basic price of its current equivalent is £416—well over three times as much. These proportionate increases should be borne in mind when assessing the value for money of today's Rolls-Royce cars.

Only by adopting advanced production methods and thereby increasing yearly output can a superlative machine like this be made today at a price its clientele can afford. The Rolls-Royce is one of very few surviving top quality cars; the maintained standard of overall excellence is rewarded by full order books, and a world reputation which has never stood higher.

ROLLS-ROYCE SILVER CLOUD II

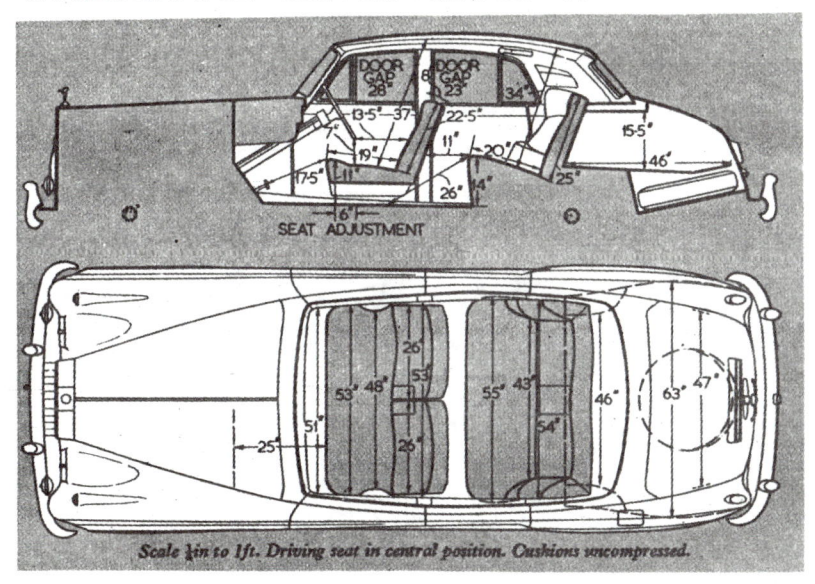

Scale ⅛in to 1ft. Driving seat in central position. Cushions uncompressed.

PERFORMANCE

ACCELERATION TIMES (mean):

Speed range, M.p.h.	Gear Ratios and Time in Sec.			
	3·08 to 1	4·46 to 1	8·10 to 1	11·75 to 1
10—30	—	—	3·0	—
20—40	—	4·9	3·5	—
30—50	7·6	5·1	4·5	—
40—60	7·8	5·4	—	—
50—70	8·1	6·8	—	—
60—80	8·5	—	—	—
70—90	11·9	—	—	—
80—100	17·4	—	—	—

From rest through gears to:
- 30 m.p.h. .. 3·8 sec.
- 40 ,, .. 5·9 ,,
- 50 ,, .. 8·3 ,,
- 60 ,, .. 11·5 ,,
- 70 ,, .. 15·5 ,,
- 80 ,, .. 21·1 ,,
- 90 ,, .. 27·4 ,,
- 100 ,, .. 38·5 ,,

Standing quarter mile 18·2 sec.

MAXIMUM SPEEDS ON GEARS:

Gear	M.p.h.	K.p.h.
Top (mean)	113·1	182·1
(best)	115·0	185·2
3rd	70	113
2nd	50	80
1st	21	34

TRACTIVE EFFORT (by Tapley meter):

	Pull (lb per ton)	Equivalent gradient
Top	270	1 in 8·2
Third	390	1 in 5·7
Second	680	1 in 3·1

SPEEDOMETER CORRECTION: M.P.H.

Car speedometer	10	20	30	40	50	60	70	80	90	100	110	120
True speed:	10	20	30	40	50	59	68	76	84	93	102	112

BRAKES: (at 30 m.p.h. in neutral):

Pedal load in lb	Retardation	Equiv. stopping distance in ft
25	0·28g	108
50	0·68g	44
75	0·81g	37·5
80	0·87g	34·7

FUEL CONSUMPTION (at steady speeds in top gear):

30 m.p.h.	19·5 m.p.g.
40 ,,	18·4 ,,
50 ,,	17·1 ,,
60 ,,	16·4 ,,
70 ,,	14·9 ,,
80 ,,	13·5 ,,
90 ,,	12·3 ,,
100 ,,	9·5 ,,

Overall fuel consumption for 1,559 miles, 11·8 m.p.g. (23·9 litres per 100 km.).
Approximate normal range 10—17·5 m.p.g. (28·2—16·1 litres per 100 km.)
Fuel: Premium grade.

TEST CONDITIONS: Weather: Dry and sunny, no wind.
Air temperature, 56 deg. F.
Model described in *The Autocar* of 25 September, 1959.

STEERING: Turning circle,
Between kerbs, L, 43ft 4in. R, 42ft 10in.
Between walls, L, 45ft 3in. R, 44ft 9in.
Turns of steering wheel from lock to lock, 4·25.

DATA

PRICE (basic), with standard saloon body £4,095.
British purchase tax, £1,707 7s 6d.
Total (in Great Britain), £5,802 7s 6d.
Extras: Electric window lifts £92 1s 8d (inc. P.T.)
Refrigeration system, £389 11s 8d (inc. P.T.) (H.M.V. radio fitted as standard).

ENGINE: Capacity, 6,230 c.c. (380 cu. in.)
Number of cylinders, 8 in vee.
Bore and stroke, 104·4 × 91·44mm (4·1 × 3·6in.).
Valve gear, o.h.v., self-adjusting tappets.
Compression ratio, 8·0 to 1.
M.p.h. per 1,000 r.p.m. in top gear, 27·8.

WEIGHT: (With 8 gals fuel), 40·4cwt (4,522lb).
Weight distribution (per cent): F, 49·5; R, 50·5.
Laden as tested, 43·4cwt (4,858lb).
Lb per c.c. (laden), 0·78.

BRAKES: Type, Rolls-Royce Girling drum. 2 T.S. front, L and T rear. Method of operation: F, hydraulic; R, hydro-mechanical. Mechanical servo assistance, and duplication of hydraulic system.
Drum dimensions: F and R, 11·25in. dia; 3·0in. wide.
Lining swept area: F, 212 sq. in. R, 212 sq. in. (196 sq. in. per ton laden).

TYRES: 8·20—15in. Dunlop tubeless.
Pressures (p.s.i.): F, 22; R, 27. (All conditions).

TANK CAPACITY: 18 Imperial gallons.
Oil sump, 13 pints.
Cooling system, 21 pints.

DIMENSIONS: Wheelbase, 10ft 3in.
Track: F, 4ft 10·5in.; R, 5ft 0in.
Length (overall), 17ft 7·75in.
Width, 6ft 2·75in.
Height, 5ft 4in.
Ground clearance, 7in.
Frontal area, 26·4 sq. ft. (approximately).

ELECTRICAL SYSTEM: 12-volt, 67 ampère-hour battery.
Head lamps, double dip; 60—36 watt bulbs.

SUSPENSION: Front: Independent, coil springs and wishbones, with anti-roll bar. Rear: Live axle on asymmetric, semi-elliptic leaf springs.

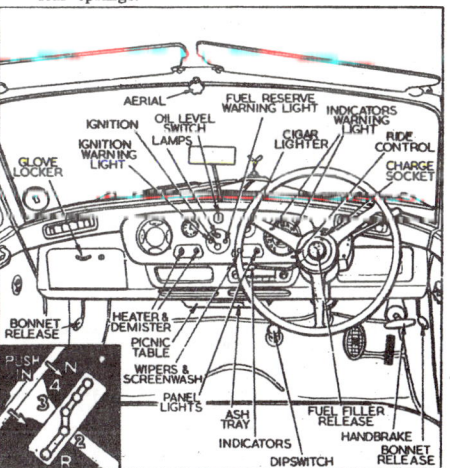

69

The Motor ROAD TESTS
Road Test No. 18/60

Make: Rolls-Royce **Type:** Silver Cloud II
Makers: Rolls-Royce Ltd., 14-15 Conduit Street, London, W.1

Test Data

World copyright reserved; no unauthorized reproduction in whole or in part.

CONDITIONS: Weather: Warm and dry with light wind. (Temperature 60°-70°F., Barometer 29.9-30.0 in. Hg.) Surface: Dry tarred macadam and concrete. Fuel: Premium-grade pump petrol (approx. 96 Research Method Octane Rating).

INSTRUMENTS
- Speedometer at 30 m.p.h. .. 6% slow
- Speedometer at 60 m.p.h. .. 7% slow
- Speedometer at 90 m.p.h. .. 3% slow
- Distance recorder .. 1% slow

WEIGHT
Kerb weight (unladen, but with oil coolant and fuel for approx. 50 miles) .. 40¼ cwt.
Front/rear distribution of kerb weight 50/50
Weight laden as tested .. .. 44½ cwt.

MAXIMUM SPEEDS
Flying Mile.
Mean of four opposite runs .. 112.3 m.p.h.
Best one-way time equals .. 113.9 m.p.h.

"Maximile" speed. (Timed quarter mile after one mile accelerating from rest.)
Mean of four opposite runs .. 104.7 m.p.h.
Best one-way time equals .. 105.9 m.p.h.

Speed in gears. (Automatic upward gearchange speeds at full throttle.)
Max. speed in 3rd gear .. .. 71 m.p.h.
Max. speed in 2nd gear .. .. 39 m.p.h.
Max. speed in 1st gear .. .. 27 m.p.h.

FUEL CONSUMPTION
- 19.5 m.p.g. at constant 30 m.p.h. on level.
- 18.5 m.p.g. at constant 40 m.p.h. on level.
- 17.5 m.p.g. at constant 50 m.p.h. on level.
- 17.0 m.p.g. at constant 60 m.p.h. on level.
- 15.5 m.p.g. at constant 70 m.p.h. on level.
- 14.0 m.p.g. at constant 80 m.p.h. on level.
- 12.0 m.p.g. at constant 90 m.p.h. on level.
- 10.0 m.p.g. at constant 100 m.p.h. on level.

Overall Fuel Consumption for 1,670 miles, 139.1 gallons, equals 12.0 m.p.g. (23.6 litres/100 km.)

Touring Fuel Consumption (m.p.g. at steady speed midway between 30 m.p.h. and maximum, less 5% allowance for acceleration). 13.0 m.p.g.
Fuel tank capacity (maker's figure) 18 gallons (warning lamp operates when 3 gallons remain).

STEERING
Turning circle between kerbs:
- Left .. 40¾ ft.
- Right .. 40 ft.

Turns of steering wheel from lock to lock 3⅔

BRAKES from 30 m.p.h.
- 0.92 g retardation (equivalent to 32½ ft. stopping distance) with 60 lb. pedal pressure.
- 0.85 g retardation (equivalent to 35⅓ ft. stopping distance) with 50 lb. pedal pressure.
- 0.46 g retardation (equivalent to 65⅓ ft. stopping distance) with 25 lb. pedal pressure.

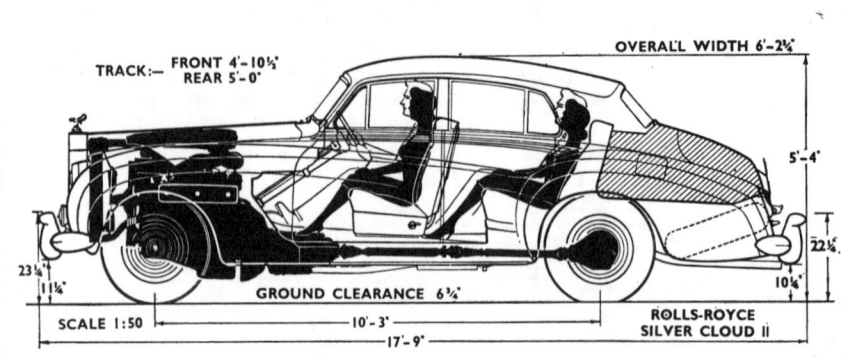

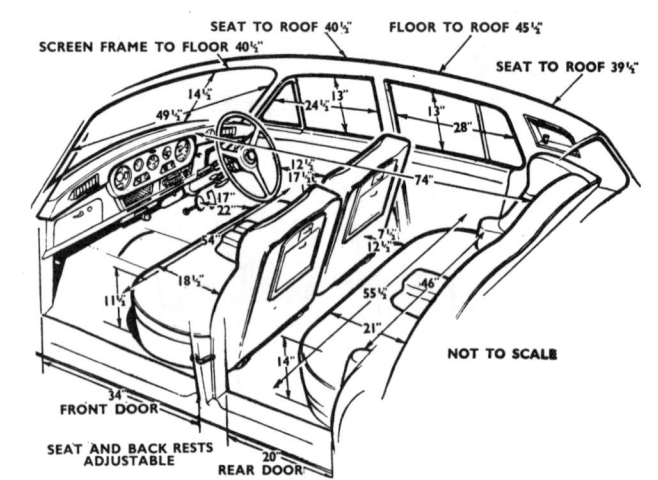

ACCELERATION TIMES from Standstill
0-30 m.p.h.	3.4 sec.
0-40 m.p.h.	5.5 sec.
0-50 m.p.h.	7.9 sec.
0-60 m.p.h.	10.9 sec.
0-70 m.p.h.	14.7 sec.
0-80 m.p.h.	20.2 sec.
0-90 m.p.h.	28.0 sec.
0-100 m.p.h.	40.1 sec.
Standing quarter mile	18.0 sec.

ACCELERATION TIMES from Rolling Start
Driving range
0-20 m.p.h.	2.0 sec.
10-30 m.p.h.	2.4 sec.
20-40 m.p.h.	3.5 sec.
30-50 m.p.h.	4.5 sec.
40-60 m.p.h.	5.4 sec.
50-70 m.p.h.	6.8 sec.
60-80 m.p.h.	9.3 sec.
70-90 m.p.h.	13.3 sec.
80-100 m.p.h.	19.9 sec.

HILL CLIMBING at sustained steady speeds
- Max. gradient on top gear .. 1 in 7.1 (Tapley 315 lb./ton)
- Max. gradient on 3rd gear .. 1 in 4.5 (Tapley 485 lb./ton)
- Max. gradient on 2nd gear .. 1 in 3.0 (Tapley 720 lb./ton)

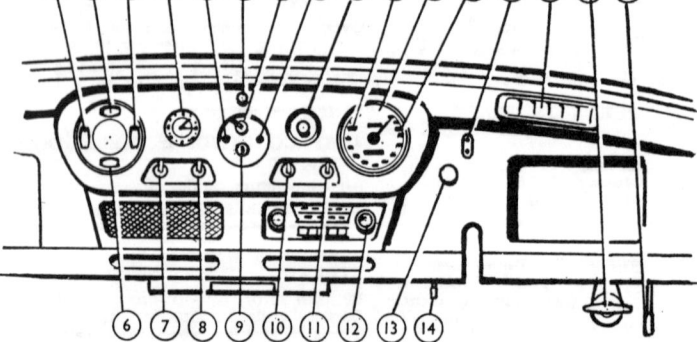

---

1. Headlamp dipswitch. 2. Ride control switch. 3. Direction indicator switch. 4. Transmission selector. 5. Horn button. 6. Fuel and oil level gauge. 7. Heater temperature control and fan switch. 8. Demister control and fan switch. 9. Ignition and starter switch. 10. Windscreen wiper and washer switch. 11. Panel light and map-reading light switch. 12. Radio. 13. Petrol cap cover release. 14. Trip adjuster. 15. Oil pressure gauge. 16. Ammeter. 17. Water thermometer. 18. Clock. 19. Dynamo charge warning light. 20. Oil level indicator button. 21. Lights switch. 22. Fuel warning light. 23. Cigar lighter. 24. Direction indicator warning light. 25. Speedometer, and distance recorder. 26. Inspection lamp plug. 27. Air ventilator (one each side). 28. Handbrake. 29. Bonnet release catch (one each side).

The Rolls-Royce Silver Cloud II

Effortless Speed and Extreme Quietness with a New V-8 Engine

UNFASHIONABLE though it may be to welcome an increase in the price of anything, we finished our recent test of a Rolls-Royce Silver Cloud II saloon convinced that its makers were 100% right in making it an appreciably more expensive car than its predecessors. When the company first marketed an all-steel saloon some eight years ago, it cost as much as nine contemporary small cars, whereas today's equivalent Rolls-Royce costs as much as 12¼ modern small cars. By spending this extra money, the designers have been able to ensure that a car of the 1960 vintage is amongst the finest in their history.

In this new version of what is proudly advertised as "the best car in the world" the prime mechanical change is replacement of a 4.9-litre six-cylinder engine by a V-8 power unit of 6.2-litre size. Redesigned ventilation for the bodywork, and standardized power steering, with some appropriate chassis modifications, are other welcome major improvements which are backed up by countless less conspicuous refinements. Whereas not so many years ago it was possible to wonder whether or not one really wanted a Rolls-Royce, now it only seems possible to wonder whether or not one can afford a Rolls-Royce.

Essentially the Silver Cloud II is laid out on traditional "big car" lines, with a luxurious five-seat body built upon a separate chassis of 10-ft. 3-in. wheelbase and having an overall length of 17 ft. 9 in. Very high performance is provided without any fuss or effort at all, the minor irregularities of ordinary roads are silently ironed out by the suspension almost as if they had been steam-rollered out of existence, and extremely complete sound insulation keeps passengers quietly apart from the turmoil of life outside the car. Beautifully proportioned, the Silver Cloud II looks a pleasantly but not exaggeratedly low-built car, yet its size is, in fact, such that you step easily into it without stooping, and sit with ample headroom in a virtually flat-floored body from which you look down on other motorists.

Silent Power

Putting their new light-alloy eight-cylinder engine into a car which it is difficult to distinguish visually from its recent predecessors, Rolls-Royce Ltd. have greatly increased the Silver Cloud's power-to-weight ratio. Top speed on the level has gone up by about 10 m.p.h., to more than 112 m.p.h.; the time needed to reach 80 from a standstill is 22½% lower than formerly, at 20.2 sec.; despite higher gearing, the new engine provides about 10% more pulling power in top gear, it now being possible to sustain a steady speed up gradients of the 1 in 7 in order without the automatic gearbox needing to engage a lower ratio. The best testimony which can be given to the quietness with which large but undisclosed amounts of power are developed is that, at any motorway cruising speed up to 100 m.p.h. or more, it proved possible for the driver to slip the gear selector into neutral and let the engine idle without his passengers noticing anything except that the car was slowing down. Modest enough in its fuel anti-knock requirement to run happily on the premium-grade petrol of any country which a tourist is likely to visit, this engine does have a considerable thirst for fuel, but consumption figures which on the road are likely to range between 10 m.p.g. and 14 m.p.g. should not worry the buyer of this car from the cost point of view; the non-stop cruising range provided by an 18-gallon tank (a warning lamp on the facia lights up when 3 gallons or so remain) is however rather limited, and the petrol filler will not always accept fuel at the maximum delivery rate of a modern pump.

Transmitting the high torque of a big engine working very well within its potential, the Rolls-Royce four-speed automatic transmission has an easier job than have comparable units fitted to less powerful cars, and in almost all conditions it does this job excellently without the driver needing to use the manual override control. In first and second ratios the power unit is just audible inside the car, but the great majority of driving is done with third or top gear automatically engaged and not only are these ratios silent, but the upward and downward changes between them are virtually imperceptible. The gap between third and second ratios is rather wide, and if jerky changes are not to take place the driver must be discreet in his use of the accelerator pedal at speeds below 20-25 m.p.h. A clumsy driver, given a car which in response simply to hard pressure on the accelerator pedal will out-accelerate most sports two-seaters, can jerk his passengers, but a sensitive driver can accelerate the Rolls-Royce very rapidly indeed with almost complete smoothness and quiet. Idling at a speed low enough to eliminate "creep" due to drag in the fluid coupling at traffic checks, the V-8 engine was not altogether its usual smooth self at tick-over, but its automatic choke eliminated all warming-up temperament in the mild spring weather of our two-week test.

Almost every element of the Rolls-Royce braking system is duplicated to guard against possibilities of failure, and as hitherto drum brakes of large size (and with very little of the self-wrapping effect

TWIN hinged panels give access to the top of the V-8 engine, the air filter also tilting upwards if required: platinum-point sparking plugs below the exhaust manifolds should seldom need attention.

In Brief

Price £4,300 plus purchase tax £1,792 15s. 10d. equals £6,092 15s. 10d.
Capacity ... ... 6,230 c.c.
Unladen kerb weight ... 40¼ cwt.
Acceleration:
 20-40 m.p.h. in driving range ... ... 3.5 sec.
 0-50 m.p.h. through gears 7.9 sec.
Maximum direct top gear gradient ... ... 1 in 7.1
Maximum speed 112.3 m.p.h.
"Maximile" speed ... 104.7 m.p.h.
Touring fuel consumption ... 13 m.p.g.
Gearing: 27.3 m.p.h. in top gear at 1,000 r.p.m.; 45.5 m.p.h. at 1,000 ft./min. piston speed.

SPACIOUS and with an almost completely flat floor, the interior is finely furnished in traditional style, and very easy to enter or leave through four wide doors. Details in the rear quarter (*above*) include a separate light, vanity mirror and cigar lighter. Below are seen the slide-out picnic table and folding armrests in the front compartment.

The Rolls-Royce Silver Cloud II

which can magnify fade) are applied with help from a gearbox-driven servo. Extremely smooth control over a very heavy and very fast car is provided by these brakes, and although in very slow stop-go traffic those who knew early versions of this system can still detect a lag amounting to a few inches of motion along the road before full servo assistance is obtained, only those who have previously found fault are likely to detect vestiges of it in the latest car. With an open window, a hard and almost metallic rubbing is audible if these fade-resistant brakes are applied on a road which echoes the sound off stone walls, but there was never any trace of squeal.

Light Steering

Steering by a cam and lever gear has power assistance from an engine-driven hydraulic pump as a standardized feature. As a town runabout or in narrow and winding country lanes this big car is transformed into quite a nimble one by the fact that its steering remains light right down to the lowest speeds—in fact, on most surfaces and with the engine running, it is quite easy to swing from lock to lock with the car at rest. In ordinary driving, however, it is not possible to detect when power assistance begins to magnify the driver's manual effort, the steering being light at all times but never embarrassingly so.

At first acquaintance, this does not seem to be the most completely stable of cars, it being easy for a driver to over-correct slight deviations from a true course caused by road camber or by wind. Full confidence is soon acquired however; and fairly sensitive response to small corrective movements of the steering wheel, but a mounting degree of understeer if corners are tackled fast, come to be appreciated as quite suitable handling characteristics for a very fast touring car which is not meant to be thrown around like a sports model. Attempts to corner inappropriately fast produce body roll and tyre squeal, but quite brisk cornering (silence and that great rarity, a pessimistic speedometer, often tend to disguise the car's true speed) and effortlessly quick acceleration back to cruising speed even if the road is quite steeply uphill, get this car along traffic-avoiding secondary roads very rapidly.

As has been indicated, the primary virtue of this model's very orthodox but very carefully engineered suspension system is its ability to eliminate small bumps. Living up to its "Silver Cloud" name, it floats along over ordinary main-road irregularities which are neither felt not heard inside the rubber-mounted body. When a bigger bump or wave in the road does penetrate the defences, the fact that the car then rises and falls just like any other is almost more noticeable because it is a surprise. A two-position ride control switch is provided on the steering column, to alter the rear shock absorber settings, but we found only a very few occasions when the "soft" setting was preferable for driving over cobblestones at less than 30 m.p.h., and rather more numerous occasions when we wondered if something harder than the "hard" setting (which we used almost continuously) would have given even greater comfort and stability at speed along an undulating road. With a good driver at the wheel and on British roads, the standard of riding comfort provided for a rear-seat passenger is at least as high as that enjoyed in the front seat, despite the considerable unsprung weight of a rigid rear axle and semi-elliptic leaf springs.

With all the windows closed, the complete quietness of this car at very high speeds is astonishing. Only one facia-mounted loud speaker is used with the standard-equipment radio, yet no more than a very slight adjustment of the volume control setting appropriate to a traffic jam is needed to let the rear-seat passenger enjoy music as he is driven along a motorway at more than 110 m.p.h. Opening any window by even the smallest amount destroys this silence, especially at any speed above about 60 m.p.h., by producing a great deal of wind noise and pressure pulsation, but the latest ventilation system goes far towards eliminating any need for open windows. Two air-circulating systems have separate variable-speed fans and air and water control valves, one circulating heated air in the front and back compartments at floor level when required, and the

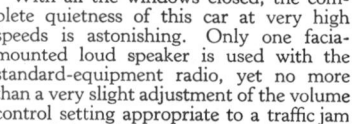

PROPORTIONED to look low although it is in fact of uncramped interior height, the Silver Cloud II is a quality car of classic form with modern cleanliness of detail design.

LONG and devoid of the tail fins which act as corner markers on fashion-following cars, the tail of the all-steel body encloses a flat, carpeted luggage floor of generous area.

other letting fresh air blow onto the windscreen interior and through adjustable louvres towards the faces of the driver and front passenger. Whilst temperature control is in steps rather than being fully progressive, this system copes excellently with cold and temperate weather, but when the outside temperatures rose above 70 deg. F. it proved necessary to open windows if rear-seat passengers were to be truly comfortable. At an extra cost of £275 plus £114 11s. 8d. purchase tax, refrigeration can be incorporated in the ventilating system to provide true air conditioning, and limited experience of this installation on another example of the Silver Cloud II suggests that it is a worth-while aid to comfort in British summer weather as well as being a "must" for tropical conditions. A very useful option which was fitted on our test car was electrically-operated windows with a four-switch master control on the driver's door.

Submitted to us for test after it had run nearly 16,000 miles, the subject of this test report was still in almost immaculately perfect condition inside and out, emphasizing that the rather vulnerable-looking polished woodwork, deep pile carpets and fine-quality leather upholstery are not unsuited to hard usage. A minor short-circuit at the switch for variable-brightness instrument lighting led to our confirming that this car offers far more than skin-deep beauty, removal of four screws allowing the walnut facia panel to be removed without further difficulty, each individual control unit and instrument then being easily and separately removable from an engineered backplate should it require attention. Hidden quality of this sort is in character with the refinements of detail which have resulted in chassis greasing being required only at 10,000 mile intervals.

A bench-type front seat is used to make three-abreast seating possible, but with a split backrest of which each half is separately adjustable for rake and carries its own folding central armrest. The rear seat has a central folding armrest, and despite intrusion of the wheel arches offers fair accommodation for three people, rear-seat legroom and headroom being very ample. Details of equipment such as folding picnic tables, multiple interior lights and vanity mirrors, separate front and rear cigar lighters, and rear-window de-misting by an electrical heater element, are in accordance with what is expected of a Rolls-Royce, but a headlamp flasher would be a welcome further item. Ample capacity for luggage is provided on the carpeted floor of an internally-lit rear locker, and inside the car two door pockets supplement twin glove-boxes and a rear parcel shelf as stowage for odds and ends. A full set of instruments is provided (not including an engine rev. counter), but apart from the big and clearly calibrated speedometer they act merely as indicators rather than giving precise readings.

There are few more satisfying ways of travelling by road than to sit behind the familiar Rolls-Royce radiator and long bonnet, with a fine view in all directions but almost completely sound-insulated from the world at large, floating gently along at an unhurried pace which is far above the speed of most other traffic. Supplemented perhaps by a very small car for errands into places where parking is difficult, the Rolls-Royce Silver Cloud II can provide its owner with a superb form of motoring.

The World Copyright of this article and illustrations is strictly reserved © Temple Press Limited, 1960

Specification

Engine

Cylinders V-8
Bore ... ... 104.14 mm.
Stroke ... ... 91.44 mm.
Cubic capacity ... ... 6,230 c.c.
Piston area ... ... 105.5 sq. in.
Valves: Pushrod-operated o.h.v. with hydraulic tappets.
Compression ratio 8.0/1
Carburetter: Twin 1¾ in. S.U. horizontal type HD6.
Fuel pump Twin S.U. electrical, rear mounted
Ignition timing control ... ... Centrifugal
Oil filter Full-flow (British Filters Ltd.)
Max. power ... ... not quoted
Piston speed at car's max. speed 2,470 ft./min. at 4,100 r.p.m.

Transmission

Clutch: Hydraulic coupling incorporated in Rolls-Royce automatic transmission.
Top gear 3.08
3rd gear 4.46
2nd gear 8.10
1st gear 11.75
Reverse 13.25
Propeller shaft: Rolls-Royce Hardy Spicer divided open.
Final drive ... ... 13/40 hypoid bevel
Top gear m.p.h. at 1,000 r.p.m. 27.3
Top gear m.p.h. at 1,000 ft./min. piston speed 45.5.

Chassis

Brakes: Rolls-Royce drum type, with gear-box-driven servo: two separate hydraulic circuits for front wheels, hydraulic and mechanical operation of rear brakes.
Brake drum diameters ... ... 11¼ in.
Friction areas: 240 sq. in. of lining area working on 424 sq. in. rubbed area of drums.
Suspension:
Front: I.f.s. by coil springs, transverse wishbones and anti-roll torsion bar.
Rear: Semi-elliptic leaf springs and single radius arm.
Shock absorbers: Rolls-Royce lever-arm hydraulic, with electrical two-position remote control of rear damper setting.
Steering gear: Rolls-Royce hydraulic power-assisted cam and roller steering.
Tyres: 8.20–15 tubeless. (Dunlop on test model).

Coachwork and Equipment

Starting handle ... ... None
Battery mounting: Under floor of luggage locker.
Jack ... ... Bevel-geared bipod screw pillar
Jacking points: Two external sockets below body sides.
Standard tool kit: Jack, tyre pump, wheel brace, wheel disc remover, box spanner, adjustable spanner, 3 open-jaw spanners, contact breaker spanner with feeler gauge, drain plug key, pliers, screwdriver, tyre pressure gauge, spare bulbs.
Exterior lights: 2 headlamps, 2 sidelamps, 2 foglamps, 2 stop/tail lamps, number plate lamp, reversing lamps.
Number of electrical fuses: 15 (including standardized radio and optional electrical windows).
Direction indicators: Self-cancelling flashers (amber lenses at rear, combined with fog-lamps at front).
Windscreen wipers: Two-speed Lucas D.R.3 electrical twin-blade, self-parking.
Windscreen washers Lucas electrical pump type
Sun visors ... ... Two, universally pivoted
Instruments: Speedometer with total and decimal trip distance recorders, clock, oil pressure gauge, coolant thermometer, ammeter, fuel contents gauge (also indicates oil level).
Warning lights: Dynamo charge, headlamp main beam, turn indicators, low fuel level.
Locks:
With master key: Ignition/starter switch, either front door, glove box, luggage locker.
With other key: Ignition/starter switch and doors only.
Glove lockers: Two on facia (one open, one with lockable lid).
Map pockets ... Two inside front doors
Parcel shelves ... One behind rear seat
Ashtrays One on facia, two behind front seats
Cigar lighters ... One front, one rear
Interior lights: One in roof with manual switch and four door-operated switches, one map-reading lamp, two lamps in rear quarters.
Interior heater: Fresh air heating, with separate volume and temperature controls for de-misting and heating air.
Car radio: H.M.V. Radiomobile 400T fitted as standard.
Extras available: Power-operated windows, full air conditioning (including refrigeration).
Upholstery material Top grain leather
Floor covering ... ... Pile carpets
Exterior colours standardized: 16 colours, or two-tone combinations of these: any other colour to order at extra cost.
Alternative body styles: On long wheelbase chassis, saloon with division, or H. J. Mulliner drop-head coupe.

Maintenance

Sump: 13 pints, S.A.E. 20W winter and S.A.E. 30 summer, or S.A.E. 10W/30 multigrade.
Gearbox: 20 pints, automatic transmission fluid
Rear axle ... 1⅝ pints, Castrol Hipress SC
Steering gear lubricant: S.A.E. 90 hypoid gear oil in steering gear, and automatic transmission fluid in power-assistance system.
Cooling system capacity 21 pints (3 drain taps)
Chassis lubrication: By grease gun every 10,000 miles to 21 points.
Ignition timing ... ... 2° b.t.d.c. static
Contact-breaker gap ... 0.019-0.021 in.
Sparking plug type: Lodge CLNP or Champion RN8.
Sparking plug gap ... ... 0.024-0.027 in.
Valve timing: Inlet opens 20° b.t.d.c. and closes 61° a.b.d.c.; exhaust opens 62° b.b.d.c. and closes 19° a.t.d.c.
Tappet clearances ... ... Not adjustable
Front wheel toe-in ... ... 1/16 in. to 3/16 in.
Camber angle ... ... Zero
Castor angle ... ... Zero
Steering swivel pin inclination ... 4½°
Tyre pressures:
Front ... ... 22 lb.
Rear ... ... 27 lb.
Brake fluid: Wakefield-Girling Crimson (S.A.E. spec. 70-R-1).
Battery type and capacity: 12 v. 67 amp. hr. (Dagenite 6HZP 11/9 62F or Exide 6.XTHZ 11/L).

ROAD TEST
ROLLS-ROYCE SILVER CLOUD II

Prestige and performance, at £1 per pound

JERRY CHESEBROUGH PHOTOS

SEVEN YEARS AGO the Rolls-Royce firm decided that an entirely new powerplant would, ultimately, be necessary. Today, we have it—a beautifully engineered all-aluminum V-8 unit of a size and potential sufficient to insure that no major changes will be necessary for a good many years.

In typical Rolls-Royce tradition, no power and torque figures are published. However, careful consideration of the acceleration and speed capabilities, as shown by this test, indicates a useful output of about 220 honest bhp. We say "honest," because certain American V-8 engines of similar size claim a great deal more power, but their actual road performance definitely proves that the true net horsepower at the flywheel and as-installed is nowhere near the claims of 300 plus.

One of the amazing things about the new Rolls-Royce is that it looks exactly like the last Rolls-Royce we tested; a 1957 Silver Cloud model powered by a 6-cyl, F-head engine of approximately 180 bhp (R&T, May 1958). While there are many minor chassis changes, as well as the major change of a new engine, a general driving impression of the V-8 model is not really much different from that of 1958. It still has power steering and a hydramatic transmission, for example. All principal dimensions, both inside and out, are unchanged.

At idle the engine is so quiet that it is difficult to tell whether it is running or not. So was the six. But press the throttle slightly and as the engine speeds up you can tell the difference. There is, in fact, a slight trace of exhaust noise that seems somewhat more noticeable than before and our general impression was and is that the V-8 isn't as quiet as the six at full throttle, at low speeds. However, you *can* hear the clock tick at 60 mph—provided all windows are closed and the air-conditioning is turned off. At 60 mph and up the V-8 is definitely smoother and quieter than the six and, of course, the primary reason for the new and larger engine was to improve the performance, particularly in the 60 to 90 mph range. This does it; 0-90 mph, for example, comes up in about the same time as was formerly required for 0-80.

Since we seldom drive cars of the luxury type, we felt that our evaluation of the Rolls-Royce should include a check of a comparable American luxury car. For the record the American car, which advertises 325 bhp and weighs about the same, gave virtually identical acceleration times to 30, 60 and 90 mph (after correction for speedometer error). Engine noise and general running smoothness were somewhat inferior to the Rolls-Royce. The ride was slightly better, as judged by several passengers on the basis of boulevard performance. In this respect the Rolls-Royce seems to suffer particularly from tire thump and, unless the ride control was on "firm," there was some pitch. We mention this because a car of Rolls-Royce stature invites criticism, though finding fault is actually very difficult.

As for roadability, the Rolls is a vehicle par-excellence for comfortable long-distance driving. It does not, however, invite sports-car-like tactics on winding roads and the tires protest loudly if so treated. Power steering is standard equipment, and necessary. It provides 50% assistance when cruising, 80% boost for parking: an admirable arrangement. The feel is a little sensitive at speed and the first high speed bend encountered causes some excitement, because it takes more force and more rim

SEPTEMBER 1960

A surfeit of plumbing.

movement to hold the desired line than one expects. This is a matter of experience and practice and also, perhaps, due to our own lack of experience with power steering. At any rate, the new model definitely has more understeer than formerly. This is readily noticeable and was done deliberately, with safety in mind.

The electrically actuated hydraulic control for the rear dampers works, although it doesn't seem to make too much difference, one way or the other. For high speed cruising we used "firm" at all times, because the car seemed to feel more stable and there was considerably less cornering roll. The softer setting might be preferable on the expressway and would certainly be used all the time on the boulevards.

For the driver, the Rolls abounds in novelties and niceties, designed to enhance comfort, convenience and safety. Foremost is the very high seating position, accompanied by a floor that is very nearly flat. Three in the front seat is, however, a little crowded, and a pair of folding center armrests in front invites driving in armchair elegance. Instrumentation is complete and readable, a minor criticism being that some of the gauges are too far away for easy checking. The array of knobs, buttons and controls is formidable and we're not sure, even yet, whether we had full control of the air-conditioning system. However, it works very well except for the fact that the rear blower fans are noisy and once, in a traffic jam, the engine thermometer went up to a temperature of nearly 200° F (which isn't even close to boiling, because ethylene glycol under pressure is used as the coolant).

The Rolls-built hydramatic transmission provides smooth, effortless driving and is, of course, extremely efficient, due to its use of a fluid coupling which slips almost not at all (about 1%) at cruising speeds. Transmission control is sensibly arranged with a N-2-3-4-R sequence. The Rolls people say their over-riding control is unique, but it functions exactly the same as on American cars. If you put the lever in "2," the car starts in first gear, but quickly shifts to 2nd—and stays there unless the speed drops to 5 mph. This is exactly the same as the usual low range. If you select "3" the car operates exactly as in "4" (or drive), with the exception that the transmission downshifts from 4th to 3rd whenever the speed falls to 60 mph, regardless of throttle position. Full throttle (3-4) upshifts occur at 70-72 mph when the control is either at "3" or at "4." This method of controlling the transmission is admirable, though one of our passengers having his first ride in a Rolls-Royce remarked that he had the same thing in his car (a 300 hp V-8) and he felt that a car such as this, designed for chauffeur driving, should have "the good old 4-speed box." Our opinion is more or less the same, and we wish the R-R engineers would make the hydraulic system completely controllable by giving the driver a full 1-2-3-4 sequence in which "4" only would give full automatic operation.

As for performance, the figures in the data panel show truly remarkable accelerative powers, particularly with a machine weighing 2.5 tons. During the tests it was impossible to get even one timed top speed run, but we rate the car as being capable of an honest 112 mph for a two-way level road average. Incidentally, the speedometer was commendably accurate up to 70 mph and then got progressively fast. The highest speed check we got was at 110 mph indicated, equivalent to exactly 99.0 mph. Two owners told us they have seen well past the last mark on the dial (120 mph), which figure, allowing for error, would probably be over 110 mph. Fuel economy isn't very good; we got 10.4 mpg on one check and a best figure of 11.5 on a 200-mile trip. However, the tank holds 21.6 gallons and the cost of re-filling it wouldn't be important to a Rolls owner.

From the right angle, you can see the engine.

Instrumentation is conservative, complete.

ROAD & TRACK

The brakes are boosted by the traditional R-R servo motor and they really are powerful, particularly at high speed. The lining area (240 sq in.) is tremendous and this is the only large car we know with completely adequate brakes, which are fully up to the car's weight and performance capability. There is, however, a minor fault which we discovered quite by accident. When putting the car in our office garage the front wheel passed over a small board, and just as it rolled over the obstruction we applied the brakes and smacked into a large wooden box. In other words, when creeping at one mph, the mechanical brake booster doesn't work and the brakes require considerably more force on the pedal.

It would take several more pages to enumerate all of the many interesting and unique features of the Rolls-Royce. A few of these are: a push-button release on the instrument panel for the gasoline filler, a similar button which causes the fuel gauge to read the oil level in the engine sump, a socket for plugging in a battery charger, a picnic table that slides out from under the instrument panel, chassis lubrication required only at 10,000-mile intervals, self-adjusting brakes, and so on. Every detail of the car, inside and out, shows careful planning and near-perfect workmanship. Anyone having the slightest knowledge of what it costs to build a car with this kind of finish and workmanship will not wonder at the price asked —actually, he will wonder why the price isn't much higher.

An explanation of the fact that the Rolls-Royce selling price has not increased in the same ratio as other cars is simple. Before the war, Rolls built only the chassis and all bodies were strictly custom built. However, since the war, Rolls has offered one standardized body type (the model shown here) and the economies effected have been passed on to the customer. The exact saving is essentially $9000, for other body types are catalogued at around $24,000 and up, these being the very limited-production custom-built types.

While the price of even this standard-type Rolls-Royce seems to be a lot of money, roughly twice as much as one of our own luxury cars, the man who can afford one also enjoys the benefits of drastically reduced depreciation.

To rephrase an old cliche, "it takes money to save money," in short, the current Rolls-Royce is an excellent value and it is a pleasure to report that there is still a company in business dedicated to the task of producing the best car in the world, regardless of cost.

What a spot for a picnic.

ROAD & TRACK ROAD TEST 258

ROLLS-ROYCE V-8

SPECIFICATIONS
List price	$15,655
Curb weight	4680
Test weight	4980
distribution, %	49.5/50.5
Dimensions, length	212
width	74.8
height	64.0
Wheelbase	123
Tread, f and r	58.5/59.9
Tire size	8.20-15
Brake lining area	240
Steering, turns	4.3
turning circle, ft	41.7
Engine type	V-8, ohv
Bore & stroke	4.1 x 3.6
Displacement, cu in	380.0
cc	6230
Compression ratio	8.00
Bhp @ rpm (est)	220 @ 4000
equivalent mph	112
Torque, lb-ft. (est)	340 @ 2200
equivalent mph	61.1

GEAR RATIOS
O/d ( ), overall	n.a.
4th (1.00)	3.08
3rd (1.45)	4.46
2nd (2.63)	8.10
1st (3.82)	11.8

CALCULATED DATA
Lb/hp (test wt)	20.8
Cu ft/ton mile	95.4
Mph/1000 rpm (4th)	27.8
Engine revs/mile	2160
Piston travel, ft/mile	1295
Rpm @ 2500 ft/min	4160
equivalent mph	116
R&T wear index	20.0

PERFORMANCE
Top speed (est.), mph	112
best timed run	n.a.
3rd (3700)	71
2nd (3700)	39
1st (3150)	23

FUEL CONSUMPTION
Normal range, mpg	11/14

ACCELERATION
0-30 mph, sec	3.3
0-40 mph	5.0
0-50 mph	8.0
0-60 mph	11.4
0-70 mph	15.5
0-80 mph	21.3
0-90 mph	28.5
0-100 mph	43.0
Standing ¼ mile	18.0
speed at end, mph	75

TAPLEY DATA
4th, lb/ton @ mph	240 @ 55
3rd	355 @ 40
2nd	520 @ 30
1st	off scale
Total drag at 60 mph, lb	220

SPEEDOMETER ERROR
30 mph	actual 29.4
40 mph	40.0
50 mph	50.0
60 mph	59.8
70 mph	68.7
80 mph	77.5
90 mph	85.0
100 mph	92.0

SEPTEMBER 1960

New Air Conditioning for Rolls-Royce and Bentley

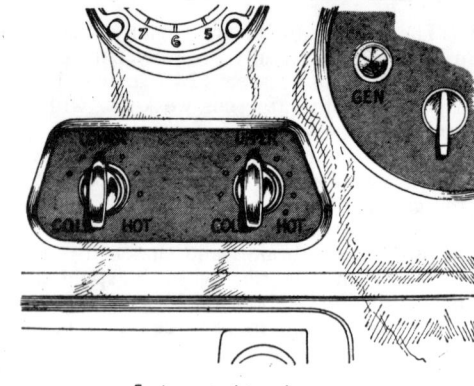

Facia control panel

IN our description, on earlier pages, of the latest S.2 series Rolls-Royce and Bentley cars, a brief reference is made to the completely new ventilation system developed for them. The refrigeration is an optional extra, and provision is made in the air conditioning unit for the evaporator matrices. The whole assembly is now mounted under the right front wing above and behind the wheel, and so the components of the refrigerator no longer occupy any of the rear boot space. Basically the refrigerator system is as described fully in *The Autocar* of 4 April 1958, operating on the vapour cycle system, with a straightforward series layout of engine-driven compressor, condenser and evaporator; Freon 12 is the gaseous cooling medium. The compressor is belt-driven through a magnetic clutch.

Apart from the fact that the basic elements do not occupy useful space in the car, the most important feature of the new air conditioning is that a split air system is used. One supplies fresh air from the front of the car, and the other recirculation of the air already inside. By means of two facia controls, the balance between these is adjusted. In very hot countries, where refrigeration is becoming increasingly popular, the air temperature inside the car may be higher than that outside when it has been standing in the sun. By using the fresh air side of the circuit only, much quicker cooling can be obtained.

In the fresh air circuit, the intake is from the grille at the right-hand side of the radiator to a nylon filter and two-speed fan; ram effect assists flow at high speeds. Through polythene trunking air passes to an electrically operated control valve which can be selected to direct the flow through the heater, or the evaporator matrix when the refrigeration unit is fitted. The air is then directed to three slots along the lower edge of the windscreen, and two on the facia panel. These are provided with adjustable deflectors for directing the air to the face if desired or, if need be, to demist the side windows.

For the recirculatory system, the intake is on the floor just behind the front right seat. The air is passed in turn through a nylon filter, two-speed motor, and, to choice, the evaporator or lower part of the heater matrix in this secondary circuit to the body sides. There are two exits on each side near the front passengers' feet, and one on the left side in the gap between the door and the seat. A pull-out drawer beneath the facia can be used to direct air over the top of the seats for the benefit of the rear passengers when refrigeration is fitted; it is not used on the standard system.

By means of the two facia controls a very wide range of conditions can be chosen. A degree of dehumidifying can be achieved by passing the recirculating air through the heater side of its system and mixing it with a cool charge on the fresh air side. With full refrigeration, 50 to 60lb is added to the weight, and cars fitted with refrigeration have additional roof and exhaust system insulation.

The Phantom V and Continental Bentleys retain the previous refrigeration layout designed for boot mounting. For the former this is because the conditioning is naturally concentrated on the rear passengers' compartment rather than that of the chauffeur, while in the case of the Continental the more compact frontal area precludes the use of the under-wing unit.

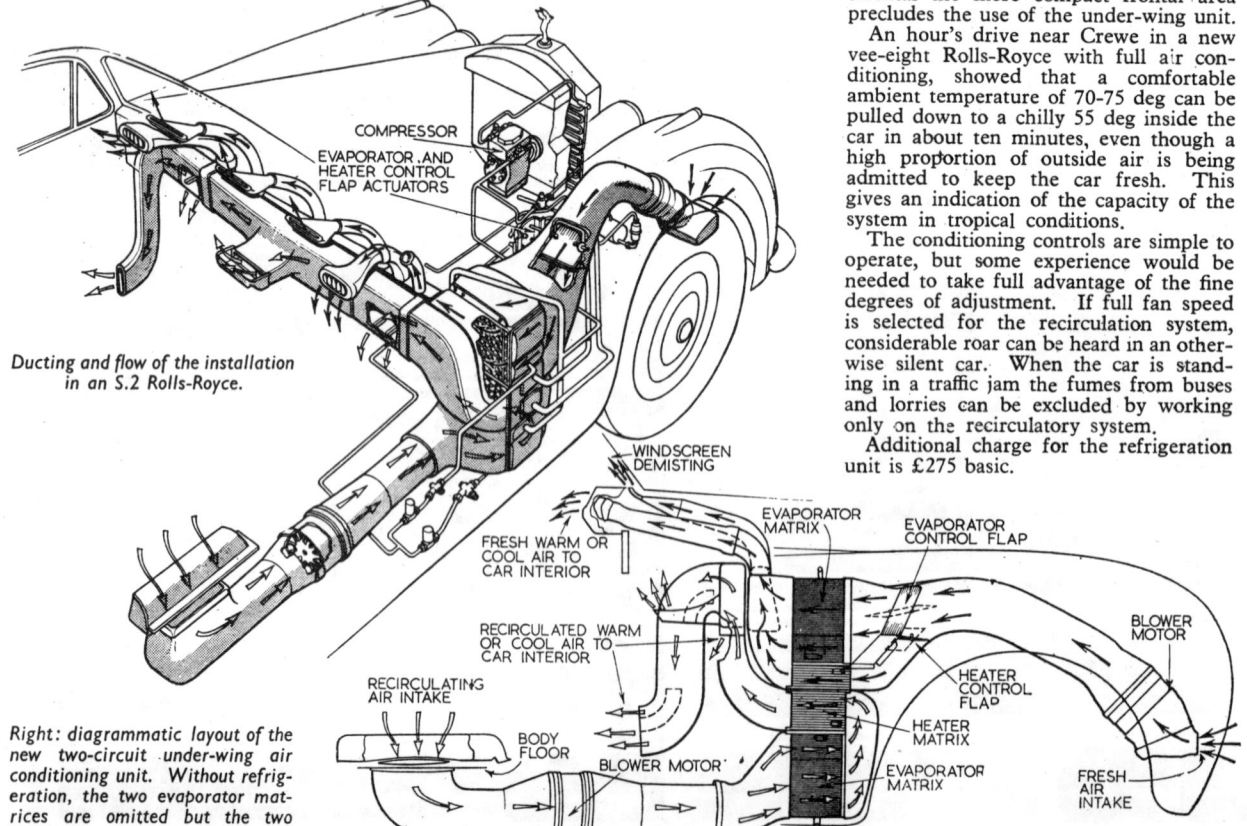

An hour's drive near Crewe in a new vee-eight Rolls-Royce with full air conditioning, showed that a comfortable ambient temperature of 70-75 deg can be pulled down to a chilly 55 deg inside the car in about ten minutes, even though a high proportion of outside air is being admitted to keep the car fresh. This gives an indication of the capacity of the system in tropical conditions.

The conditioning controls are simple to operate, but some experience would be needed to take full advantage of the fine degrees of adjustment. If full fan speed is selected for the recirculation system, considerable roar can be heard in an otherwise silent car. When the car is standing in a traffic jam the fumes from buses and lorries can be excluded by working only on the recirculatory system.

Additional charge for the refrigeration unit is £275 basic.

Ducting and flow of the installation in an S.2 Rolls-Royce.

Right: diagrammatic layout of the new two-circuit under-wing air conditioning unit. Without refrigeration, the two evaporator matrices are omitted but the two sections of the heater remain

THE AUTOCAR, 21 OCTOBER 1960

No. 164 1956 ROLLS-ROYCE SILVER CLOUD I

PRICE: Secondhand £3,875: New—basic £3,385, with tax £5,079

Petrol consumption 13-15 m.p.g. Mileometer reading 31,234
Oil consumption negligible Date first registered 20 June 1957

USED CARS
on the Road

UNTIL thought is given to prices and depreciation rates, it comes as a gentle shock to realize that even a Rolls-Royce can still command an asking price of nearly £4,000, when more than four years old. This implies depreciation at a rate of £300 a year—which, although considerable, is equivalent to only 6 per cent of the capital outlay. In justification is the fact that the car itself is still in splendid condition, offering all the comfort and refinement to be expected of the model.

Indeed, the mechanical side of the car has gained in some respects as a result of the distance travelled, and the engine is, if anything, quieter and smoother than that of the Silver Cloud, of which we carried out a full Road Test in May, 1958; performance is also better. Starting is instantaneous, and when the engine is cold the automatic choke behaves efficiently.

Automatic transmission is standard, of course, on the Silver Cloud, and the changes of ratio are scarcely detectable on small throttle openings. A small fault is that whether it is selected with the manual control or not, the change down to second gear lags slightly; if the driver remembers this, and pauses for a moment, the delay passes unnoticed, but if he accelerates before the changeover to the lower gear is completed an undignified lurch can result. In normal use the manual selector is convenient between fourth and third ratios, giving the driver effortless control over the transmission.

Slightly more tremor and road wheel reaction than one would expect are felt through the steering, but the control is precise. Power-assisted steering, now standard on the Rolls-Royce, was an optional extra when this model was made, and it was not fitted to the test car. Considerable effort is needed in manœuvring but at normal speeds the steering is light. The ride provided by the suspension is notably level and free from pitching; road surface irregularities are sensed or detected rather than noticed. Cruising at 90 m.p.h., which the Silver Cloud will sustain effortlessly, the directional stability is superb, and deteriorates little on a windy day.

Slight vibration occurs between 70 and 75 m.p.h.—probably the result of the all-too-common fault of lack of balance on one of the wheels.

To anyone who has not experienced the smooth progression of Rolls-Royce brakes, those on the Silver Cloud would probably seem excellent: they are capable of powerful retardation, but in fact they do not come up to the high standards of the model. Heavier pedal pressures are needed than should be, and there is less than the normal response to extra pressure on the pedal. The hand brake is powerful.

It is reasonable to expect very high standards of exterior and interior condition on a car of this make, because one so rarely sees a Rolls-Royce which is anything but immaculate. As far as the interior is concerned, this Silver Cloud is absolutely up to the required standard, and even for a Rolls-Royce the mild degree of deterioration in four years is commendable. The beige leather of front and rear seats shows some slight creasing, but they remain extremely comfortable and have not sagged. The polished wood facia and door trim and the cloth roof linings are all unmarked; only the slightly soiled beige carpets show anything like the signs of use which may be expected.

Externally the car bears no evidence of rust, but there are visible a number of tiny blemishes. Worst of these is a long and broad scratch on the driver's door, and there is a substantial dent in the rear bumper. Otherwise the dark blue—almost black in appearance—coachwork, and the glistening chromium, give an appropriately impressive appearance to the car. Slight evidence of respraying on the right wing, and scarcely detectable out-of-line of the radiator grille, suggest that there has been damage at some time in the past, but examination underneath showed that this could not have been anything major.

The car is not without minor failures in the electrical and mechanical equipment: the window winder on the right rear door is not working, nor are the windscreen washer and the cigarette lighter, situated in the quarter panel to the right of the rear seat.

Only the tyre-pressure gauge is missing from the magnificent tray of fitted tools in the luggage locker, all of which appear practically unused. On the costly aspect of tyres (for a car of this size), there are four Avon tyres less than half-worn, and the spare is an unused Dunlop.

Experience of this Rolls-Royce shows its price to be fully justified, and the mileometer reading above may be believed. Most buyers would probably insist on proper attention to the small points mentioned, and the car may then offer almost the appearance and superb luxury, with more than the initial performance, of the Silver Cloud I when new.

There are no additional accessories on the car—heater, radio windscreen washer and clock were all included in the specification. There are even filaments embedded in the glass to demist the rear window

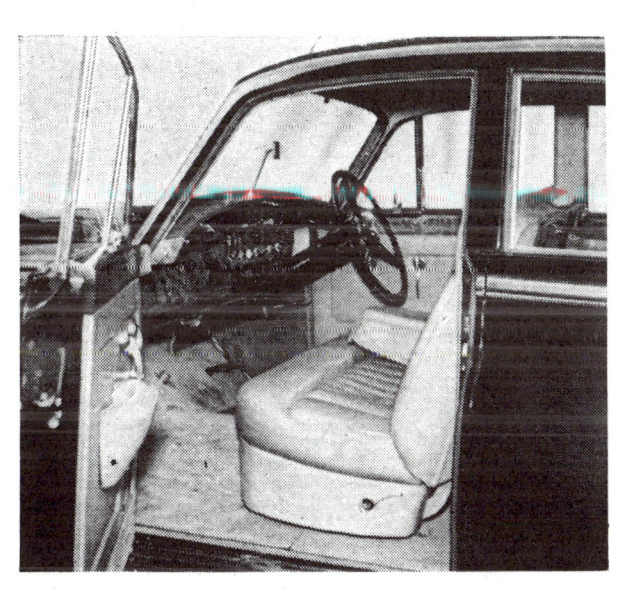

PERFORMANCE CHECK

(Figures in brackets are those of the original Road Test—16 May 1958)

0 to 30 m.p.h.	3·7sec (4·1)	0 to 70 m.p.h.	16·6 sec (18·4)
0 to 50 m.p.h.	8·7sec (9·4)	0 to 80 m.p.h.	22·3 sec (25·0)
0 to 60 m.p.h.	12·1sec (13·0)	0 to 90 m.p.h.	29·4 sec (34·1)

Standing quarter-mile 17·9sec (18·8)

Provided for test by Weybridge Automobiles, Ltd., Weybridge, Surrey. Telephone: WEYbridge 2233.

LIFE ON A

Incredible four-speed automatic drive and new 6-litre V8 engine put latest Rolls-Royce further than ever out of this crude world, says Bryan Hanrahan

modern MOTOR ROAD TEST

IF I had the sort of bank balance that would buy a Rolls-Royce, I'd reckon the money itself would give me sufficient insulation from the world's cares.

Call me a renegade if you like, but two tons and nearly 18 feet of Silver Cloud S II V8 is not my brand of champagne.

You sit up high in a four/five-seater body that is relatively narrow by today's standards. Your backside is insulated by magnificent leather from any feel of road surface, the power-assisted steering takes all effort out of guiding the car, the automatic box trips up and down through four ratios with less disturbance than a flea hiccupping — all in a deathly silence.

I'm not suggesting that I prefer train travel. But I do prefer a light, sensitive car that gives you its messages and expects its answers.

That S II was like the cloying essence of a perfect scent, rather than the blended, inviting after-product.

There is no fault to be found in engineering or craftsmanship. It's just that the whole concept seems unreal to me (how poverty has restricted the refinement of my tastes!).

Anyway, let's to it.

A Cloud that Flies

They never tell you what power a Rolls car engine develops. My guess is about 280 b.h.p. for the firm's new 6.2-litre aluminium V8.

ROLLS never quote the output of their engines, but Hanrahan estimates that the new 6230c.c. V8 churns out around 260 b.h.p. It's well and truly buried and servicing isn't easy — but would a Rolls owner care?

MODERN MOTOR — July 1961

THICK walnut, leather and carpets swamp you with luxury — but that wheel would look better in a bus and instruments are still central.

Only clue given in the specifications is that it churns out 340ft./lb. of torque at unspecified revs.

So you will hardly be surprised to hear that, despite a test weight of 46cwt., the car would go from 0 to 30 m.p.h. in 3.8 seconds, from 0 to 60 in 12, and from 0 to 100 in 38.3.

All this Gran Turismo class tyre-burning is achieved without wheel-spin, axle tramp or any sort of mechanical noise.

Top gear gives a whopping 27.3 m.p.h. per 1000 revs. So you let road conditions be your judge of cruising speed. Anything between 70 (2600 r.p.m.) and 105 m.p.h. (4000 r.p.m.) is just peanuts for this mass of machinery.

Best average top speed I could get was 116.8 m.p.h. — only because I didn't have enough road for more.

(SILVER) CLOUD

A best one-way run put the needle past 120.

Maximum gear speeds were: first, 30 m.p.h.; second, 58; third, 80. All were achieved with no more fuss from the engine than greasing the palm of a Grand Vizier.

And this Cloud nonchalantly knocked off the standing quarter-mile in 18.8 seconds.

Suspension, Steering

The only unusual thing about the suspension is that Rolls-Royce designed it.

Front end has trailing wishbones with radius rods, coil springs and stabiliser bar; the back has a solid axle with long semi-elliptic springs.

The set-up works like no other of its kind. And it is helped by variable settings on the back shockers.

Flick a switch, and either hard or soft setting is brought in immediately.

I used the hard setting for all acceleration runs, fast corners and rough going. The effect was to stop any wheel hop.

The Rolls flowed over all average surfaces like hot butter; in the rough, with the hard setting she levelled potholes like a bulldozer blade.

Understeer is constant — unless you overdo things (I found it very

CONTINUED ON PAGE 93

MODERN MOTOR — July 1961

MAIN SPECIFICATIONS

ENGINE: V8, o.h.v.; bore 104mm., stroke 91.4mm., capacity 6230c.c.; compression ratio 8 to 1; maximum b.h.p. not quoted; maximum torque 340ft./lb. at unspecified r.p.m.; twin SU horizontal carburettors, twin SU electric fuel pumps; 12v. ignition.
TRANSMISSION: Rolls-Royce adapted Hydra-Matic drive with torque-converter and four forward speeds; overall ratios — 1st, unquoted; 2nd, 8.1; 3rd, 4.4; top 3.0 to 1; hypoid bevel final drive, 3.0 to 1 ratio.
SUSPENSION: Front independent, by trailing wishbones, coil springs and radius rods; semi-elliptics at rear; Rolls-Royce shock-absorbers all round, with adjustable type at rear.
STEERING: Recirculating-ball with servo assistance, 4¼ turns lock-to-lock; 40ft. 6in. turning circle.
BRAKES: Servo-assisted hydraulic drum-type; lining area, 240 sq. in.
WHEELS: Pressed-steel discs, with 8.40 by 15in. tyres.
CONSTRUCTION: Separate box-section chassis.
DIMENSIONS: Wheelbase 10ft. 3in.; track, front 4ft. 10½in., rear 5ft. 0in.; length 17ft. 7½in., width 6ft. 2¾in., height 5ft. 4in.; ground clearance 7in.
DRY WEIGHT: 40½cwt.
FUEL TANK: 18 gallons.

PERFORMANCE ON TEST

CONDITIONS: Fine, cool, longitudinal breeze; dry bitumen; two occupants, premium fuel.
BEST SPEED: 120.7 m.p.h.
FLYING quarter-mile: 116.8 m.p.h.
STANDING quarter-mile: 18.8s.
MAXIMUM in indirect gears: 1st, 30; 2nd, 58; 3rd, 80.
ACCELERATION from rest (all four gears engaging automatically in selector position 4): 0-30, 3.8s.; 0-40, 5.9s.; 0-50, 8.7s.; 0-60, 12.0s.; 0-70, 15.9s.; 0-80, 22.1s.; 0-90, 30.4s.; 0-100, 38.3s.
PASSING ACCELERATION (position 4): 20-40, 4.0s.; 30-50, 5.3s.; 40-60, 5.9s.; 50-70, 7.0s.; 60-80, 9.1s.; 70-90, 11.9s.; 80-100, 14.8s.
BRAKING: 31ft. 10in. to stop from 30 m.p.h. in neutral.
FUEL CONSUMPTION: 12.8 m.p.g. overall for 92-mile test.
SPEEDO: 1 m.p.h. slow at 30 m.p.h.; accurate throughout rest of range.

PRICE: £8750 including tax

ROLLS-ROYCE SILVER CLOUD III
BENTLEY S3

NEW MODELS

Crewe-cut, 1963 style: Immediate recognition points for the new Rolls-Royce Silver Cloud III are the paired headlamps, bolder wings with recessed parking-signalling lamps, and more discreet bumper overriders. A second glance reveals that the classic radiator shell has been lowered by about 1¼in

MORE POWER, MORE PASSENGER SPACE, BETTER LIGHTING, EASIER STEERING

VERY high grade cars such as the Rolls-Royce and Bentley are made in such relatively small numbers that the fruits of constant research and development can be embodied quickly on the production line. This is a more or less constant process, but modifications of a somewhat basic nature, in particular those which have a sufficiently marked effect on performance or appearance to justify a new series or model designation, are naturally introduced together.

Both the Rolls-Royce Silver Cloud and the S-series Bentley—identical in all except radiator shell and bonnet—advance a major step and are now designated the Silver Cloud III and S3 respectively. In brief, they have more power, an improved lighting system with paired headlamps, reduced steering effort, and increased passenger space in the back.

Firm to their tradition, Rolls-Royce do not reveal power output figures, but claim a seven per cent increase to this unknown quantity for the latest version of their 6,230 c.c. vee-8 engine. This results from a rise in compression ratio, now 9·0 to 1 instead of 8·0 (the lower ratio being optional to suit markets where 100 octane fuels are not available); and from the replacement of 1¾in. S.U. HD6 carburettors by 2in. HD8s. A new vacuum advance-retard mechanism for the ignition distributor, supplementing the centrifugal weights, together with the raised compression, are said by the makers to have lowered the specific fuel consumption.

To meet the higher power, the chrome molybdenum steel crankshaft is now nitride-hardened, and larger diameter gudgeon pins are fitted. These are now offset in the pistons, the better to withstand the increased thrust and contributing also to quieter running. Crankcase breathing is drawn into the engine intake through sealed ducting, to avoid the possibility of fumes escaping.

Extra power-assistance has been provided for the cam-and-roller steering gear, reducing the maximum load at the wheel rim to 6lb. While this brings about an appreciable saving in effort for normal driving, its effect is most marked during low-speed manœuvring and parking. Plenty of "feel" is retained in this system, and the hydraulic power valves remain closed up to ¼lb load. There are

Above: This is the Bentley S3, which shares the senior partner's amendments. Driving vision is improved over the steeply sloping bonnet. From the rear only the overriders are new, and the earlier type is retained for export cars. Left: Sealed crankcase breathing into the induction is new, and the ignition distributor now has a supplementary vacuum advance-retard control

Autocar, 19 October 1962

Above: Interior innovations include a padded capping rail above the facia. Right: Less prominent corner bolsters save space, allowing three abreast in greater comfort. There is extra legroom, too

no further main modifications to the mechanical components, the complex drum braking system with all-trailing shoes, a gearbox-driven friction servo and three independent systems being retained.

A four-headlamp system is now standardized, all being Lucas sealed beam units of 5¾in. dia. The inner, long-range lamps each have a single filament, the outer units two filaments—one slightly out of focus to add spread to the main beams, the other for dipped beam. Wattage for the main beams is up from 120 to 150. Parking and direction-signalling lamps are now combined in neat oval units recessed in the wings; the twin foglamps are retained. Except for the North American market, smaller and simpler overriders are fitted to the bumpers.

More Seat Room

Inside the cars are important changes in that individual front seats, each with its own centre-folding armrest, have become standard equipment, although the one-piece bench remains optional. In the back two valuable extra inches of legroom have been found, the seat cushion being moved back by that amount. Most of this has resulted from making the backrest more upright—and certainly no less comfortable. Also, the previously rather "aggressive" corner padding has been reduced, this allows greater effective width for seating three abreast, as well as being more comfortable to lean against in any circumstances.

There are five new standard colours—Astral blue, Antelope, Dusk grey, Pine green and Garnet. The four discontinued colours (Velvet green, Midnight blue, Opal and blue-grey) can still be specified, but now at extra cost. Other listed supplementaries include electric window-lifts, a refrigeration system, tinted glass and fitted suitcases for the boot.

While this completes the particular innovations for the Silver Cloud III and Bentley S3, the later Silver Cloud IIs and S2s had quite a few features which had been progressively incorporated without any fanfares, and which are continued on the new cars. Those which follow are especially noteworthy. The heating and ventilating installation has two ventilating matrices for independent fresh air and recirculating systems, the latter with improved delivery to the middle of the rear floor. A ram position has been added to the fresh air circuit. A headlamp flashing button has been incorporated in the direction indicators lever, and there is a tell-tale lamp in circuit with the ignition switch to warn the driver when the handbrake is on.

For the standard Radiomobile receiver (there is a choice between three) a second speaker has been added on the rear parcels shelf, with a balance control for adjusting the front and rear outputs. Footrests are provided for those riding in the back, who also have ashtrays of an improved design. There is a map reading lamp on the facia, the instrument panel lighting is more powerful, the switch for the electrical demisting element in the rear window is on the facia instead of in the back compartment, and the driving mirror is better supported to resist vibration. In the luggage boot there is a detachable metal panel covering the spare wheel, which lies horizontal below the boot floor.

Other Rolls-Royce and Bentley types—the long-wheelbase Silver Cloud III and S3, the huge Phantom V and the S3 Continental—naturally all incorporate the revised engine and other changes. Of the coachbuilt cars in series production, there is a new two-door sports saloon by Park Ward on the Continental chassis, and the Phantom V seven-passenger limousine (Park Ward and H. J. Mulliner) has a modified rake to the screen as well as a completely redesigned boot.

Price increases have brought the Silver Cloud III and Bentley S3 standard saloons up to £6,277 17s 9d and £6,126 12s 9d respectively, the previous figures being £5,913 10s 3d and £5,769 2s 9d. Current chassis prices for these two cars are £3,135 and £3,035.

During an afternoon spent on the Oulton Park racing circuit in Cheshire, when examples of the latest products were tried in direct comparison with a Rolls-Royce Silver Cloud II, we were able to confirm the gain in performance and to appreciate the sensitive control provided by the unusually light steering. These fine cars have always felt smaller than they look to drive, and the increased close-range field of view over the lowered bonnet adds to this impression. Further remarks concerning the handling of the Silver Cloud III and S3 must await a full Road Test.

Autocar's representatives were given the run of the Oulton Park racing circuit for an afternoon, to sample the increased power and lightweight power-steering of the new cars

THE MOTOR August 21 1963

Extended Road Test No. 33/63

MAKE Rolls-Royce ● **TYPE** Silver Cloud III
● **MAKERS** Rolls-Royce Ltd., 14-15 Conduit St., London, W.1

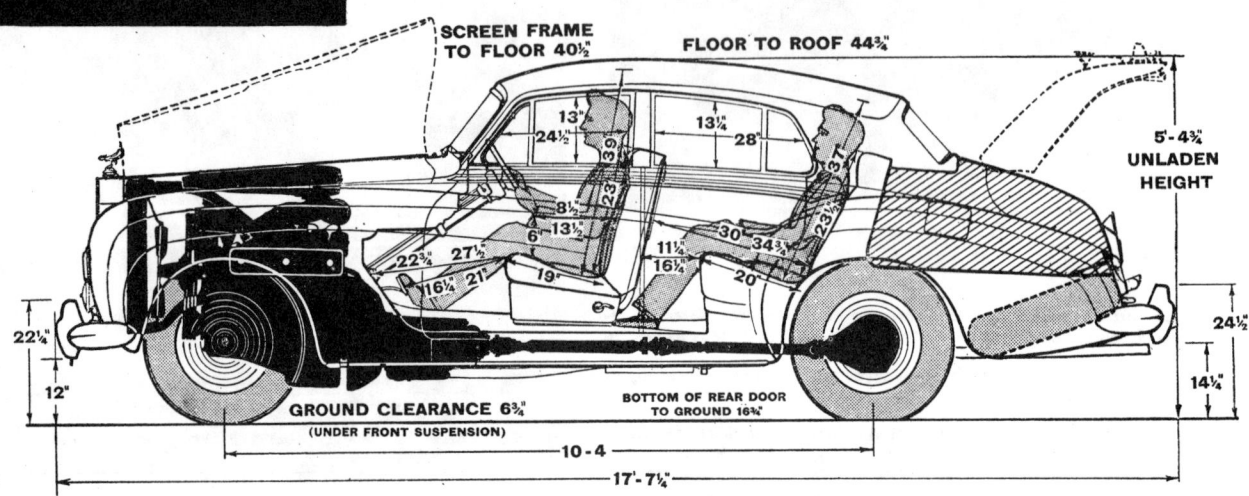

Test Data

World copyright reserved; no unauthorized reproduction in whole or in part.

Conditions: Weather: Dry, light wind (Temperature 60°–66°F., Barometer 29.5–29.6 in Hg.) Surface: Concrete, dry tarmacadam. Fuel: Super Premium grade pump petrol (101 Octane by Research Method).

MAXIMUM SPEEDS
Flying Mile
Mean of opposite runs	114.3 m.p.h.
Best one-way time equals	115.4 m.p.h.
"Maximile" Speed: (Timed quarter mile after one mile accelerating from rest)	
Mean of four opposite runs	106.3 m.p.h.
Best one-way time equals	108.9 m.p.h.

Speed in gears (automatic change-up speeds)
Max. speed in 3rd gear	70 m.p.h.
Max. speed in 2nd gear	38 m.p.h.
Max. speed in 1st gear	24 m.p.h.

ACCELERATION TIMES
from standstill
0–30 m.p.h.	3.2 sec.
0–40 m.p.h.	5.3 sec.
0–50 m.p.h.	7.6 sec.
0–60 m.p.h.	10.1 sec.
0–70 m.p.h.	13.5 sec.
0–80 m.p.h.	18.1 sec.
0–90 m.p.h.	25.5 sec.
0–100 m.p.h.	34.7 sec.
Standing quarter mile	17.8 sec.

on upper ratios
	Top gear	"Kick-down" range
10–30 m.p.h.	— sec.	2.2 sec.
20–40 m.p.h.	7.1 sec.	4.3 sec.
30–50 m.p.h.	7.6 sec.	4.4 sec.
40–60 m.p.h.	8.1 sec.	4.8 sec.
50–70 m.p.h.	9.1 sec.	5.9 sec.
60–80 m.p.h.	9.6 sec.	8.0 sec.
70–90 m.p.h.	11.9 sec.	11.9 sec.
80–100 m.p.h.	16.6 sec.	16.6 sec.

Overtaking
Starting at 40 m.p.h. in direct top gear, distance required to gain 100 ft. on another car travelling at a steady 40 m.p.h.=545 ft.

FUEL CONSUMPTION
Overall fuel consumption for 2,134 miles, 190 gallons, equals 11.2 m.p.g. (26 litres/100 km.)

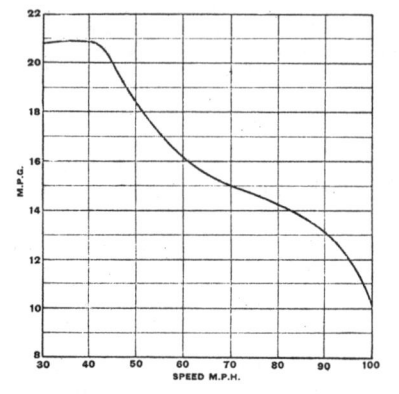

Touring Fuel Consumption (m.p.g. at steady speed midway between 30 m.p.h. and maximum, less 5% allowance for acceleration) ... 14.1 m.p.g.
Fuel tank capacity (maker's figure) 18 gallons

BRAKES
Deceleration and equivalent stopping distance from 30 m.p.h.
0.34 g with 25 lb pedal pressure	(=88 ft.)
0.85 g with 50 lb pedal pressure	(=35 ft.)
0.90 g with 60 lb pedal pressure	(=33 ft.)

Handbrake
0.25 g deceleration from 30 m.p.h. (=120 ft.)

Brake Fade
TEST 1. 20 stops at ½ g deceleration at 1 min. intervals from a speed midway between 30 m.p.h. and maximum speed (=72 m.p.h.)
Pedal force at beginning = 40 lb.
Pedal force for 10th stop = 50 lb.
Pedal force for 20th stop = 50 lb.

TEST 2. After top gear descent of steep hill falling approximately 600 ft. in half a mile, increase in brake pedal force for ½ g stop from 30 m.p.h. = 8 lb.

PARKABILITY
Gap needed to clear a 6 ft obstruction

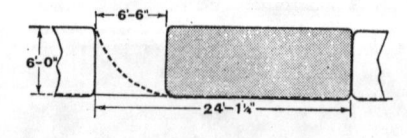

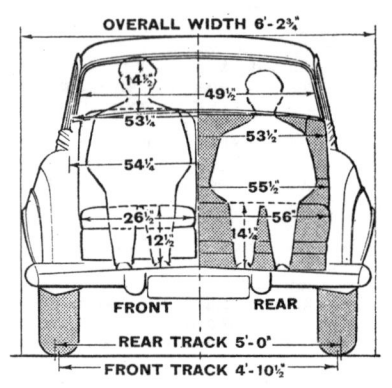

Waterproofing
Increase in brake pedal force for ½ g stop from 30 m.p.h. after two runs through shallow watersplash at 30 m.p.h. = 0 lb.

STEERING
Turning circle between kerbs:
Left	41 ft.
Right	39½ ft.
Turns of steering wheel from lock to lock	4
Steering wheel deflection for 50 ft. diameter circle	1.75 turns
Steering force (at rim of wheel) to move front wheels at rest	7 lb.
Steering force to hold car on 100 ft. diameter circle at 15 m.p.h. (=0.3g approx.)	5 lb.

INSTRUMENTS
Speedometer at 30 m.p.h.	1½% slow
Speedometer at 60 m.p.h.	1% slow
Speedometer at 90 m.p.h.	2½% slow
Distance recorder	¾% slow

WEIGHT
Kerb weight (unladen, but with oil, coolant and fuel for approximately 50 miles) 40¼ cwt.
Front/rear distribution of kerb weight 50/50
Weight laden as tested 44 cwt.

THE MOTOR, August 21 1963

ROLLS-ROYCE SILVER CLOUD III

IT is unlikely that any motorcar will ever quite enjoy the prestige of the Rolls-Royce. For many years it has carried an aura which in some ways flatters it but in others fails to do it justice. The car, as sold to the public has rarely been in the forefront of technical development; rather does it exemplify well-tried techniques and refined, if apparently conventional engineering. It has seldom been amongst the fastest of its contemporaries, or even the biggest or roomiest. And the latest Silver Cloud III raises doubts about any claim to be the quietest.

For £5,517 for the basic car (including purchase tax, but without extras) the Rolls-Royce offers elegance and good taste rather than fashionable flamboyance, a high standard of luxury for five people, brisk performance, and very good handling. Light alloys are used extensively in the 6,230 c.c. V-8 engine, which has the refinement of hydraulic tappets, and the massive chassis frame is carried on independent front suspension by coil springs and wishbones with a live axle and semi-elliptic leaf springs at the rear. The ride may be varied by a damper control at the discretion of the driver. Identification features which distinguish it from the Silver Cloud II are four headlights and a lower bonnet line which has meant shortening the famous radiator by 2½ in.

On the debit side, the noise level does not seem exceptionally low by present-day standards and adherence to drum brakes will be difficult to justify for faster Rolls-Royces in view of a tendency to fade which our prolonged tests showed up. The maximum speed and swift acceleration however, are remarkable for a car weighing more than two tons and with a shape hardly ideal for air penetration. The size and grandeur of the car (the view along the bonnet is a very commanding one) deceive one about the performance which would do justice to many a car of more sporting pretensions. The finish, both in detail and the broader sense of equipment and trim is superb. The excellence of the interior is the sort which only skilled craftsmen can achieve using traditional materials with modern manufacturing aids to durability. Heating and ventilation are excellent and typical of the car's character. Care and deep thought have obviously gone into its assembly, traditional features have not been discarded lightly and the richness of the car, its long life expectancy, and the modest depreciation which can be expected go a long way to justifying its cost.

Performance

AN automatic choke operates on depression of the throttle so that after letting the S.U. electric pumps push petrol into the two 2 in. S.U. carburetters, starting was instantaneous. The choke remained in operation for only a few hundred yards during which some creep with the automatic transmission was apparent during stops. The big V-8 engine demands 100 octane (Super Premium) fuel, a little label inside the filler flap reminding pump attendants. There was never any pinking or running-on but the engine was slightly uneven during idling. It vibrates little, but there is a remote rumble when the car is stationary. Opening the throttle wide produces really rapid acceleration up to well over 100 m.p.h. to the accompaniment of quite a pronounced power roar from under the bonnet. At speed this is replaced by a low hum which, although not unpleasant is louder than expected in a Rolls-Royce. With gentler driving, the engine is practically inaudible. The 1 in 3 test hill was treated easily, restarts being accomplished on less than full throttle.

Transmission

ROLLS-ROYCE have put many years of development into an automatic transmission which drives a four-speed epicyclic gearbox through a fluid coupling and in its latest form they have

In Brief

Price £4,565 plus purchase tax £951 12s. 1d. equals £5,516 12s. 1d.
Capacity .. .. 6,200 c.c.
Unladen kerb weight .. .. .. 40¼ cwt.
Acceleration:
  20-40 m.p.h. in top gear .. .. 2.3 sec.
  0-50 m.p.h. through gears .. .. 7.6 sec.
Maximum speed .. .. 114.3 m.p.h.
Overall fuel consumption .. .. 11.2 m.p.g.
Touring fuel consumption .. .. 14.1 m.p.g.
Gearing: 27.3 m.p.h. in top gear at 1,000 r.p.m.

Traditional materials and layout (*above*). The absence of a clutch pedal is one of the few clues which identify this interior as a current Rolls-Royce. (*Right*) Small tables fold down from the adjustable backs of the front seats.

brought an older design to a level of performance which is competitive with much newer gearboxes. All the gears are quiet except first which makes a slight whine; the automatic changes are in general very smooth, although on full throttle the second to third shift is accomplished with a curious double surge.

There is, however, a very wide gap between second and third gears and the resulting change in engine speed makes shifts between these ratios sound fussy. With the present power/weight ratio, first and second gears are little needed and all normal driving is done in top and third. For overtaking, the throttle kick-down switch recalls third very smoothly indeed at any speed and it can also be engaged on the overrun by means of the rather stiff steering column selector lever if the driver prefers to override the automatic mechanism. Neutral, 4, 3, 2, and Reverse positions are offered by the selector quadrant. "2" holds second gear for maximum engine braking when descending very steep hills or for traffic crawling, "3" normally confines the box to first, second, and third gears only although it will change into top if the safe rev. limit is reached, and the Reverse position offers a safety lock for parking on hills.

The best of the modern torque converter transmissions bridge ratio changes less perceptibly but the Rolls-Royce offers more control to its driver.

Running costs

NOT unexpectedly with six-and-a-bit litres, economy is not a strong point. The cost of luxury is indicated by an overall consumption of 11.2 m.p.g. although gentler drivers would reach nearer our touring figure of 14.1 m.p.g. The Super Premium fuel requirements raise the cost per mile of the Silver Cloud but the test car consumed hardly any oil during 2,000 miles. Specimen fuel consumptions under various conditions were as follows:

Motorway: Cruising at 100-110 m.p.h. when possible, average speed 96 m.p.h.—9.6 m.p.g. **Undulating main road:** 50-60 m.p.h. cruising seldom using full throttle, average speed 52 m.p.h.—14.1 m.p.h. **Cold start:** Choke in operation for approximately ½ mile, heavy city traffic, average speed 12½ m.p.h.—6.5 m.p.g.

The bonnet is split along the centre axis and released by handles on each side of the front compartment. These work a complicated but beautifully smooth and effective system of rods and levers which allow the two sides to swing upwards where they are held by automatic stays. The dipstick is easy to get at; the massive air cleaner hinges upwards where it is held by a wire to clear the oil filler. High wings hamper maintenance work on the engine and the opening is in any case small. The hydraulic reservoirs are convenient alongside the engine where they can be easily inspected and topped-up.

Although the boot looks shallow, it held a surprising amount of dummy luggage. The total capacity was 9.2 cu. ft., made up of two 2.0 cu. ft. (24×18×8 in.), one 1.3 cu. ft. (21×15×7 in.), three 0.8 cu. ft. (17½×13×6 in.) and three 0.5 cu. ft. (14×11×5 in.).

ROLLS-ROYCE Silver Cloud III

THE MOTOR   August 21 1963

The Rolls-Royce is an elegant car from any angle. Tinted glass in the test car gave a sense of privacy without impairing outward vision. The cover over the petrol filler is released electrically by a button on the facia. Bumper overriders are smaller than those on the Silver Cloud II.

Handling

EXCELLENT power steering contributes to handling which is surprisingly good for such a large car. It is much more agile than it looks and can negotiate winding country roads at a cracking pace. The lightness of the steering makes manoeuvring easy and at speed there is just the right amount of feel transmitted to the driver's hands once he is acclimatized to its light response. Only on unusually rough surfaces is any reaction felt through the column and there is no feed-back on ruts. Castor action seems to suit everybody; after a tight corner the wheel will spin back at the correct rate without the necessity to "catch up" with the fairly low (four turns lock-to-lock) gearing. This is not so low-geared, however, as to be an embarrassment in correcting a skid.

Despite its near 50/50 weight distribution, the Silver Cloud understeers but there is not much roll on corners. The characteristics are very consistent and there is never any feeling that, when pushed to its limit, the car will react violently. In the wet, it can be cornered fast under power to the point where the back breaks away gently to be corrected by unwinding the appropriate amount of steering. Far from feeling a large car, it can be handled with great delicacy. At speed, it is usually better to set the suspension control to Hard, to avoid uncomfortable floating; mild tyre squeal on sharp corners becomes rather loud if the car is driven in sporting fashion. On motorways, wet or dry, fast cruising seems natural and the car remains secure and stable; cross-winds can make it wander slightly at high speeds. Unusually, Rolls-Royce consider their tyres adequate for high speed motoring without raising the pressures.

Brakes

BOTH our fade tests showed the brakes up in a poorish light. They have large drums with two trailing shoes but at the bottom of the test hill, the pedal pressure for a ½g stop had risen by 8 lb. Reference to the other brake data will show that normal operating pressures are light. When fading did occur, there was no pulling to one side or grabbing. The handbrake stopping power was also poor and it failed to hold the car facing down a 1 in 3 test hill, but would hold it successfully pointing up. Care is necessary at low speeds with the gearbox driven mechanical servo; pedal pressures rise sharply during backing-and-filling manoeuvres, and there is a small distance-lag before the servo is operative. Separate hydraulic systems guard against total brake failure.

Comfort and control

WITH the suspension control at Normal, the low-speed ride in both front and rear is good over all but the roughest surfaces. At the Hard setting, the handling improves and the ride does not deteriorate materially until rough roads are encountered. This includes long wave bumps taken fast which can produce pitching and short bumps which can jolt the occupants. The very size of the car seems to absorb bumps but they can be heard quite distinctly. Cat's eyes make a resounding thump and there are even occasions, such as under hard acceleration, when the live back axle judders.

The seats in front and rear are superbly comfortable, giving good support all the way up the spine and across the shoulders. When the car is moving, however, the width of the seats and the slipperiness of the leather make them less appealing. The driver has to use the steering wheel or the very well-placed adjustable armrests to locate himself. The seats are big, with a high, upright commanding position and backs which can be made to recline to about 40 degrees from the vertical. The cushions are large and wide, coming close together at the front where it would be practical (but uncomfortable) for a third occupant. Drivers of all sizes found they could easily accommodate themselves behind the large wheel, but some would have preferred a thicker rim. A rear seat passenger with size 10 or larger shoes has insufficient room for his feet on the toeboard.

No car is silent, but the Silver Cloud is certainly very quiet indeed, particularly when driven gently. Wind noise is exceptionally low in spite of the turbulence which the bluff front must cause, and it is only necessary to turn up the radio substantially over 95 m.p.h. Certain road surfaces produce quite a loud body drumming and opening a window even a little (which the makers do not recommend under cruising conditions) promotes wind roar. The cool-air ventilator increases road noise perceptibly when it is open.

Heating and ventilating equipment is substantial and complex but works well. There are two control systems, one for the upper part of the car, another for the lower and it was never really necessary to drive with a window open. The upper system directs hot or cold fresh air (which collects unwanted fumes through its low-mounted intake in heavy traffic) to the front seat occupants' faces and/or the windscreen and side windows, while the lower warms or cools front and rear with recirculated air as required. Tiny wires in the rear window glass demist the inside, taking about ten to fifteen minutes from switching on to clearing condensation which had formed into small droplets. The blowers for the interior systems are quite noisy at the higher of their two speeds. We tried an alternative full air-conditioning system (which both refrigerates and dries the air) on a Bentley. After a stationary spell in hot sunshine, the temperature dropped from 110 deg. F to 90 deg. F on a warm evening within about 15 minutes of switching on the engine and letting it idle with the windows closed and the car standing still—the most testing condition.

Those who habitually drive automatics with two feet (i.e. left foot braking) will find this impossible in the Silver Cloud because the brake pedal, which can be double width to order, is on the right of the steering column. The handbrake has to be reached for on the right and the minor switchgear is also rather far away from the driver except for the less often operated items like the ride control, electric petrol filler flap, and rear window demister switch which are near the wheel. Visibility is assisted by the high driving position and the new, lower bonnet line but the imposing radiator shell often conceals low-mounted brake or indicator lights of a car ahead. The screen pillars are not thick for the size of car and the convex mirror fills the rear window precisely and does not vibrate. A convenient flasher works all four headlights which give a good spread of light for 100 m.p.h. motoring on good roads. Reversing lights work automatically with selection of reverse on the quadrant.

Fittings and furniture

SUPERB workmanship in traditional materials gives the interior an air of well-bred quality expected of a Rolls-Royce. All the switchgear works with a satisfying click and such items as glove-box lids, picnic tables in the backs of the front seats, and ashtrays work smoothly. Less satisfactory are the electric window winders which buzz. The doors needed a firm slam to close them, even with a window open to release the air pressure inside. The speedometer proved slightly pessimistic; its calibrations are easily understood once one becomes used to starting at the top of the scale and working clockwise round the dial. Four segments in a distant dial on the passenger's side supply information on fuel contents (a warning light supplements this and the gauge also acts as an indicator of engine oil level with the car stationary) oil pressure, battery charge, and coolant temperature. A reasonably accurate clock is standard equipment.

There is plenty of space for small items, with glove compartments each side of the facia, the passenger's having a lockable lid. Small compartments in the thickness of the doors have smoothly sliding covers and a substantial shelf behind the rear seat is useful but there is little accommodation for large parcels. The boot is long and wide, but shallow, the 18-gallon fuel tank, battery and the spare wheel in a separate compartment underneath taking up much of the room in the tail. Few other manufacturers provide both a trap door for topping up the spare tyre and a tailored carpet for reaching the trap. Next to it is a similarly covered, fitted tool kit. An H.M.V. radio, fitted as standard, has a control to balance front and rear speakers, and reproduction and tone were good.

The World Copyright of this article and illustrations is strictly reserved © Temple Press Limited, 1963

Data and performance comparison overleaf ▶

ROLLS-ROYCE Silver Cloud III

THE MOTOR August 21 1963

Coachwork and Equipment

Starting handle .. .. None
Battery mounting .. .. Under boot floor
Jack .. .. Bevel-geared bipod screw pillar
Jacking points .. Two external sockets below body sides
Standard tool kit: Jack, tyre pump, wheel brace, wheel disc remover, box spanner, adjustable spanner, 3 open-jaw spanners, contact breaker spanner with feeler, gauge, drain plug key, pliers, screwdriver, tyre gauge, bulbs.
Exterior lights: 4 headlamps, 2 side, 2 fog, 2 stop/tail, 1 number plate, 2 reversing.
Number of electrical fuses 8
Direction indicators .. Self-cancelling flashers
Windscreen wipers 2-speed electrical, self-parking
Windscreen washers .. .. .. Electric
Sun visors 2
Instruments: Speedometer with total and decimal trip mileage recorders, clock, ammeter, combined fuel and oil contents gauge, oil pressure gauge, coolant thermometer.
Warning lights Generator, fuel, and handbrake
Locks:
  With ignition key .. .. Both front doors
  With other keys .. .. Glove box and boot
Glove lockers .. .. .. Two, one lockable
Map pockets .. .. One in each front door
Parcel shelves .. .. One behind rear seat
Ashtrays .. .. 5; 3 front, 2 rear
Cigar lighters .. .. .. 2; 1 front, 1 rear
Interior lights: Roof light with courtesy switches, map light in facia, 2 reading lights in rear, 1 in boot with courtesy switch.
Interior heater: Fresh air and recirculating standard. Rolls-Royce refrigeration unit extra.
Car radio .. .. Radiomobile 620T standard
Extras available: Tinted glass (standard with refrigeration unit), electrically operated windows, safety belts, refrigeration unit, plus range to customer's choice.
Upholstery material .. .. .. Leather
Floor covering .. .. .. Wilton carpet
Exterior colours standardized 16
Alternative body styles .. Limousine on long wheelbase chassis

Key to facia

1. Airstream direction control. 2. Oil pressure gauge. 3. Ammeter. 4. Water temperature gauge. 5. Clock. 6. Generator warning light. 7. Lights switch. 8. Oil level indicator switch. 9. Fuel warning light. 10. Cigar lighter. 11. Speedometer. 12. Total mileage recorder. 13. Trip mileage recorder. 14. Ride control. 15. Fuel filler flap switch. 16. Handbrake warning light. 17. Gear selector lever. 18. Bonnet release. 19. Fire extinguisher. 20. Combined fuel and oil contents gauge. 21. Radio speaker balance. 22. Heater control (lower). 23. Heater control (upper). 24. Ignition/starter switch. 25. Screen wiper/washer switch. 26. Panel light switch. 27. Radio. 28. Trip mileage recorder reset. 29. Direction indicator/headlamp flasher control. 30. Horn. 31. Handbrake. 32. Bonnet release. 33. Ventilator. 34. Electric window switches.

Specification

ENGINE

Cylinders	V-8
Bore	104.14 mm.
Stroke	91.44 mm.
Cubic capacity	6,230 c.c.
Piston area	105.5 sq. in.
Valves	Pushrod o.h.v.
Compression ratio	9.0/1
Carburetter	Two S.U. HD8
Fuel pump	Two S.U. electric
Ignition timing control	Centrifugal
Oil filter	Full flow
Maximum power	Not disclosed
Maximum torque	Not disclosed

TRANSMISSION

Clutch: Fluid coupling and automatic 4-speed epicyclic gearbox.

Top gear	3.08
3rd gear	4.46
2nd gear	8.10
1st gear	11.75
Reverse	13.25
Propeller shaft	Divided
Final drive	Hypoid spiral
Top gear m.p.h. at 1,000 r.p.m.	27.3
Top gear m.p.h. at 1,000 ft./min. piston speed	45.5

CHASSIS

Brakes: Rolls-Royce drum type with gearbox driven servo: two separate hydraulic circuits for front wheels, hydraulic and mechanical operation of rear brakes.
Brake drum diameters .. .. 11¼ in.
Friction areas: 240 sq. in. of lining area working on 424 sq. in. rubbed area of drums.
Suspension:
  Front: I.F.S. by coil springs, transverse wishbones and anti roll torsion bar.
  Rear: Semi-elliptic leaf springs and single radius arm.
Shock absorbers
  Rolls-Royce lever arm hydraulic with electrical two-position remote control of rear damper setting.
Steering gear: Rolls-Royce hydraulic power-assisted cam and roller steering.
Tyres .. .. .. 8.20-15 tubeless Dunlop

Maintenance

Sump: 12 pints, S.A.E. 20 winter, 30 summer 10 w/30 approved.
Gearbox .. .. 20 pints, transmission fluid
Rear axle .. .. 1¾ pints, S.A.E. 90
Steering gear lubricant .. Transmission fluid
Cooling system capacity 21 pints (3 drain taps)
Chassis lubrication: By grease gun every 12,000 miles to 21 points.
Ignition timing .. .. 2° b.t.d.c. static

Contact breaker gap .. .. .014 in.-.016 in.
Sparking plug type .. .. Champion RN8
Sparking plug gap .. .. .024 in.-.027 in.
Valve timing: Inlet opens 20° b.t.d.c. and closes 61° a.b.d.c. Exhaust opens 62° b.b.d.c. and closes 19° a.t.d.c.
Tappet clearances .. Hydraulic, self-adjusting
Front wheel toe-in .. ¹⁄₃₂ in.-⅛ in.
Camber angle .. .. Vertical to ½° positive

Castor angle .. .. 1°
Steering swivel pin inclination 4½° at 0° camber
                                  4° at ½° positive
Tyre pressures:
  Front .. .. 22 lb.
  Rear .. .. 27 lb.
Brake fluid .. Castrol-Girling 6293 Crimson
Battery type and capacity: Dagenite or Exide 12 volt, 67 amp.-hour.

Visibility
180° from the driving seat. Shaded areas show one-eye visibility.

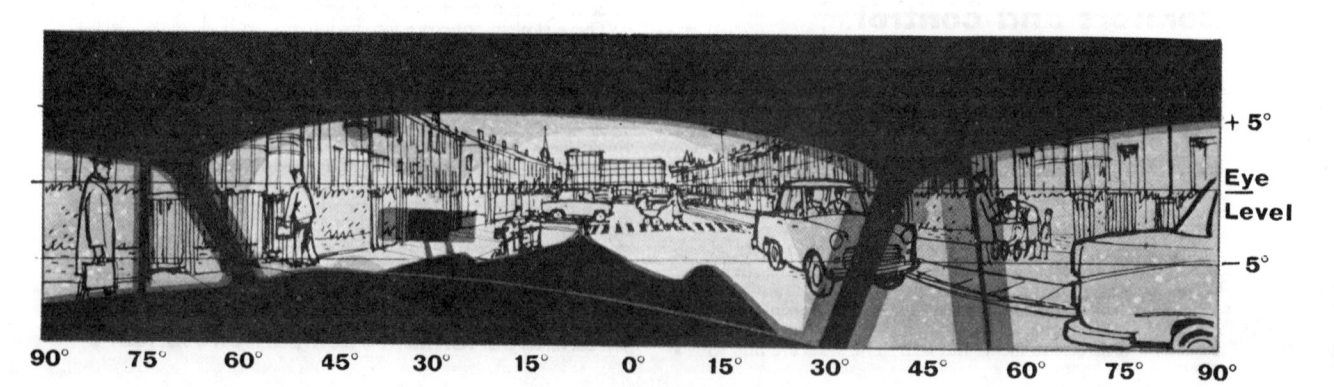

+5°
Eye Level
-5°

90° 75° 60° 45° 30° 15° 0° 15° 30° 45° 60° 75° 90°

THE MOTOR August 21 1963

89

PERFORMANCE COMPARISONS
with four competitive cars previously tested by THE MOTOR

MAXIMUM SPEED, m.p.h.

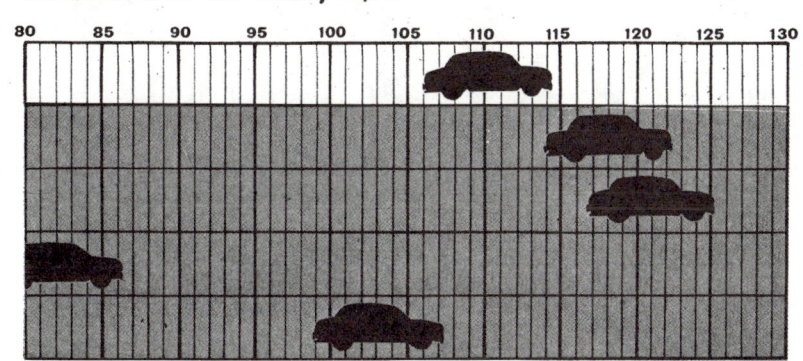

Car	mph
Rolls-Royce	~111
Daimler Majestic Major	~120
Bristol 407	~122
Vanden Plas 4-litre	~82
Mercedes-Benz 300SE	~102

FUEL CONSUMPTION, m.p.g.
TOURING (open) — OVERALL (filled)

Car	Touring / Overall
Rolls-Royce	13.5 / 10.5
Daimler Majestic Major	17 / 15
Bristol 407	16.5 / 14.5
Vanden Plas 4-litre	15 / 12
Mercedes-Benz 300SE	15.5 / 13

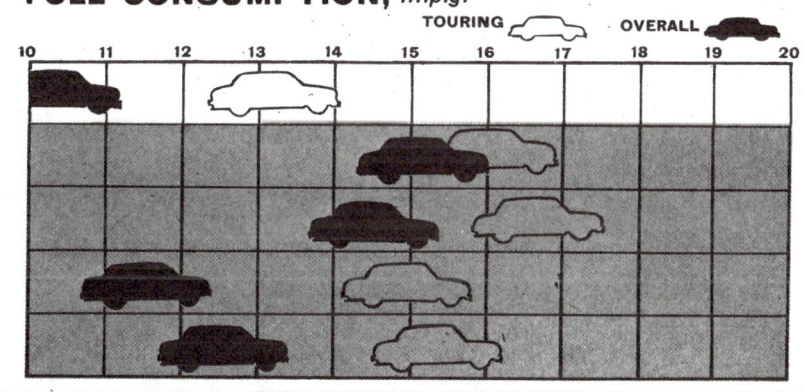

ACCELERATION, 0-50 m.p.h. in seconds

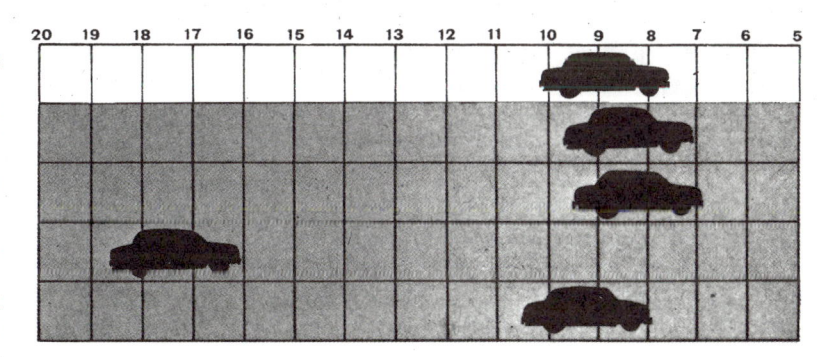

Car	seconds
Rolls-Royce	~9
Daimler Majestic Major	~8.5
Bristol 407	~8.5
Vanden Plas 4-litre	~17
Mercedes-Benz 300SE	~9.5

ACCELERATION, kick down, 20-40 m.p.h. in seconds

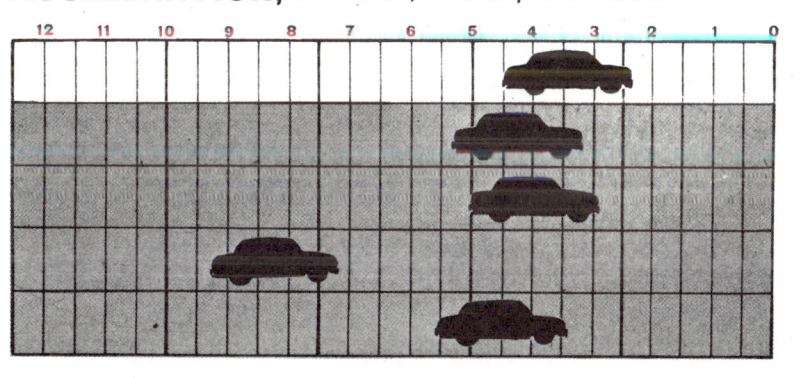

Car	seconds
Rolls-Royce	~4
Daimler Majestic Major	~5
Bristol 407	~4.5
Vanden Plas 4-litre	~8.5
Mercedes-Benz 300SE	~5.5

THE MOTOR March 4 1964

FULL STOP BY ROLLS-ROYCE

PERSONAL COLUMN

by Charles Bulmer

WHEN we road tested the Rolls-Royce Silver Cloud III last August we criticized the brakes for various reasons, notably for appreciable pedal pressure rise during our long hill descent and the fade test of 20 stops at minute intervals from 72 m.p.h. These criticisms were received with some horror at Crewe, where they are particularly proud of their brakes and also rather sensitive to un-informed criticism suggesting that a less reactionary firm would have adopted discs by now.

The road test car was examined, and it was found that there was good reason for indifferent fade results and also for some high speed vibration and rumble which we noticed. The shoes had not been properly centred in the drums during assembly, and they had been re-lined with loose linings instead of lined shoes ground accurately to the drum radius, thus concentrating all the braking on a few small friction areas.

Sixty stops in one hour

Rolls-Royce were anxious that we should repeat the fade tests when the car was free, and this we have done now. It can be said straight away that, within the limits of experimental error, our repeat tests show no pedal pressure rise at all, as they forecast. "But," they said, "why stop your fade test after twenty minutes; why not go on for an hour or even until the linings are worn out?" We thought we might be worn out as well by then, but we accepted the challenge of 60 stops from 72 m.p.h. in one hour, which is a severe test for the observer as well as for the brakes. This had no effect on the brakes at all, the pedal pressure remaining at 37½ lb. ($\pm$ 2½ lb. experimental error) from start to finish. There was no pulling, no vibration or rumbling at any time.

In suggesting this test Rolls-Royce made no claim that it could not be paralleled by some disc-braked cars. They did say, however, that they thought they could name some which would fail due to the hydraulic fluid boiling in these conditions of intense prolonged heat build-up, a much more catastrophic occurrence than fade because it results in immediate total brake failure. Rate of wear is another vital consideration with a two-ton car capable of 115 m.p.h., and so far it has proved necessary to replace the friction linings at least twice as often with discs as with the drum brakes which have an enormous lining area; the friction materials which wear well with discs tend to squeak. All these factors and others have been weighed in retaining drums, although the engineers at Crewe have been experimenting with discs for over 10 years now and are still doing so.

The servo

The whole drum layout is unorthodox. Two trailing shoes are used at the front, an arrangement which has a negative self-servo effect and therefore has an abnormally low sensitivity to changes of friction coefficient with rising temperature; the shoes are self-adjusting and run in very heavily ribbed drums to guard against distortion. At the rear the shoe layout is orthodox, but the two are linked for equal wear. The whole system is energized by the unique R-R mechanical servo driven from the output-shaft of the gearbox which was first introduced by Sir Henry Royce in 1925. This has one disadvantage—lag in response which represents a constant distance of about 18 in. Translated into time this means 1/50 sec. at 50 m.p.h. or 1/5 sec. at 5 m.p.h.; in other words it is unnoticeable at speed but observable in heavy town traffic and when manoeuvring.

Its advantage is an ability to provide assistance of a magnitude beyond that of a reasonable-sized vacuum servo—it multiplies the driver's efforts by seven and applies the resultant force to three independent linkages, two hydraulic and one mechanical. One master cylinder operates one shoe in each front brake and applies 76% of the rear braking, the remaining 24% being transmitted by the mechanical linkage which also serves the hand brake; the second master cylinder operates the second shoe in each front brake. The failure of a single component in any part of the braking system therefore leaves enough braking on both front and rear wheels to achieve braking efficiencies between 65% and 85% of the maximum possible without resort to the handbrake.

OPINION

♦

TOPICAL

TECHNICAL

DIARY

WONDERS will never cease, and that alone is explanation enough for using a flying Rolls-Royce action shot to illustrate a point for the second month running! The fact is that this venerable old firm suddenly decided to wake up and join in the fun, spurred to it no doubt by the recent spate of Good-Better-Best verbiage arising out of the arrival of the Mercedes 600.

Anyway they handed out a Silver Cloud III to Tony Brooks . . . most II of the motoring journalist set apparently despite his associations with the London *Observer*, and invited him to drive it round Goodwood in the manner born. Then, by way of a thank-you, the makers stood him a trip from Le Touquet to Cape Ferrat (740 miles) in the same car. The " wonders " part of the arrangement is that R-R subsequently encouraged Brooks to discuss the car with the Chief Engineer in quite the frankest manner that any R-R has ever been discussed in public and what is more, put it all down in a sales brochure for everybody to read and inwardly digest.

Brooks: My first impression was one of the car being extremely light to handle in relation to its size, and I was very surprised at its *controllability*. You could do more or less what you wanted with it and there was no tendency to oversteer or understeer. It had absolutely neutral characteristics with the tyre pressures used. It always slid bodily, and it could actually be drifted quite easily in the wet if you wanted to. Even when the circuit was wet, the car felt extremely safe at all times. It rolls so very little too. I drove the car through the corners as hard as I could and the roll for this class of car was very low indeed — and this was commented on by the track manager. We also put in some fast laps with four people up and even with this sort of load there wasn't very much roll. I was particularly surprised at the way the car could be thrown around and yet behave in such a gentlemanly way. The impression was not of a large heavy car at all.

The steering of the car was a little too low-geared for my taste but this didn't present any real problem at Goodwood other than at the chicane where it was necessary to cross my arms to get through in a really fast manner; I rarely had a need to correct more than about one inch or two inches of steering wheel movement and therefore I didn't worry about the low-geared steering from this point of view. In day to day motoring this is a very small problem.

S. H. Grylls (Chief Engineer): We think our steering ratio is the best compromise. You *can* drive in a sporting way with this, as you've discovered. And at the same time less continuous attention is needed under normal circumstances. We describe it as having a " low sneeze-factor."

Brooks: I should really have prefaced all these comments with the point that I have d r i v e n this car not as an average Rolls-Royce owner would — but in a decidedly sporting manner.

From that point of view I found the change down to second on our particular car not very smooth. If I was behind traffic and pottering along at about 20 m.p.h. in top, then all of a sudden the road became clear, I would press the accelerator to the floor and the car would lurch forward.

And there's another criticism here. Between second and third from the point of view of ultimate performance, there's too big a gap — particularly if you happen to hit the change-up on a hill. The fall off in acceleration is very noticeable as you get into third.

S. H. Grylls: This is the biggest unsolved problem in our particular form of automatic transmission. We have retained a stepped transmission having the least possible slip because we think this is most suited to motoring in Europe. Of course, any transmission needs a little time to bed down. The car you tested, I think, was a new one.

The fall in acceleration you mention would only be " very noticeable " I would say, if you are particularly brutal with the throttle."

Brooks: It was deliberate, of course. But is there no possible way of closing down the second to third gap?

S. H. Grylls: There's no reasonable way of closing down the gap. You see you've got these four ratios in a total of three epicyclic trains and the fluid flywheel actually moves up and down the box. Although geographically it's in the front, it's in the middle of the box in some of these ratios.

Brooks: But surely a torque-convertor overcomes this, doesn't it?

S. H. Grylls: It does in that with three ratios you get almost as good a coverage. But you can't drive at full throttle in top gear at low speeds as you can with our type of transmission. We can get down to 900 revs per minute in top gear before changing down, but we can manually select third gear for acceleration or braking when desired. A torque-convertor transmission necessarily burns more fuel.

Brooks: It seems a pity that the design of the new engine has produced rather a noisy tick-over. Couldn't this be reduced — even though it is quickly lost on moving off from rest?

S. H. Grylls: This is inherently more difficult with an aluminium engine and so

— OPINION

far we have not found a satisfactory answer.

Brooks: **I found the brakes wonderful. But they do bring up the obvious question of discs and their even greater resistance to fade. Hasn't it been possible yet to overcome the incidental drawbacks of discs?**

S. H. Grylls: As you know, discs do have advantages over conventional drum brakes. They have better heat dissipating qualities and so show less fade under continual high speed use. But our drum brakes are far from conventional. The Rolls-Royce arrangement of trailing shoes at the front and equal wearing shoes at the rear is less temperature sensitive than disc brakes. When you combine this with heavy ribbing of the brake drums, a very large brake lining area and specially developed linings, brake fade has ceased to be a problem. Our tests in fact have shown that we can stop once a minute from 70 mph until the linings are worn out. And, of course, pedal pressure is no problem in a Rolls-Royce because we give you servo-assistance that makes you seven times the man you are.

Brooks: **Yes, but what are the disadvantages of discs?**

S. H. Grylls: Disc braking systems are usually heavier than drum brake systems. They are inclined to be noisy in operation at certain speeds and their performance often varies in different climatic conditions. And their rate of wear is at least twice. In other words, you would have to reline them twice as often as our brakes. And the discs that wear comparatively well squeak like blazes. All of which we'll overcome one day. But we won't fit discs until they are as silent and smooth and progressive as the brakes we have now.

Brooks: **Right. Then let's turn to suspension. I found it very good. But there was a slight tendency for the car to wander at speed, particularly noticeable on the rough French roads, in spite of our 5 lb pressure differential between front and rear tyres when cold. Was this perhaps partly due to the large suspension movements?**

S. H. Grylls: Self-centring which is essential in a motor car is incompatible with complete freedom from wander. Incidentally the tyre pressure differential when hot is about 8 lbs.

Brooks: **Surely independent rear suspension would further improve the ride and roadholding?**

S. H. Grylls: Well, it's simply not true that irs is the answer to every ride problem. Take the swing axle type which is the commonest of all. It can lead to some very tricky handling characteristics.

Brooks: **I agree there. I've never liked ordinary swing axle layouts myself. You can get a sudden change to vicious oversteer when you're cornering fast.**

S. H. Grylls: Of course, there are other forms, but they all pose problems — of wear and noise — which just aren't acceptable when you're designing to Rolls-Royce standards. Our cars are heavy, they're capable of travelling at very high speeds, and they have to be very comfortable. Consequently the suspension has to be fairly soft and you have to cater for considerable vertical movements. No present independent rear system could handle all this as well as our system does.

Brooks: **I was very impressed with the handling of the car in the wet — particularly at Goodwood. It rolled moderately easily up to a certain point but after that it didn't seem to go any more.**

S. H. Grylls: We were the first people in England to have a skid pan. We made ours in 1935. You can find out in five minutes there all about the handling of a motor car. I think we know what to look for now and we know what we want to make a car handle.

Mr. Brooks (concluding) had a number of other minor points to make, some of which he agrees are a matter of personal preference. He felt that amendments to the dash layout, lack of full crash-padding, rearwards visibility under given circumstances, heavy-to-shut doors, long travel of the hand-brake lever, performance of the screen-wipers at very high speeds, strength of the horn-note and noise made by the scuttle air-scoop and ventilation fan could all do with attention.

LIFE ON A CLOUD

CONTINUED FROM PAGE 81

hard to take liberties with £8750 worth of someone else's machinery).

Power assistance helps to overcome it in the hands. I was never quite sure how the forces were balanced when trying hard on a corner.

You see, the power build-up is proportional to the effort at the wheel. Normally it takes about 45 percent out of the effort, increasing to 80 percent when hauling the car around or parking.

The wheel is low-geared and not the handiest at 4½ turns lock-to-lock. But the 40ft. 6in. turning circle is magnificent for such a monster.

The best I can say about this power system is that I wouldn't care to try parking without it.

But you don't have to worry. When the tail goes, there's plenty of time to correct. The Rolls is beautifully balanced.

There's really no need to fool around the corners, anyway; go in relatively slow and just let the clamoring torque shoot you out like the second stage of a moon rocket.

Mighty Stoppers

Obviously, the next thing the Rolls needs is brakes. Brakes there are aplenty—and the arrangement must be the safest yet designed.

Three separate systems work the big drums, with 240 sq. in. of lining area:
- A separate master-cylinder does for the front brakes;
- Another master-cylinder works the rear brakes;
- A separate mechanical linkage from the handbrake goes to the rear brakes.

I can't imagine all three failing at the same time.

Power assistance is provided by a gearbox-driven pump — a system that Rolls have used for more than 30 years. Even if you coast down a hill with the engine switched off, the gearbox will keep driving the servo pump, so you'll still have some power assistance for the brakes.

All other brake-assisting power systems that I know of work off the engine manifold vacuum and don't operate unless the engine is running.

However hard I tried, I couldn't fade these anchors: stopping distance from 30 m.p.h. in neutral was always under 32 feet.

Unique Transmission

But perhaps the most fascinating part of the car is its automatic transmission.

Basically it is a General Motors design — but not, of course, in its original crudity. Rolls-Royce bought manufacturing rights for it about ten years ago, when it was a three-speed transmission with torque-converter.

They set to work and put in a fourth gear. They modified the control valves. Then they gave it a system of control like no other automatic transmission has.

In addition to the usual park, neutral and reverse, the steering-column quadrant has separate selector positions — 1, 2, 3, 4 — for all four forward gear ratios.

For mucking around and admiring other people's houses (Rolls-Royce owners would be hardly likely to do this, of course) you can just lock out top by pushing the selector to 3. Unless you exceed maximum revs in third, the transmission won't go above that gear.

The same applies to positions 2 and 1. In fact, position 4 is the equivalent to Drive on less "refined" transmissions.

Lowest gears can also be engaged by a kick-down switch on the throttle. Maximum speeds for kick-down are: 4 to 3, 60 m.p.h.; 3 to 2, 20 m.p.h. First is not available on the move above 10 m.p.h.

Upward changes on a light throttle are: 1 to 2, 10 m.p.h.; 2 to 3, 19 m.p.h.; 3 to 4, 28 m.p.h.

If the throttle is used normally, neither up nor down changes can be detected without a stethoscope on the gearbox. All the indirect gears are as quiet as top.

But if you put her at a winding bit of road at full throttle in 4, there's inevitably a switching-down to third and back again. That's why Rolls provide lock-out selection for each gear.

Park and reverse are blocked out from accidental engagement by a gate. The engine will start only with the selector at neutral.

On a test of less than 100 miles, all this jiggling with the transmission —exploiting its almost endless and breathtaking variations—was paid for at the rate of 12.8 m.p.g. But what's a bit of petrol? If you pinch a Rolls with a full 18 gallons aboard, you can have at least 200 miles of fun.

And for anyone who's dickering on the edge of buying a new Rolls (I don't expect any correspondence on this point) there is a sort of celestial trade-in plan for cars "not more than two years old" (at this point I almost feel like embracing Communism).

The Rolls is very high by today's standards at 5ft. 4in. But you get a good view over the long bonnet. Only bad part of visibility is backwards—both driving mirror and rear window are too small.

I didn't like the instruments being set in the middle of the dash, spread out like a hand of cards. But, given time, you can read almost anything you want—including the engine-oil level.

Going through all the "bits" in detail would fill half the magazine—but the more intriguing odds-and-ends are:
- A dash button that opens the petrol-cap flap;
- An astoundingly efficient twist-and-pull handbrake, carefully concealed under the dash;
- Separate heating and ventilating arrangements for the back compartment;
- Rear-window demisting by electric wires buried in the glass;
- Picnic tables, vanity mirrors and wing-chair-type seating in the back;
- Front bench seats with split squabs that are individually adjustable for rake;
- A very small transmission hump which gives plenty of room for a third party in front;
- Real pile carpets and real walnut trim on dash and door cappings.

Bodywork is built on a separate box-section chassis stuffed with fibreglass insulation, which is one of the reasons the car is so quiet.

The boot sports a usable 17 cubic feet of space. The spare is housed in a separate compartment below—plus all hand tools necessary for routine adjustments, stowed neatly in a rubber-padded drawer.

Engine accessibility is not good (as if any Rolls owner would care about that!). The alligator bonnet is long and narrow. But the air-cleaner hinges upward to let you in at the two fat SU carbies and valve gear. I wouldn't like to have to change plugs, though.

Unusual points about the aluminium block, with oversquare 104mm. by 91.4mm. cylinder bore and stroke, are:
- Light weight—same as for the old six-cylinder 5-litre unit;
- Wet steel cylinder liners;
- Hydraulic tappets that need no periodic adjustments (these have been used before by Rolls-Royce but dropped for a while, for some reason).

Compression ratio is 8 to 1. I couldn't make the engine ping on our premium fuel.

Chassis lubrication is needed only at 10,000-mile intervals—a year's driving, on the average. And a radio set is thrown in as standard equipment.

Well, there it is—the peculiar Rolls-Royce mixture of ancient and modern. Whether you like it or not, the engineering and finish are THE best, and ownership is a sign that you have arrived. Which, my dear Sir or Madam, is where we came in...

FOOTNOTE for graziers: you can't get more than three or four sheep in the back. ● ● ●

MODERN MOTOR — July 1961

BUYING SECONDHAND

Rolls-Royce Silver Cloud
Bentley S-series

AUTOCAR, w/e 27 December 1975

FOR MOST PEOPLE the ownership of a Rolls-Royce is an unattainable dream, yet a close look at what is available on the secondhand market shows that Rolls-Royce quality is available for the same sort of money as a new Rover 3500, Volvo 244GL or Mercedes-Benz 230 and for less than the cost of any new Jaguar. The snag, of course, is that a Rolls-Royce for this sort of money (£4,000) means going back to the Silver Cloud series prior to the introduction of the current Silver Shadow.

It was as long ago as April 1955 when the Rolls-Royce Silver Cloud was publicly announced and as usual it represented a considerable step forward in Rolls-Royce motoring. Particularly striking was the all-new steel saloon body combining flowing lines with aerodynamic efficiency. There was still the traditional Rolls-Royce chassis, of course, upon which the standard steel body or one of a number of coachbuilt bodies were mounted. This model was designed very much with the owner-driver in mind and although a large car — it was 17ft 7¾in long, almost 6ft 3in. wide and had a 10ft 3in. wheelbase — the excellent commanding driving position and such refinements as automatic transmission made it seem considerably smaller from the driving seat. Rear seat passengers were well looked after with generous leg room, while a comprehensive heating and ventilation system was provided. For the first time its counterpart Bentley, known simply as the S-series, was identical in all except badging—though the Continental Bentleys continued to differ in a number of details, not least of which were a 3½in lower radiator, higher 2.92 rear axle ratio, narrower 8.00-15in. tyres and twin leading shoe front brakes as well as a higher compression and more power, and were all coachbuilt.

Power for the Silver Cloud was provided by the ultimate stretch of the six-cylinder engine which had powered the R-R Silver Dawn/Bentley R-type now bored out to 4,887 c.c. and with a new aluminium cylinder head, and allied to the Rolls-Royce built Hydra-Matic automatic gearbox. No manual gearbox was available, although a few buyers of Bentley Continentals were permitted the option of a manual box up to 1957. Power assisted steering was not available for the first Silver Clouds.

Rolls-Royce have pursued a policy of continuous development which still continues today, so changes to the basic design have been introduced on an "as and when" basis rather than lumped together under the guise of a facelift. The original Silver Cloud, now known as the Silver Cloud 1, and its Bentley counterpart the S-series (known now as S1), were to continue with very few modifications until the introduction of the Cloud II/Bentley S2 in September 1959. As far as Cloud Is were concerned 1957 was when some changes likely to influence a secondhand choice occurred. In February summer/winter heater taps, a worthwhile refinement allowing the heater to be turned off for the summer (thus increasing cold air throughput) were introduced, in April power assisted steering became optional, and in September the compression ratio was raised from 6.6 to 8.0 to 1 and larger 2in. SU carburettors were fitted giving a modest power increase.

The advent of the Silver Cloud II in September 1959 brought with it a number of improvements and a bigger engine. Although outwardly unchanged — Clouds I and II are identical externally — the Silver Cloud II boasted a completely new aluminium alloy V8 engine of 6,230 c.c. under the bonnet. The power increase that this brought — it is not Rolls-Royce policy to reveal actual power output figures — gave considerable improvements in acceleration shortening the car's 0-60 mph time by 1.5sec to 11.5sec and the 70-90 mph time by 3.8sec to 11.9sec. All this was not obtained for nothing and the V8 was certainly not as quiet or smooth as the six even though hydraulic operation had been adopted for the tappets. From a maintenance point of view the bonnet was more tightly packed, the wet-linered V8 less straightforward for the owner to work on — a feature that mattered little to the original owner but is more relevant now — with the necessity to remove the offside front wheel to change the sparking plugs on that side of the engine. Chassis improvements included the replacement of the Rolls-Royce pedal operated centralized chassis lubrication system by 21 grease nipples, duplicated hydraulic systems serving the massive 11¼ x 3in. drum brakes which continued to use a gearbox driven mechanical servo, standardization of power assisted steering (previously an extra), pre-engaged starter, increased capacity heating and ventilation systems with fresh air outlets on the facia capping rail and a separate recirculating system for the rear compartment, and a higher 3.08 final drive ratio. There were no major engineering changes during the Silver Cloud II's three year life, though there were a host of minor changes. In October 1960 fresh air ducting was added under the nearside front wing allowing air to enter the car under the facia, and chrome handrails were fitted to the facia capping rail. A year later came more changes to the interior including blue instrument lighting, a combined headlamp flasher/indicator switch, provision of a handbrake warning lamp, central heater duct to the rear compartment and sliding doors to the front door pockets. Final changes to the series came with the fitting of a larger rear lamp assembly in May 1962 and sealed beam headlamps for the final cars from August 1962.

The Silver Cloud III and Bentley S3 models were announced in October 1962 and were heralded as a major advance in Rolls-Royce motoring. With the benefit of

1963 Rolls-Royce Silver Cloud III retained the classic shape of earlier Clouds but had a lowered bonnet and radiator, sidelamps and flashers recessed in the front wings and a paired four-headlamp system. The twin foglamps and roof mounted radio aerial were standard fittings

hindsight it is now clear that they were very much better cars than the Cloud II V8s — quite apart from their changed appearance and higher performance. Immediate visual changes were the adoption of a paired four-headlamp system, using 5¾in. Lucas sealed beam units, the forward slope of the bonnet to a radiator 1½in. lower, recessed sidelamps/indicators and smaller overriders. On the mechanical side the power of the 6,230 c.c. V8 had been increased by the adoption of a 9.0 to 1 compression ratio (8.0 compression was optional for some markets) and larger 2in. SU HD8 carburettors in place of the S2's 1¾in. HD6s. This resulted in the S3 requiring 100 octane Super petrol — a disadvantage when buying secondhand now, though fuel consumption at around 12 mpg is slightly better than Silver Clouds I and II. Engine changes to cope with the increased power included a nitride-hardened crankshaft (broken crankshafts were not unknown on early S2s) and larger diameter gudgeon pins in the pistons. A sealed crankcase ventilation system was also new, while extra power assistance for the steering reduced the load at the wheel when parking to 6lb. Inside the S3 there were changes too with the adoption of individual front seats (replacing the one-piece bench which remained optional) and a redesigned rear seat with more upright backrest position giving 2in. more leg room as well as proving more comfort for three-abreast seating.

Again major changes during the S3's life were few. Stainless steel wheel discs replaced chromium-plated ones in April 1963 and a more effective rear window demister was introduced in November of that year. 1964 saw several changes: wider front seats in January, a restyled headlamp surround incorporating the RR or B monogram in March, and wider rims on the rear wheels for standard and long wheelbase cars (making fronts and rears no longer interchangeable) in May. Production of Silver Cloud III and Bentley S3 standard and long-wheelbase cars stopped just prior to the introduction of the completely new Silver Shadow in October 1965, though production of H. J. Mulliner, Park Ward coachbuilt S3s continued until well into 1966.

Coachbuilt cars

There were a great many more examples of the Silver Cloud and Bentley S-series produced than many people imagine, the majority being the standard cars emanating from Crewe. Coachbuilt cars were therefore relatively few in number though Freestone and Webb, Hooper, H. J. Mulliner, James Young and (after 1961) the combined firm of H. J. Mulliner, Park Ward all produced coachbuilt versions of Silver Cloud and Bentley S-series chassis. Production figures for the various models help to put the matter in perspective: Silver Cloud 1 2,238, long wheelbase 121; Bentley S1 3,009, long wheelbase 35, Continental 431; Silver Cloud II 2,417, long wheelbase 299; Bentley S2 1,863, long wheelbase 57, Continental 388; Silver Cloud III 2,044, long wheelbase 253, coachbuilt 79; Bentley S3 1,286, long wheelbase 32, Continental 312. All Bentley Continentals were coachbuilt though by the time the S3 arrived on the scene they offered no extra performance and had even lost their high axle ratio, in effect they were just coachbuilt saloons. The standard long wheelbase car had 4in. longer wheelbase arrived at by cutting the standard body and adding in an extra piece behind the central pillar; this was carried out by Park Ward. H. J. Mulliner produced a convertible version of the standard saloon and there were standardized H. J. Mulliner, Park Ward two-door coupé and convertible bodies on Silver Cloud III and Bentley S3 models. The adoption of new design features and mechanical changes on coachbuilt cars could be up to 12 months behind the standard saloons and individual cars varied considerably in the equipment fitted.

Finest coachbuilt bodies were by H. J. Mulliner. This is a Bentley Continental S3, but the same body was used on S2 Continentals and also on Silver Cloud IIIs

Rolls-Royce Silver Cloud III with H. J. Mulliner, Park Ward steel body is a car to be wary of as the rust problems can be serious. Convertible version is basically similar

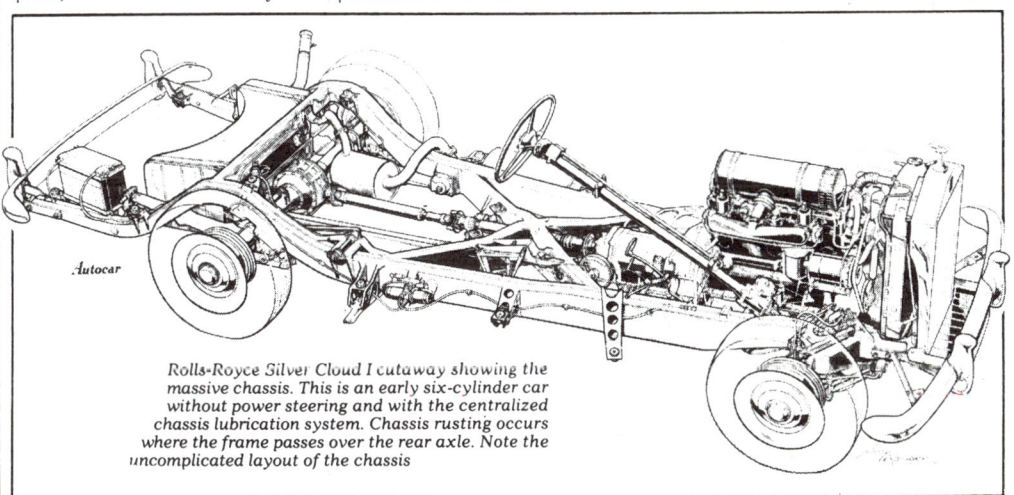

Rolls-Royce Silver Cloud I cutaway showing the massive chassis. This is an early six-cylinder car without power steering and with the centralized chassis lubrication system. Chassis rusting occurs where the frame passes over the rear axle. Note the uncomplicated layout of the chassis

Interior of the Silver Cloud III standard saloon. Trimming was to a high standard in leather. Individual front seats were a feature of the model and the rear seat was moved further back to give increased legroom

What to look for

When it comes to assessing a used Rolls-Royce Silver Cloud or Bentley S-series model, the old myths about Rolls-Royces never wearing out and that a former titled owner makes the car a better buy should be forgotten. The plain truth is that — although exceptionally well built in the first place — even Rolls-Royces wear out if maintenance has been

BUYING SECONDHAND
Rolls-Royce Silver Cloud
Bentley S-series

neglected as may often have been the case with cars we are considering. A car which has been properly maintained in accordance with Rolls-Royce recommendations and had jobs done when they became necessary is likely at a mileage of around 100,000 to be good for as much mileage again without major problems. A car that has been neglected could be in need of a major rebuild' by 100,000 miles — remember the youngest of the models we are considering is 10 years old and the oldest could be 20 — which will be very, very expensive. It is far better to pay a top price for a car in first class bodily and mechanical condition than to buy a cheap poor condition (say) Silver Cloud I or II for £2,000 which then costs a further £5-6,000 to bring up to scratch whereas a good example could have been found for around £4,000.

All three series had their individual weaknesses, the Silver Cloud I/Bentley S1 being probably the easiest to work on thanks to its six-cylinder engine which gave little trouble. Early SCI/S1 models up to 1957 are perhaps best avoided because they do not have twin brake master cylinders (a worthwhile safety feature on a car weighing around 2 tons) and also lack power steering. Neither of these features can be added retrospectively. The centralised lubrication on these cars often failed at the front suspension points, resulting in wear of the swivel pin on the top wishbone. Wear here can only be detected when the suspension is not under load, but has been the cause of MoT failures. At around £150 a side this is not a cheap fault to remedy and combined with attention to a power steering leak could mean expenditure of about £400 to pass an MoT. By and large, though, Silver Cloud I front suspension gives less trouble than series II and III cars and once rectified should be troublefree for around another 100,000 miles. Gearbox problems are few on the six-cylinder cars with a life of over 100,000 miles.

Early Silver Cloud II/Bentley S2 models of 1959-60 are best avoided. Some of these had crankshaft breakage problems due to lack of lubrication to the bearings while others suffer from corrosion of the aluminium alloy cylinder block. This occurs around the base of the pressed-in steel cyclinder liners and allows them to move causing piston ring breakage. Be very wary of any S2 (or even S3) which is badly down on compression, using a lot of oil (consumption is normally negligible) and smoking from the exhaust as this could be due to liner movement breaking a piston ring. Once the engine has been taken down, it will be impossible to put it back as it was if this does prove to be the case. The remedy is a major engine rebuild with a new block — this could cost as much as £2,000! Noisy tappets on the six-cylinder engine can be adjusted, but the same problem on the V8 engine (hydraulic self-adjusting tappets) means camshaft replacement, involving removal of the engine from the car, and a bill of around £300 including fitting. There are also sometimes cylinder head leaks from the steel gasket used on S2s, and the flaps at the bottom of the front wings controlling the face-level ventilation are prone to seize. Gearbox life on S2 and S3 is around 80,000 miles; there is provision for adjusting the brake bands. Other problems common to all models include distorted front brake drums (about £48 each), wear on brake servo plates and drive gear (about £60), weak rear suspension dampers (leaks), leaks from the pinion seal, and power steering leaks (not good for the rubber engine mountings). Rusting of the underwing heater matrices can be a problem — replacement means removal of one of the bolt-on front wings on SC II and III and the driver's door as well on SCI. It is not recommended to replace the original 8.20-15in. cross-ply tyres with radials. Rolls-Royce dealers can supply the correct tyres at about £35 a tyre and these have a life of 12-15,000 miles, so any example with worn tyres is going to result in a big bill early on. Brake lining life, too, is short and renewal will be needed at about the same interval as tyres.

Corrosion

Rust attacks even Rolls-Royces, though of course the aluminium bodied coachbuilt cars are less affected. Doors, bonnet, and boot on all Silver Clouds/Bentley S-series are aluminium and therefore rust free. Look for rust on the vertical spot welds at the rear of the front wings and around the sidelamps on S1/S2. Front wings on S3s are made in two parts and joined along the swage line — a rust point as is the area under the paired headlamp units. On all models the sills rust (usually right underneath where it does not show). Other rust spots are where the centre stay joins the front wing (above the wheel), the trailing edge of the rear wheelarches and below the rear bumper. The chassis is of massive box section construction but nevertheless rusts right through where it passes over the rear axle. However, this can be plated easily. Silver Cloud IIs are somewhat better than Is and IIs where this chassis rust is concerned. Fuel tanks also rust where the securing straps pass round them — only remedy is a new tank.

Among the coachbuilt cars by far the worst offender is the H. J. Mulliner, Park Ward two-door

Performance data

	Rolls-Royce			Bentley			
	Silver Cloud I	Silver Cloud II	Silver Cloud III	S1	S1 Cont PW	S2 Cont JY	S2 Cont HJM
Road tested in *Autocar*:	6 May 1958	13 May 1960	9 Aug 1963	7 Oct 1955	21 Dec 1956	30 Dec 1960	16 Aug 1973
Mean Maximum Speed (mph)	106	113.1	115.8	101	119.2	112.7	—
Acceleration (sec)							
0-30 mph	4.1	3.8	3.5	4.4	4.3	4.0	3.8
0-40 mph	9.4	5.9	5.4	—	—	6.3	6.2
0-50 mph	9.4	8.3	7.7	10.3	9.3	8.9	8.7
0-60 mph	13.0	11.5	10.8	14.2	12.9	12.1	11.8
0-70 mph	18.4	15.5	14.2	19.9	17.1	15.9	15.9
0-80 mph	25.0	21.1	19.2	28.0	21.3	20.5	21.0
0-90 mph	34.1	27.4	25.4	39.4	29.5	26.9	28.9
0-100 mph	50.6	38.5	34.2	—	40.2	37.1	41.5
0-110 mph	—	—	—	—	51.8	—	—
Standing ¼-mile (sec)	18.8	18.2	17.7	19.7	18.8	18.6	18.5
Top Gear (sec)							
30-50 mph	8.9	7.6	7.9	—	—	7.6	—
40-60 mph	9.4	7.8	8.0	—	—	7.8	—
50-70 mph	11.1	8.1	8.5	—	—	8.7	—
60-80 mph	13.0	8.5	9.5	—	—	10.2	11.3
70-90 mph	—	11.9	11.7	—	—	12.2	14.1
80-100 mph	—	17.4	15.0	—	—	16.6	22.5
Overall fuel consumption (mpg)	12.0	11.8	12.3	14.0	15.2	13.1	—
Dimensions							
Length		17ft 7¾in.			17ft 8in.		
Width		6ft 2¾in.			6ft 0in.		
Height		5ft 4in.			5ft 1in.		
Weight (lb)	4,144	4,522	4,578	4,242	3,976	4,460	—

Bentley Continentals: Dimensions are approximate. PW = Park Ward Drophead Coupé; JY = James Young 4-door fitted with low 3.08 axle ratio; HJM = H. J. Mulliner 4-door Flying Spur tested after 115,000 miles as a used car.

Spares prices

	SCI/S1		SCII/S2		SCIII/S3	
	New	Exchange	New	Exchange	New	Exchange
Short engine	N/A	N/A	N/A	£1,241.24	N/A	£1,241.24
Automatic gearbox (R-R Hydra-Matic)	N/A	£314.80	N/A	£314.80	N/A	£314.80
Propellor shaft centre joint	£3.02	N/A	£3.02	N/A	£3.02	N/A
Final-drive assembly	N/A	£314.80	N/A	£314.80	N/A	£314.80
Brake shoes — front (set)	£15.33	N/A	£15.33	N/A	£15.33	N/A
Brake shoes — rear (set)	£15.33	N/A	£15.33	N/A	£15.33	N/A
Suspension dampers — rear (pair)	N/A	£86.05	N/A	£86.05	N/A	£86.05
Suspension dampers — REAR (pair)	N/A	£86.05	N/A	£86.05	N/A	£86.05
Radiator matrix assembly	N/A	£76.14	N/A	£76.14	N/A	£74.07
Dynamo	£100.04	£64.80	£100.04	£64.80	£100.04	£64.80
Starter Motor	£101.70	£82.08	£108.86	£82.08	£108.86	£82.08
Front wing panel	£133.11	N/A	£133.11	N/A	£189.00	N/A
Bumper, front	£35.64	†	£35.64	†	£69.33	†
Bumper, rear	£35.64	†	£35.64	†	£35.64	†
Windscreen, laminated	£47.13	N/A	£47.13	N/A	£47.13	N/A
Exhaust system complete	£219.65	N/A	£248.46	N/A	£280.70	N/A
Bentley radiator shell	£245.32	†	£245.32	†	£258.02	†
Rolls-Royce radiator shell	£260.65	£108.00	£260.65	108.00	£258.02	£108.00

N/A — Not available 'Rolls-Royce Silver Cloud I and Bentley S1 six-cylinder engines are not offered complete but parts are available for engines to be rebuilt. † Bumpers and radiator shells can be reconditioned on exchange basis by independent specialists

AUTOCAR, w/e 27 December 1975

Silver Cloud III in either coupé or (worst of all) drophead coupé form. These cars have steel (not aluminium) front and rear wings and the rear of the car is a mass of welded steel boxes. Also bad are early Park Ward bodied cars while James Young bodied cars leak water through the rear window and boot seals. By far the best of the specialist bodies were those carried out by H. J. Mulliner, particularly the Bentley Continentals.

Interior condition

It may come as a great surprise to some to learn that the instrument surrounds and facia capping rails on standard S-series cars are plastic — it wears well. The woodwork is not costly to refurbish unless it is badly chipped or has split and the veneer lifted off. Headlinings and carpets (except on James Young cars) clean up well. Both front seats on a Bentley Continental would cost about £300 to re-trim in leather — a standard saloon would be rather more because the seats are larger. Early cars with electric windows can be a problem as these were never very reliable.

Spares for Silver Clouds are readily available even bodyshells and chassis frames. Body panels for coachbuilt cars are mostly not obtainable — and the cost of making them on a one-off basis is prohibitive, so a bad accident is likely to result in the car being written off. We are indebted to Rolls-Royce distributors Lex Mead Maidenhead for their help in preparing these notes.

Silver Clouds and Bentley S-series models have now joined the ranks of Bentley Continentals and are appreciating. Best value for money from the series has to be a Bentley S1 saloon — bearing in mind that the equivalent Rolls-Royce version is about £1,000 more — while a good Silver Cloud I or III, an H. J. Mulliner bodied Bentley Continental or a James Young-bodied long wheelbase Silver Cloud should prove a worthwhile investment.

Approximate selling prices

Condition not year of manufacture is the basis of deciding the value of a used Silver Cloud or Bentley S-series. Coachbuilt cars and Bentley Continentals are worth up to £2,000 more than the equivalent Rolls-Royce standard saloon. Exceptional cars command higher prices.

Rolls-Royce Silver Cloud I — pre 1957 (single circuit brakes and no power assisted steering) — poor condition £2,000 / good condition £3,000

Rolls-Royce Silver Cloud I — 1957-62 (twin brake master cylinders and power assisted steering) — poor condition £2,300 / good condition £3,500

Rolls-Royce Silver Cloud II (1959-62) — poor condition £2,300 / good condition £4,000

Rolls-Royce Silver Cloud III (1962-65) — poor condition £3,000 / good condition £6,000

Bentley S1, S2, S3 standard saloons are worth about £1,000 less than the equivalent Rolls-Royce.

Bentley Continental S1 (H. J. Mulliner) — poor condition £3,000 / good condition £5,500

Bentley Continental S2 (H. J. Mulliner) — poor condition £3,750 / good condition £5,500

Bentley Continental S3 (H. J. Mulliner, Park Ward) — poor condition £4,000 / good condition £6,000

Chassis identification

April 1955: S-Series cars announced. First chassis numbers—
— Silver Cloud I — SWA2
— Bentley S1 — B2AN

September 1955: Bentley Continental S1 introduced — BC1AF
April 1956 Twin brake master cylinders — Silver Cloud I — SYB116
— Bentley S1 — B245BC
— Bentley Continental S1 — BC16BG

March 1957: Last Bentley Continental with manual gearbox — BC79BG
September 1957: 8.0 to 1 compression, 2in. SU carburettors
— Silver Cloud I — SFE23
— Bentley S1 — B257EK
— Bentley Continental S1 — BC21BG

November 1957: Long wheelbase cars
— Silver Cloud 1 — ALC1
— Bentley S1 — ALB1

1959: Final chassis numbers six-cylinder cars
— Silver Cloud 1 — SNH262
— Silver Cloud 1 lwb — CLC47
— Bentley S1 — B50HA
— Bentley S1 lwb — ALB36
— Bentley Continental S1 — BC31GN

September 1959. First chassis numbers series 2 cars with 6,230 c.c. V8 engine
— Silver Cloud II — SPA2
— Bentley S2 — B1AA
— Bentley Continental S2 — BC1AR

January 1960: Long wheelbase S2 cars
— Silver Cloud II — LCA1
— Bentley S1 — LBA1

October 1961: Revised facia with blue instrument lighting
— Silver Cloud II — SZD347
— Bentley S2 — B416DV
— Bentley Continental S2 — BC66CZ

1962: Final chassis numbers series 2 cars
— Silver Cloud II — SAE685
— Silver Cloud II lwb — LCD25
— Bentley S2 — B376DW
— Bentley S2 lwb — LBB33
— Bentley Continental S2 — BC139CZ

October 1962: First chassis numbers series 3 cars with four headlamps and lowered bonnet
— Silver Cloud III — SAZ1
— Silver Cloud III lwb — CAL1
— Bentley S3 — B2AV
— Bentley S3 lwb — BAL2
— Bentley Continental S3 — BC2XA

May 1964: Wider rear wheel rims
— Silver Cloud III — SFU675
— Bentley S3 — B484EC

August-September 1965: Production of standard and long wheelbase cars ceases. Final chassis numbers
— Silver Cloud III — CEL105
— Silver Cloud III lwb — CGL27
— Bentley S3 — B40JP
— Bentley S3 lwb — BCL22

January 1966: Final Bentley S3 Continental chassis no — BC120XE
March 1966: H. J. Mulliner, Park Ward bodied Silver Cloud III discontinued. Final chassis number — CSC19C

Note: Neither Rolls-Royce nor Bentley chassis numbers run in continuous sequence but instead run in short letter series (i.e. SAZ1-61). Series starting with the number 1 use odd numbers only (but not 13) and series starting with 2 use even numbers only. All Bentley chassis numbers start with the letters B, all Bentley Continental numbers with BC, while L preceding the chassis series letter indicates left hand drive.

Milestones

April 1955: Rolls-Royce Silver Cloud I and Bentley S1 announced with Pressed Steel body mounted on substantial box-section chassis. Six-cylinder 4,887 c.c. engine and automatic gearbox standard. Chassis available for specialist bodies by various coachbuilders. Independent front suspension by coil springs and wishbones, live rear axle and semi-elliptic leaf springs and drum brakes all round with mechanical servo were to remain chassis features for the life of the model.
September 1955: Continental version of S1 Bentley introduced with high compression version of six-cylinder engine and lowered lightweight body by H. J. Mulliner or Park Ward. Offered improved performance and true 120 mph thanks to higher 2.92 axle ratio.
April 1956: Twin master cylinders give increased safety to Rolls-Royce and Bentleys.
April 1975: Power assisted steering available as an option on home market.
September 1957: Compression ratio raised to 8.0 to 1 and 2in SU carburettors
November 1957: Park Ward long wheelbase saloon with 10ft 7in. wheelbase available. Fitted with electrically-operated glass division.
April 1958: Underwing refrigeration unit available as optional extra.
October 1958: Electrically-operated windows available as optional extra on Crewe-built cars.
September 1959: Rolls-Royce Silver Cloud II and Bentley S2 introduced with all-new 6,230 c.c. V8 engine, power assisted steering standard and improved heating and ventilation system. Continentals also available.

January 1960: Long wheelbase standard saloons announced.
October 1961: Revised facia with blue instrument lighting, handbrake warning lamp, combined headlamp flasher/indicator switch, doors to door pockets, provision for ram air demisting
May 1962: Larger rear lamp assembly.
August 1962: Sealed beam headlamps with no monogram in lamp.
October 1962: Rolls-Royce Silver Cloud II and Bentley S3 with 9.0 compression version of 6,230 c.c. V8 giving more power, lower bonnet and four headlamps and interior seating changes. Long wheelbase case and Bentley Continentals available at same time.
April 1963: Stainless steel wheel discs fitted instead of chromium plated ones.
August 1963: Improved handbrake efficiency.
November 1963: More effective rear window demister — width decreased to 2ft 6in.
January 1964: Wider seats, heat shield for right hand exhaust manifold.
March 1964: Restyled headlamp surround incorporating RR or B monogram.
May 1964: Wider rear wheel rims for standard and long wheelbase cars. Front and rear wheels no longer interchangeable.
September 1964: New Nemag speedometer with 0 at bottom left.
October 1965: Silver Shadow replaces standard and long wheelbase cars, but H. J. Mulliner, Park Ward 2-door bodied Rolls-Royces and Bentleys Continentals continue intil early 1966.

SILVER CLOUD III

PHOTOGRAPHS: MICHAEL COOPER

"IF you are feeling pimply and your knees are turning blue" Spike Milligan and the Goons have one remedy for you. We have another—take a Rolls-Royce Silver Cloud III for a day's drive in the country. It is perhaps expecting too much of the National Health Service to make such cars available on prescription, but we suggest it hopefully to the Minister of Health with the certainty that it is a real tonic likely to lead to a feeling of general well-being.

The Mulliner Park Ward coupé shown on these pages is not only one of the most expensive and most exclusive British cars but also one of the most beautiful. The length to height ratio of a large car permits superb proportions and a long, low look without any of the inconveniences of a low, small car. In this case the designers have taken full advantage of their opportunities, and the colour they have chosen is a rich red, appropriately called regal red.

The performance of a Silver Cloud is not altered appreciably by the type of coachwork. Our road test of the standard saloon showed that these cars are capable of nearly 120 m.p.h. and that they are remarkably quick to accelerate from rest. No one should imagine that a big car like this must be ponderous; it is content to match its mood to that of the driver. If there is need to hurry it will cruise without fuss at 90-100 m.p.h. on main highways. It can also be guided very precisely through minor roads in a way, and at speeds, that would not shame a sports car. Treated in this uncharacteristic way it may become conspicuous but never alarming.

Rolls-Royce do not say much about the progressive development work and minor improvement that goes on, but, for example, the automatic transmission is now appreciably better than it used to be. Changes up and down are practically imperceptible. The selector lever, of course, can be used like a normal gear lever to engage or hold second or third.

It takes a little time to get used to the power steering and take a perfect line through a bend, yet, even so, a driver very soon finds out that he is making smooth and rapid progress without conscious effort or thought for any of the controls.

BROOKLANDS BOOKS

Rolls Royce
SILVER CLOUD
and Bentley S series
Gold Portfolio

Part 2

BENTLEY INTRO

News of the month is the announcement of two entirely new cars from Bentley and Rolls-Royce, the "S" and the "Silver Cloud". Both machines are identical in respect of bodywork and mechanical specification, and, for the first time, twin SU carburetters are to be found on a Rolls-Royce. These cars represent a lengthy period of development, entailing thousands of miles of testing on the Continent. Both can comfortably exceed 100 m.p.h., in uncanny silence, and road-holding is a revelation for vehicles in the luxury-car category.

The new welded, box-section chassis provides a 3-ins. longer wheelbase than the models it supplants, and permits a lower body line to be used. Unequal-length wishbones are used for the independent front suspension, with large helical springs controlled by Rolls-Royce hydraulic dampers and an anti-roll torsion bar. Long, semi-elliptic springs are employed at the rear, with a Z-type anti-roll bar and electrically-controlled dampers. The ride-control for the latter is a two-way switch mounted on the steering column. Steering is by cam and roller, with three-piece track-rod linkage. The 15 ins. steel disc wheels have five-stud fixing, and carry 8.20 ins. tyres.

The brakes are remarkably smooth in operation with tremendous stopping power. Peripheral cooling fins are fitted to the 11 ins. x 3 ins. cast-iron drums. The speed of the Rolls-Royce friction servo on the gearbox has been doubled. The front brakes are entirely hydraulic in operation, and zero-clearance shoes replace the earlier type. In actual fact, the brakes comprise two trailing shoes, and thus there is no reliance placed on any self-wrapping effect. This is claimed to eliminate any tendency to fierceness by ensuring consistency in operation, under all

---

Some Theo Page sketches of the new Rolls-Royce and Bentley models. (Top) Front suspension and steering linkage details; the brakes have two trailing shoes. (Below) Plastic inserts are fitted behind the rear seat squab to prevent interior body drumming. The rear springs are gaitered and a Z-type anti-roll bar is fitted. (Centre) Layout of the well-planned heating and ventilation equipment, with its dual intakes and separate ducting for the rear compartment. (Inset) Take-off points for defrosting and de-misting. (Bottom) Spring mounting and method of locating the battery; rear brake and hydraulic damper details.

UCES THE "S" –
and ROLLS-ROYCE THE "SILVER CLOUD"

100 m.p.h. Cruising in Perfect Silence—Automatic Gearchange Standardized—New Standards in Comfort and Luxury on Production Cars

conditions. A combined hydraulic and mechanical system is used for the rear, and, with the aid of the servo system, the braking proportion is automatically equalized between front and rear—even in reverse gear.

The power-unit has been developed from the successful o.h.i. and side exhaust valve, six-cylinder. Capacity has been increased to 4,887 c.c. (95 x 114 mm.), and the new aluminium-alloy cylinder head has six separate inlet ports. Great care has been taken with the nitrided crankshaft to ensure perfect balance; it is supported on seven main bearings and is provided with integral balance weights. Carburation is by the latest twin SU instruments of the diaphragm pattern, with Rolls-Royce standards of silencing. This search for extreme silence is also carried to the exhaust system, and the result readily explains why this particular engine is now used for both Rolls-Royce and Bentley. Compression ratio is 6.75 to 1.

The automatic gearbox comprises a fluid coupling and compound planetary gears. Of American origin, it has been developed to a high degree of efficiency by Rolls-Royce technicians. An overriding control is included, enabling the driver to exercise his (or her) own judgment in the selection of ratios. The automatic change is delightful in operation, and is practically 100 per cent foolproof. Transmission is via a divided propeller shaft to a hypoid bevel rear axle, ratio 3.42 to 1.

Equipment is particularly lavish, and the heating-cum-ventilation systems attain new standards of efficiency for British-made cars. Both are independent, with separate control switches, picking up fresh air through intakes above the front bumper. There is a separate duct for rear compartment ventilation. Two-speed windscreen wipers are used, and the switch incorporates the control for the washers.

Bodywork is mainly of steel, although aluminium has been used for bonnet, doors and luggage locker lid. On the Bentley, the frontal aspect is reminiscent of the "Continental". Careful design has minimized wind noise. Comfort for all occupants has been studied. The wide front seat is adjustable both for travel and rake. Provision of cutaway portions behind the seat ensure plenty of leg-room for rear passengers. Luggage space is extremely generous, the spare wheel being carried in a separate compartment.

Plenty of opportunity was given to try out the new Bentley "S", Rolls-Royce Ltd., having arranged long trips over a wide variety of countryside for several journalists. This was capably organized by the concern's P.R.O., Mr. Hugh Vesey, who was in the car in which AUTOSPORT'S representative travelled.

Most impressive quality of the "S" was its ability to put up quite remarkable average speeds on even the most winding roads. It does not handle like a big car; it does not understeer, nor does it oversteer. Perhaps the steering gear ratio might feel a trifle low on first acquaintance, but after a few miles, one realizes that this machine has been designed to reduce driver fatigue to the minimum. The very lightness of the steering, combined with the automatic gearbox, and road-holding *par excellence*, make the Bentley ideal for fast, long-distance motoring. Despite "soft" springing, no roll whatsoever is apparent, and the ride control can be ignored for the majority of British road surfaces. The suspension must be right. Although very large-section tyres are employed, it is virtually impossible to cause tyre squeal on even the tightest of roundabouts. This freedom from squeal is also a feature of the braking, emphasizing the smooth, progressive action of the powerful servo-controlled, hydraulic system.

As regards acceleration, the Bentley will leave many so-called sports cars standing. What is most impressive is the ability of the car to shoot up to over 100 m.p.h., from 75-80 m.p.h. cruising, proving that the engine must develop a high b.h.p., with the excellent torque

THE "SILVER CLOUD": (Above) The coachwork on the latest Rolls-Royce does not slavishly follow fashion dictates. The gently sloping bonnet, the curved screen and large rear window provide excellent visibility.

THE "S": (Left) The new 4.9-litre Bentley is superb from any aspect. Beneath its elegant bodywork is a new, very rigid chassis, with redesigned suspension and steering.

which is essential for middle-range acceleration. Maximum speed must be in excess of 105 m.p.h., and the car is so effortless that one feels quite confident that 100 m.p.h. cruising on Continental roads will not be exceptional. In point

CONTINUED ON PAGE 131

Cleanliness of line is emphasized by the two-tone finish. The appearance is essentially modern, but it remains restrained in the Bentley tradition. The simple grace will not quickly become dated

The Autocar ROAD TESTS

BENTLEY SERIES S SALOON

A COMPANY as conservative as Bentley, known to develop and adopt changes in design only after prolonged experiment and consideration, produces new models infrequently. When the new Series S was announced, therefore, it was awaited by the motoring public with more than ordinary interest, even though the formidable price places that interest in an academic class for the majority of people.

The Series S, which succeeds the Mark VI, is, with the exception of the traditional radiator, of entirely modern appearance, incorporating the best of current standards of line and grace, yet excluding ultra-modern styling details that may fall suddenly from fashion. This approach is illustrated by the head lamps, which are faired in to the smoothly flowing wings but not cowled. Similarly the rear wings have a cleanly sloping line but are not swept up to prominent fins at their extremities. The car, longer by 1ft 1¾in than the Mark VI Bentley, has a delightfully balanced appearance. The Bentley tested was constantly admired by drivers and non-motorists alike in England and on the Continent, where much of the testing took place.

Performance matches the appearance, for the new model is a 100 m.p.h. car, reaching 50 m.p.h. from a standing start in 10.3sec and 80 m.p.h. in 28sec. After 80 appears on the car's exceptionally accurate speedometer, the speed can still be increased by nearly 1 m.p.h. per second to 90 m.p.h., after which it rises at an understandably slower rate, a considerable length of clear road being required for 100 m.p.h. to be achieved. In normal driving, speeds much in excess of a genuine 80 m.p.h. are unlikely to be seen, but this speed is as easily held as it is reached, its maintenance frequently made practicable by the reserve of safety provided by the incomparably good brakes.

These mechanical-servo-assisted brakes are the magic hand that grasps the car and strips it of its speed in a twinkling in almost any conditions. On one occasion the car, quite heavily laden, was being driven at 85 m.p.h. down a completely open but quite steep hill when it was realized that a side turning almost reached would provide a better route. The result of firm, but not heavy, pressure on the brake pedal provided sheer joy even for a driver already familiar with the high standard of Bentley braking. The two tons of car and contents slowed smoothly to a crawl in a moment without noise from brakes or tyres, without the slightest deviation from course, and without even a tremor being detected. All that could be felt was the immensely powerful retardation.

During brake testing at lower speeds, it was found that only 25 lb pressure on the pedal produces a powerful braking effect, and that there is no advantage in exerting more than approximately 50 lb, which represents very little effort, for under this pressure deceleration is at its maximum at little short of 1g on smooth concrete. This deceleration was measured as 30ft per second per second, close indeed to the maximum of 32.2 permitted by the natural laws governing the braking by friction of a body subject to the normal pull of gravity. Incidentally, the brakes are not so efficient in reverse, although still effective. The pull-out hand brake lever holds the car in all conditions and the shortcoming of it not being ideally placed is of little consequence because automatic transmission and its attendant two-pedal control virtually reduces the hand brake lever to a safety device for parking.

Automatic transmission is standard equipment on the new Bentley and Rolls-Royce models; and it is now more refined than that used previously. As a safety precaution the engine cannot be started unless the quadrant lever is in the neutral position. Next to neutral is position 4, engagement of which provides the driver with fully automatic operation of the four forward gears in all conditions. If the throttle pedal is kept fully depressed, the gears change from first to second

102

BENTLEY SERIES S SALOON...

at 19 m.p.h., second to third at 32 m.p.h., and third to top at 60 m.p.h. When the car is travelling in top gear at speeds below 60 m.p.h., full depression of the throttle results in an automatic change down into third, or into second if the speed is in the twenties or below.

Sudden depression of the pedal in this way results in a rather fierce initial acceleration in either of the lower gears because they are engaged under full throttle. This technique of sudden, full throttle opening is primarily intended for emergency, however, for there are alternative methods of changing gear. In the ordinary way a driver simply opens the throttle smoothly and the gears will change almost imperceptibly according to load. But if below 60 m.p.h. the control lever is moved to position 3, third gear is automatically engaged, the change being made virtually imper-

The side lights high in the wings have red tell-tales, both of which are visible from the driving seat. Fog lamps incorporating winking indicators are fitted as standard on each side of the traditional radiator grille

ceptible if an appropriate touch of throttle is given simultaneously. With the lever in this position third gear is always held until 60 m.p.h. is reached. Above 60 m.p.h. top engages. In all other respects first, second and third gears operate as if the lever were in position 4.

For certain conditions—primarily the negotiation of prolonged, very steep hills—position 2 is provided to keep the car permanently in second gear. Position 2 is adjacent to reverse and, as a safety precaution, these two gears are in a different plane on the quadrant so that they cannot be engaged accidentally. When the lever is set at 2, first gear is by-passed and it is possible to hold second gear to something over 40 m.p.h., the gear being held up to maximum r.p.m. Position 2 is also useful for rocking the car, in conjunction with reverse, should the road wheels have sunk while stationary on a soft surface.

When the car is slowing gently, as when running up to red traffic lights, the changing down of the gears is imperceptible to the passengers. The changes are so smooth that they frequently pass unnoticed even by the driver. There is no tendency to creep when the car is stationary with the engine ticking over and the gear lever in one of the driving positions; the take-up when the throttle is opened in these conditions is delightfully smooth, even when maximum acceleration is required. When starting

The french polished woodwork of the facia and window surround is very beautiful and, with the fine hide upholstery, helps to provide the luxurious appearance expected in such a car. Instruments and switches are laid out neatly, and the ride control switch is mounted conveniently on the steering column. Adjustable armrests are fitted to the doors. The backrests of the bench type seat can be adjusted independently. A pull-out table is fitted under the facia; it contains the ashtray drawer

from cold the smoothest take-off is achieved by letting the engine warm up sufficiently for low r.p.m. to be held while a driving range is engaged.

The performance provided by the new model calls for outstandingly good suspension and steering if it is to be used safely, and in a car bearing such a heavy price purchasers naturally expect very high standards indeed. Yet, in a car of such considerable weight, and in which comfort and silence are at least as important as high performance, the designers are faced with almost insuperable problems in trying to provide the best of two worlds. The Bentley engineers have been markedly successful in their compromises.

The steering has 4¼ turns of the wheel from lock to lock, and as a result it is light to operate, although the driver has to work rather hard when manœuvring in crowded streets. On the open road directional control is commendably precise, and the activities of the front wheels can be felt through the steering to a nicely chosen degree. When cornering fast on indifferent surfaces, movement of the wheels is translated into a slight reaction at the steering wheel that indicates pleasantly what is happening at road level.

It is easy to be misled by the softness of the ride into thinking that the steering and suspension fall short of the standards set by high-performance sports cars, such an

Two S.U. carburettors are fed with air through a massive cleaner and silencer. The six-cylinder 4,887 c.c. engine has an overhead inlet and side exhaust system, and the plugs and double-contact distributor are readily accessible

103

assumption being not unnatural having in mind the character of the car. In this connection it is interesting to note that on the decidedly poor surface of a straight road in northern France, the Bentley was kept to a speed of 85 m.p.h. in safety and with supreme comfort, the poor surface being indicated to the driver only through the reaction in the steering wheel and a lightness in the feel that indicated how hard the front suspension was being made to work. When recently the same road was used in similar conditions by two high-performance sports cars the equally safe speed was some 5 m.p.h. less and the effects of the rough surface were pronounced. Thus, in the new model is achieved a wonderful degree of comfort, combined with a standard of control that falls short only of the best of super sports cars. The only aspect of the suspension in which the car proved disappointing concerned the control of the rear spring dampers. Instead of the lever fitted to previous models to give the driver an adjustment at will of the rear dampers, there is a switch marked normal and hard, but virtually no difference in the ride could be felt when this was operated.

One fast run in the new car is sufficient for the driver to realize that the designers have surpassed themselves in making the interior silent. On previous models the effect of closing all the windows in town traffic was to encase the occupants in a soundproof room, but at high speed there was still appreciable wind noise. The interior of the Series S is utterly silent at 100 m.p.h. when the windows are closed. Should a window be opened the cracking noise of a hundred-mile-an-hour gale shatters the calm, but otherwise one is conscious of the speed only by watching the speedometer or the rapid approach of the horizon. On a Belgian *autoroute* speeds of more than 90 m.p.h. were held for some miles, the occupants of the

BENTLEY SERIES S SALOON

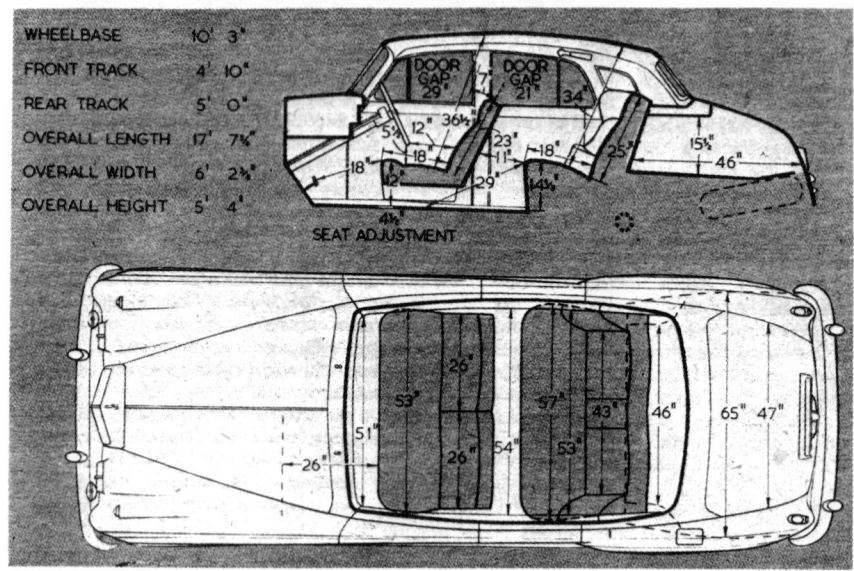

WHEELBASE	10' 3"
FRONT TRACK	4' 10"
REAR TRACK	5' 0"
OVERALL LENGTH	17' 7¾"
OVERALL WIDTH	6' 2¾"
OVERALL HEIGHT	5' 4"

Measurements in these ½in to 1ft scale body diagrams are taken with the driving seat in the central position of fore and aft adjustment and with the seat cushions uncompressed

PERFORMANCE

ACCELERATION: from constant speeds.
Speed Range, *Gear Ratios and Time in sec.

M.P.H.	4 range
10—30	3.5
20—40	4.5
30—50	5.9
40—60	6.9
50—70	9.5

From rest through gears to:

M.P.H.	sec.
30	4.4
50	10.3
60	14.2
70	19.9
80	28.0
90	39.4

* Gear ratios 3.42; 4.96; 9.00; and 13.06 to 1.
Standing quarter mile, 19.7 sec.

SPEEDS ON GEARS:

Gear	M.P.H. (max.)	K.P.H. (normal and max.)
Top (mean)	101	163
(best)	101	163
3rd	62	100
2nd	32	51
1st	19	31

(In 4 range maximum on 2nd is 40 m.p.h., 64 k.p.h.).

SPEEDOMETER CORRECTION: M.P.H.

Car speedometer	10	20	30	40	50	60	70	80	90	100
True speed	10	20	30	40	50	60	70	81	91	101

TRACTIVE RESISTANCE: 19 lb per ton at 10 M.P.H.

TRACTIVE EFFORT:

	Pull (lb per ton)	Equivalent Gradient
Top	253	1 in 8.5
Third	380	1 in 5.75
Second	550	1 in 4

BRAKES:

Efficiency	Pedal Pressure (lb)
75 per cent	25
96 per cent	50

FUEL CONSUMPTION:
14 m.p.g. overall for 261 miles (20 litres per 100 km.).
Approximate normal range 13–16 m.p.g. (21.7–17.6 litres per 100 km.).
Fuel, First grade.

WEATHER:
Air temperature 68 deg F.
Acceleration figures are the means of several runs in opposite directions.
Tractive effort and resistance obtained by Tapley meter.
Model described in *The Autocar* of April 29, 1955.

DATA

PRICE (basic), with saloon body, £3,295.
British purchase tax, £1,374 0s 10d.
Total (in Great Britain), £4,669 0s 10d.

ENGINE: Capacity: 4,887 c.c. (298.2 cu in).
Number of cylinders: 6.
Bore and stroke: 95.25 × 114.3 mm (3¾ × 4½in).
Valve gear: o.h.v. inlet, s.v. exhaust.
Compression ratio: 6.6 to 1.
M.P.H. per 1,000 r.p.m. on top gear, 25.

WEIGHT: (with 5 gals fuel), 37¾ cwt (4,242 lb).
Weight distribution (per cent): F, 51; R, 49.
Laden as tested: 40½ cwt (4,542 lb).
Lb per c.c. (laden): 1.08.

BRAKES: Type: Bentley/Girling.
Method of operation: Hydro-mechanical, servo operated.
Drum dimensions: F, 11¼in diameter; 3in wide. R, 11¼in diameter; 3in wide.
Lining area: F, 212 sq in. R, 212 sq in (209 sq in per ton laden).

TYRES: 8.20—15in.
Pressures (lb per sq in): F, 19; R, 26 (normal).

TANK CAPACITY: 18 Imperial gallons.
Oil sump, 16 pints.
Cooling system, 30 pints.

TURNING CIRCLE: 41ft 8in (L and R).
Steering wheel turns (lock to lock): 4⅛.

DIMENSIONS: Wheelbase: 10ft 3in.
Track: F, 4ft 10in; R, 5ft 0in.
Length (overall): 17ft 7¾in.
Height: 5ft 4in.
Width: 6ft 2¾in.
Ground clearance: 7in.
Frontal area: 26.4 sq ft (approximately).

ELECTRICAL SYSTEM: 12-volt; 57 ampère-hour battery.
Head lights: Double dip; 60–36 watt bulbs.

SUSPENSION: Front, independent, coil springs and wishbones. Rear, semi-elliptic. Anti-roll bar position front and rear.

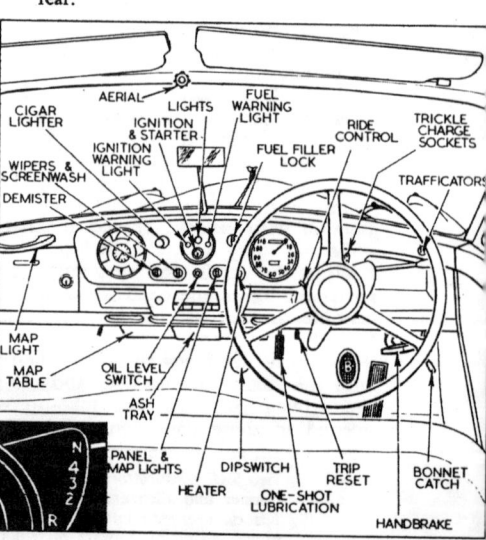

BENTLEY SERIES S SALOON...

car conversing in drawing-room tones, marvelling at the almost eerie silence.

Much of the credit for the comfort and silence is owed to the design and construction of the coachwork. A bench seat at the front has separate backrests that are adjustable within the range from bolt upright to semi-reclining. The backrests of all the seats are tall and, particularly at the rear, passengers can go to sleep if they wish with heads rested on upholstery that is as comfortable as the softest pillow. The backrest of the rear seat sweeps right round the corners to add still further to comfort, and this seat has a central and side armrests. Armrests are fitted only to the doors for the front occupants. Central armrests would be a useful addition. The car is supremely comfortable for four people, and six of average stature can be accommodated.

Detail fittings are abundant. There is a grab handle in front of the front passenger, and in addition to spring-loaded straps in the rear compartment that swing up above the line of the windows when not in use, there are grab handles on the insides of the doors. There is a pull-out table placed centrally under the facia, in which there is a subsidiary ashtray drawer. The drawer has a curved floor at the rear, and as the slide on the car tested was a little loose, ash was spilt. Fold-down tables are incorporated in the backs of the front seats with ashtrays, whose exteriors are replicas of the tables, built into the same polished wood surround. There is a cigar lighter on the facia and another beside one of the pair of vanity mirrors at the rear. Each mirror has its own light and light switch.

Apart from the conventional heater-demister unit in front, electric elements are moulded into the rear window pane to keep this clear also. However, it is difficult to understand why the switch for this device should be fitted below the rear window itself. Unless the driver has a rear passenger, he must stop the car should he wish to clear the window when his rearward vision is affected by moisture. There is a lockable glove compartment in the left side of the facia and an open pocket on the driver's right. Pockets are also fitted in the front doors, but there is none at the rear.

Controls on the instrument panel have been reduced to a minimum, most of those provided being plain and functional, and designed for more than one purpose in the pull and twist planes.

Instruments on the facia are as comprehensive as would be expected, but the rev. counter common to previous Bentleys has been omitted, it having no further useful purpose following the adoption of automatic transmission as standard. The starter is now operated by clockwise pressure on the ignition key beyond the point at which the ignition is turned on. Windscreen washers are operated by the windscreen wiper switch, and an additional switch connects a solenoid that releases the panel covering the fuel filler cap at the rear. The panel can also be released manually by a control in the luggage locker. Fine hide is used for the upholstery, and the facia and the window surrounds are finished in superbly french polished woodwork. On the car tested the doors had to be slammed fairly hard. It may be presumed that the coachwork as a whole was still tight pending final adjustment after more prolonged "shaking

The luggage locker, being very long, is more capacious than it appears at first sight. A tray of hand tools and spare bulbs is housed under the carpet in the locker. Larger tools are in with the spare wheel, and there is an inspection lamp included in the equipment

down." (Bentleys do not, of course, have to be run-in by their purchasers.)

The longer tail of the new car has enabled the luggage room to be increased. The locker is not very deep, as the spare wheel is laid horizontally in a special compartment below the floor. However, it is long, and capable of accommodating a useful amount of luggage. The larger tools are fitted in the spare wheel compartment, and there is a rubber-lined tray of small tools tucked under a corner of the carpet of the luggage locker itself. Provision is made for strapping the luggage locker lid securely should so much luggage be carried that the lid cannot be fully closed. On it are mounted the number plate and a unit containing the particularly effective reversing lights and a light for number plate illumination. The bulb of the last mentioned lamp failed during the test and its replacement—by a bulb included in the tray of tools—was an unnecessarily long job. Headlamp bulbs were easier to change (a change to yellow bulbs being undertaken when the car was taken to France) and the beams were adequate for fast touring at night.

For those able to afford a car offering such comfort, quality and performance, fuel consumption is of little account. However, it is of academic interest to record that 13 to 16 m.p.g. is not unreasonable when the size of the engine (the largest British car unit), performance, automatic transmission and weight are taken into account. Only the modest size of the fuel tank calls for criticism. A range of less than 200 miles (excluding the reserve supply) is not in accord with the long-distance touring capabilities of the car.

When a model is offered at such a high price the smallest detail that is not of the very highest standard must be criticized. The Bentley was all the more impressive, therefore, for proving to be remarkably free from the minor defects that are frequently overlooked by even the most thorough manufacturers. Indeed, the contrary was the case, praise being won for many small details as well as for the major components. Typical examples are the way in which the sun vizors are fashioned to fit the exact shape of the windscreen, and the superb craftsmanship with which even the smallest piece of piping on the upholstery is terminated with its own little hide capping.

The latest Bentley model offers a degree of safety, comfort and performance that is beyond the experience and perhaps even the imagination of the majority of the world's motorists.

Tables and ashtrays are fitted to the backs of the front seats. Operation of the windows and door locks is conventional

A cigar lighter is fitted beside one of the rear vanity mirrors, each of which has its own light switch. The hand strap swings out of the way

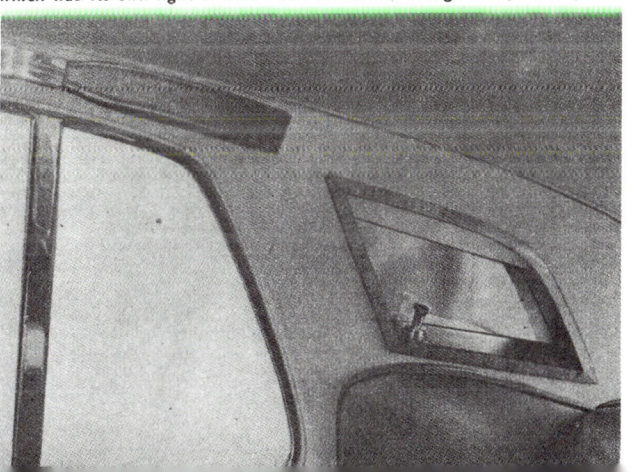

drivescription '56

Bentley

DODGING PLATOONS OF TAXIS and elevated pillars in a $12,000 Bentley saloon (especially someone else's) was frankly a novel and somewhat worrisome experience. But for once I can say I drove a cloud and really mean it since the new Bentley "S" sedan is mechanically and dimensionally identical to the equally new Rolls-Royce Silver Cloud. For the 1st few minutes, I drove as gingerly as you would sit down in a genuine Louis XIV chair.

Within minutes, tho, I had every confidence that I could do anything with the car. It is not so much what it does but how it does it that is so amazing. Factory representative Norman Miller (who was with me), stated flatly that Rolls will not tolerate noise from any component of their cars.

He was not exaggerating; the relatively small (297-cubic-inch), F-head, 6-cylinder engine is one of those rare ones you think has stalled when it is idling. Without being able to time it on city streets, my guess is that the approximately 140-horsepower (Rolls never has published this specification) unit will move the 4100-pound car from 0 to 60 mph in about 13 to 14 seconds.

Altho not publicized as such, Silver Clouds and Bentley S's have standardized on Hydra-Matic, built under license by

106

Rolls in England. What they have done to make their Dual-Range version smoother than GM's much-touted new one (as used in current Cadillacs, Oldsmobiles and Pontiacs) must remain a mystery, but smoother it is. Rolls hides production and engineering techniques in a shroud of secrecy.

The convenient selector lever carries an override control which permits manual selection of ranges from 1 thru 4 if so desired, regardless of engine rpms. Using this, the shifting technique can be developed into a fine art by owners who mourn the passing of the standard box from the option list.

Power steering is not available because it is not needed. The 212-inch car parks like a baby carriage, despite the big 8.20 x 15 tires. Cornering even at unreasonable speeds is almost devoid of lean, and totally devoid of tire squeal. Ride is soft by English standards, fairly firm by ours, but sports car stiffness is optional at a flick of the solenoid switch controlling the dual-range shock absorbers.

No switch on a Rolls or Bentley can be so crude as to click; therefore, they are all vacuum actuated unless the mechanism is located beyond earshot like the automatic lid covering the gas filler. One can (and I did) spend undue time admiring the attention to detail. One key in the set provided fits everything, the other only doors and ignition, on the long shot premise that the parking lot might close down for the night before you return. Without the ignition key in the lock the headlights can be turned off, but not on. The woodwork is solid, hand-polished African mahogany, rather than the troublesome veneer used on some lesser imports.

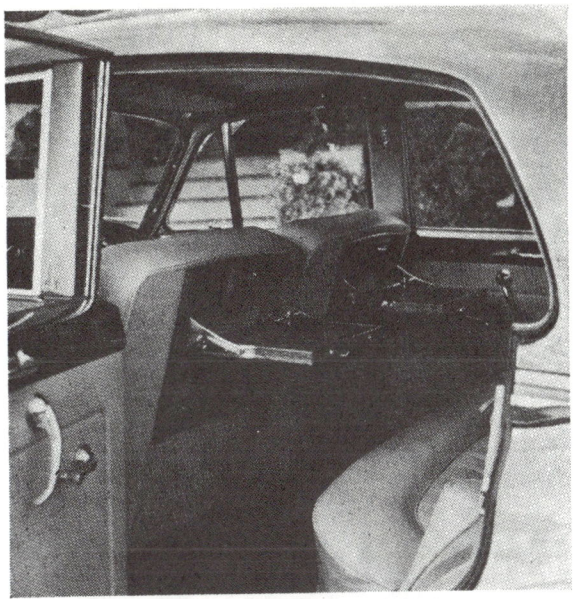

The hydraulic brakes have a unique mechanical servo, consisting of a metal disc that rotates with the driveshaft and uses its own inertia to multiply pedal effort, regardless of whether the engine is on or off, or whether the car is going forward or backward. I have never encountered a smoother, more positive system and can complain only about the English penchant for a tiny pedal.

All Rolls and Bentleys are covered by a flat 3-year warranty. It is said that the break-in period lasts for the first 100,000 miles. If only the FHA or Veterans Administration could be talked into really long-term financing arrangements, these cars would be of widespread interest as they depreciate percentage-wise much less than even a Chevrolet. As it is, about 100 people a year buy them in the U.S. —**Don MacDonald**

Editor's Note: Our thanks to J. S. Inskip (304 E. 64th St., New York) for making this one-of-the-1st Bentleys available to us for a driving impression. Westerners interested in new Bentleys and Rolls-Royces can see them at British Motor Cars, San Francisco, and Peter Satori, Pasadena, Calif.

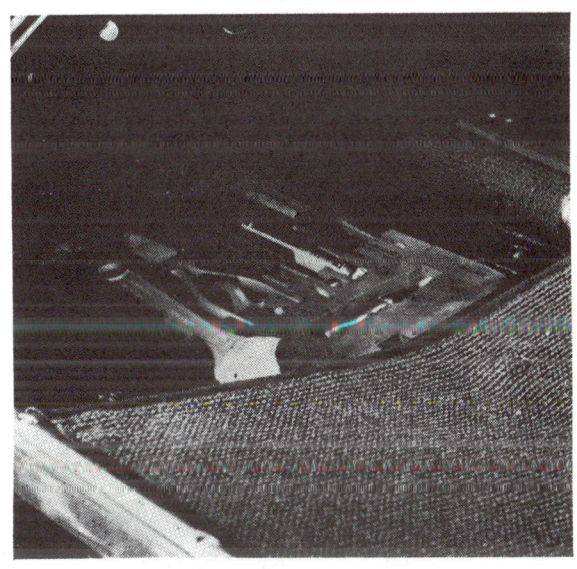

photos by Colin Creitz

Produced and built for the perfectionist— something near the ultimate in hand-made luxury— in fact, the

BENTLEY "S" SERIES

FLAWLESS modern lines have been successfully allied to the traditional Bentley radiator in the latest of the line.

THERE is something extremely satisfying about sheer quality. One can be fascinated by the ingenuity of modern small car designers, who get a quart out of a pint pot. It is also entrancing to study mass production technique, as hundreds of identical saloons roll off the lines in a day. Yet, the hand-made quality car, built to a standard with mere cost an afterthought, probably exerts a greater attraction than ever it did.

I have had the good fortune to try most of the world's fine cars, but I have never seen anything to compare with the workmanship and finish of the new "S" Series Bentley. Particularly in the interior of the body, the absolute perfection of detail, and the excellent taste shown throughout, make almost any other car seem cheap and tawdry by comparison. This is a much bigger vehicle than any previous Bentley of the Rolls-Royce line. Among its ancestors, one must go back to the almost apocryphal 8-litre Bentley or the Phantom II Rolls-Royce for comparison. There is, in fact, a good deal to remind one of the "PII" about this car, though the similarity is difficult to put into words.

The design is entirely conventional, with helical springs and wishbones in front and a hypoid axle on semi-elliptic springs behind. Yet, there is much novelty in the detail work. The rear axle has a torque member above it which also acts as an anti-roll bar. It permits the use of long, supple springs without the embarrassments occasioned by "winding up" on acceleration, while functioning as a torsion bar against roll. It is displaced towards the offside of the axle to combat propeller shaft torque, and a two-way switch gives an electrical control to the rear dampers, the "hard" setting being useful when the body is heavily laden.

The box section frame, of great depth, has cruciform bracing, and the front suspension is reinforced by a normal anti-roll bar. The cam and roller steering box is coupled by a short, transverse link to the three-piece track rod with two slave arms. The propeller shaft is divided, and the usual Rolls-Royce centralized chassis lubrication system is fitted.

The engine is a remarkably big six-cylinder unit, of nearly 5-litres capacity. The exhaust valves are in the cast-iron cylinder block, and the push rod operated inlet valves are in the detachable head. The twin SU carburetters are of the new type with diaphragm seals, and there is an automatic cold-starting arrangement.

A fully automatic gearbox, with four forward speeds, is standardized. If desired, under exceptional conditions, the driver can override the mechanism and select a lower gear manually.

On taking one's seat, the Bentley feels a very big car. It has a long bonnet, but visibility is good. The engine starts by an extra movement of the ignition key, in line with current Continental and American practice. All the controls are as usual, except that the direction indicator is mounted on the facia instead of projecting from the steering column.

If the gear lever is placed in the normal driving position, a gentle pressure on the accelerator will cause the car to glide away. First and second speeds, being very low, are disposed of almost at once, but third is held for longer before the direct drive is engaged. Driven in this manner, the changes in and out of the two lower gears can be felt, but one seldom knows, or cares, whether third or top is being employed.

Violent use of the accelerator gives a very rapid getaway indeed. The big engine can then be felt and heard at work to some extent, till third speed is found, and all mechanical sensation virtually disappears. In comparison with the best American V8s, the six-cylinder Bentley engine is not quite so smooth in initial acceleration, but it has a gloriously long

DO NOT TOUCH: Not designed for owner-driver maintenance, the highly finished engine is not too accessible.

QUALITY of finish extends to the instrument and control layout, impeccably functional. There are now only two pedals.

ENTRANCE HALL: The big car is luxuriously furnished, and dignity of entry and exit is assured. Note the folding tables.

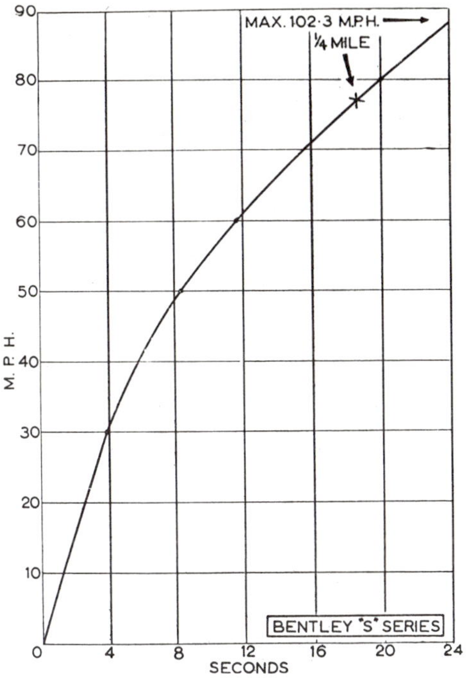

Acceleration Graph

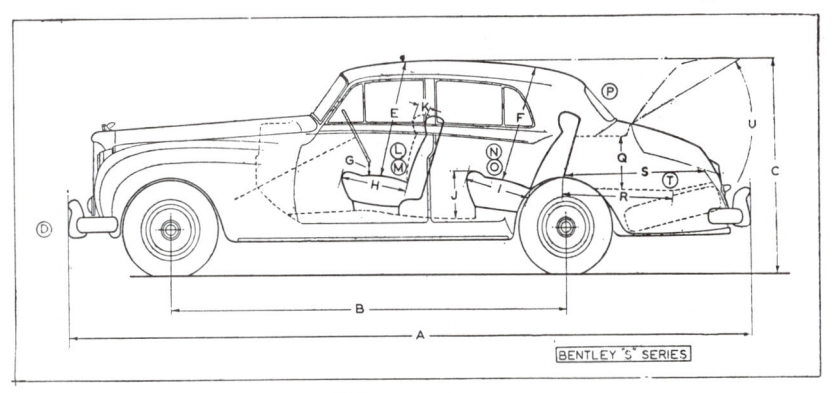

Dimensions

A Overall length, 17 ft. 7¾ ins.
B Wheelbase, 10 ft. 3 ins.
C Overall height, 5 ft. 4 ins.
D Overall width, 6 ft. 2¾ ins.
E Front head room, 3 ft. 1½ ins.
F Rear head room, 3 ft. 1 in.
G Steering wheel to seat cushion, 5 ins.
H Front seat depth, 1 ft. 7½ ins.
I Rear seat depth, 1 ft. 7 ins.
J Height of rear cushion, 1 ft. 3 ins.
K Front seat squab adjustment, 8 ins.
L Front seat width between arm rests, 4 ft. 1 in.
M Front seat width over arm rests, 4 ft. 5 ins.
N Rear seat width between arm rests, 3 ft. 9 ins.
O Rear seat width over arm rests, 4 ft. 6½ ins.
P Width of rear window, 3 ft. 10 ins.
Q Minimum depth of luggage compartment, 1 ft. 3½ ins.
R Length of floor of luggage compartment, 2 ft. 9 ins.
S Maximum length of luggage compartment, 3 ft. 10 ins.
T Overall width of luggage compartment, 5 ft. 5 ins.
U Door opening of luggage compartment, 6 ft. 2 ins.

stride which gives the most effortless high speed cruising imaginable.

The suspension is, if anything, on the firm side. The ride is very comfortable, and there are no sharp up and down movements, but there is not the slightest suspicion of transatlantic "float". I found that the softer position of the ride control gave the best roadholding, and I would only employ the harder setting if the large boot were full of luggage.

Very high marks indeed must be given to the new Bentley for its handling on wet roads. It is a fundamentally safe machine, and the method of locating the back axle has greatly improved the rear end behaviour compared with the previous model. Even on really greasy surfaces, there is a remarkable absence of wheelspin, because the springs are relieved of undesirable torque effects which would tend to promote axle patter or tramp.

As one covers the miles, there is a great sense of well-being. The seats, with their exceptionally high backs, may be set in a moment to any angle desired. The steering is not heavy, though no power assistance is provided. It is only when manœuvring or parking in a confined space that more work at the wheel is needed than in the case of a smaller car. A very elaborate heating and ventilating system is built in, so the windows may be kept closed under all normal conditions. The touch of a switch opens the flap over the petrol filler cap, so removal of the ignition key at once safeguards the fuel. There is a master key which will unlock the dashboard locker and the luggage boot as well as fitting the doors and ignition. An ordinary ignition key is also provided, with which the garage man can start the car, but he cannot gain access to your documents or baggage. The car is full of thoughtful ideas for the comfort and convenience of the occupants.

It is impossible to choose any "best" speed for the Bentley. Thus, there is no difference in mechanical sound if one cruises at 30 m.p.h. or 80 m.p.h. At the timed maximum speed of over 102

CONTINUED ON PAGE 162

ROUND THE CLOCK WITH A BENTLEY S-SERIES

A Fast Return Run from London to Scotland Underlines the Silent High-Performance, Powerful Braking and Luxurious Specification of the Latest Model of this Famous Make. Every Conceivable Comfort and Convenience, including Servo Brakes and Automatic Transmission.

BEAUTIFULLY - BALANCED LINES render the Bentley S-series standard metal saloon a very elegant fast car.

BENTLEY is not a name to be taken lightly, even by Bugatti fanatics! Indeed, the development of the Rolls-Royce-built cars has been steady, sensible and continuous. From the original 3½-litre "silent sports car" came the more powerful 4¼-litre cars, later endowed with overdrive gearboxes and developed as the war-time Mk. V with independent front suspension. Other, similar but always slightly better, Bentleys followed, leading to the recent B7 and culminating in today's S-series 4.9-litre model, of which MOTOR SPORT was recently able to conduct a road test extending over more than 1,360 miles.

The modern Bentley has a specification in which traditional engineering is blended with the requirements of the present. The six-cylinder engine has a bore and stroke of 95 by 114 mm. (4,887 c.c.) and uses a modest compression ratio of 6.6 to 1 and twin S.U. HD6 carburetters with automatic starting control. The power output remains locked in the bosoms of the Rolls-Royce technicians. Piston speed at 2,500 f.p.m. is equivalent to 3,330 r.p.m. Push-rod actuated o.h. valves are set above side-by-side exhaust valves. The pistons each have three compression rings, the top one chromium plated, and one scraper ring, and the nitrided crankshaft runs in seven copper-lead-indium-lined steel-shell main bearings and possesses integral balance weights. The lubrication system is conventional, except that the connecting-rods are drilled for lubrication of the little-ends. The sump holds 16 pints of oil. A light-alloy cylinder head with six separate inlet ports is employed. The camshaft is driven by single helical gears, with a drive for the oil pump in the centre. It runs in four plain bearings. The cooling system contains 3½ gallons of a 25 per cent. anti-freeze solution, and incorporates a belt-driven five-bladed fan and a thermostat. The latter can be supplied to open at a coolant temperature of 75-77 deg. C. or at 84-86 deg. C., as required. Fuel is supplied from an 18-gallon tank by two independent electric pumps. The 12-volt electrical system incorporates a Dagenite or Exide battery, Lucas C-47 dynamo, Lucas M-45G starter motor and a main fusebox containing eight fuses, with separate horn fuse. Lodge or Champion plugs are specified, and twin Lucas ignition coils fitted. The drive is taken *via* an automatic gearbox comprising a fluid coupling and a set of compound planetary gears through a divided propeller-shaft to a semi-floating hypoid-bevel back axle with four-star differential, giving a final-drive ratio of 3.42 to 1. The axle holds 1½ pints of lubricant.

The chassis is of welded-steel closed-box-section construction with cruciform centre bracing, a steel front pan carrying the suspension and steering units and box-section and tubular rear cross-member. Centralised chassis lubrication is fitted, supplied from a scuttle-mounted reservoir. Front suspension is by unequal-length wishbones and coil-springs, and a normal back axle is sprung on ½-elliptic leaf-springs and located by a Z-type anti-roll bar. At the front opposed-piston dampers of Rolls-Royce construction and a torsional anti-roll bar are used, while at the back the piston-type shock-absorbers are electrically controlled by the driver. The back springs are enclosed in gaiters and the shackles possess rubber bushes.

Steering is cam and roller to a transverse link and three-piece track-rod linkage, and 15-in. steel disc wheels are carried on five studs. The brakes have 11 in. by 3 in. cast-iron drums with peripheral cooling fins, and are applied hydraulically at the front and by combined hydraulic and mechanical operation at the back, through the famous Rolls-Royce gearbox-driven friction servo motor introduced in 1924, which runs at approximately one-fifth propeller-shaft speed. The proportion of braking at the back is 60 per cent. hydraulic, 40 per cent. mechanical. The hand-brake applies the back-brakes only, through part of the pedal linkage. The front brakes have twin trailing shoes.

"Static" Observations

In a car of this price and reputation one expects every conceivable luxury appointment, and in this the Bentley does not disappoint the most fastidious. The car we tried was the standard four-door, 5/6-seater saloon of stressed-skin pressed-steel construction.

It is endowed with seats upholstered in deep high-grade leather, with every possible kind of armrest, those on the front doors being adjustable for height. The front squabs are adjustable for inclination, aircraft fashion, by means of little levers beside the cushions, and their backs carry ashtrays and folding tables for the convenience of rear-compartment travellers. Upholstery is in English hide and foam-rubber overlays on spring cases, and dash and garnish rails are finished with french walnut veneer. The floor is covered with deep carpets. Incidentally, although the body is of steel, luggage-boot lid, bonnet and doors are made of aluminium in the interests of keeping the kerb weight under two tons.

Reverting to the luxury of the "static" appointments, the high-backed front seats, lacking only a headrest, can be adjusted separately or set as a three-passenger bench seat. The front doors possess normal windows, calling for 2¼ turns of the handles to fully raise or lower, ventilator windows with unpleasantly stiff catches, and well-pockets, while the armrests act as "pulls." The rear doors have rigid metal "pulls" and the window handles call for almost 2¾ turns, up-to-down. Fixed quarter-lights match the front ventilator windows. External push-button handles are used. Recessed mirrors, with a cigar-lighter adjacent to the off-side mirror, flank the back seat. Locks are fitted in both front doors.

The front-seat passenger has a very large, lined cubbyhole before him, closed with a lid which matches the dashboard and possesses a Yale lock. A similar cubbyhole, unlidded but with a useful step to retain small objects, is provided in front of the driver. Behind the rear seat is a very wide parcels shelf. Under the dash is a pull-out table, with the H.M.V. radio above it, the radio using a roof aerial elevated by turning an interior knob.

The air of high quality and refinement conveyed by the beautiful upholstery and veneered instrument panel is enhanced by the sense, from contemplation of small details, that this is a car in the true Rolls-Royce-built Bentley tradition. The large non-sprung three-spoke black steering wheel, unencumbered save for a sedate plated horn-push in its boss, the switch-gear in separate panels, the type of direction-"flashers" control set on the right of the dashboard sill, and the shape of the brake-pedal in this now two-pedal car are typically Bentley.

The instrument panel contains a Smith's 110-m.p.h. speedometer with trip and total mileometers, matched by a dial incorporating fuel gauge, water thermometer, oil gauge and ammeter. These gauges have only superficial calibrations such as "Hot" and "Cold," "High" and "Low," and, in the centre, a Smith's clock. Between these main dials is the panel containing the detachable ignition key, which is turned to start the engine (after the gear-lever has been set to neutral) and its removal locks various circuits, according to its setting, rendering the car tamper-proof, generator and fuel-warning lights (the latter indicating when less than three gallons of petrol remain), and a switch for selecting side and tail-lamps, head, side and tail-lamps or fog, side and tail-lamps, as required. On the left of this panel is a cigar-lighter, on the right a switch for releasing the flap over the petrol-tank filler, which is in the near-side back wing and has a screw cap, rather small and secured by a wire—not a filler for your chauffeur to approach with dignity and a can. Two further panels carry switches for controlling demisting and ventilation, wipers and washer, panel and map lights and ventilation and heating, these

110

being cleverly arranged to perform several functions by either turning or pulling them; thus hot or cold air is selected by pulling out to one of two positions the appropriate knob, fan speed by turning it, and the wiper switch controls the two-speed, self-parking screen-wipers by turning, the screen-washer by pulling it out, etc. A button enables sump oil level to be read on the petrol gauge. There is provision for an inspection-lamp.

A little switch on the left of the steering column selects hard or soft ride control, the left foot operates the automatic chassis lubricator, which requires two strokes every 200 miles, and a lever on the right of the steering column provides a degree of control over the automatic transmission, a button on the end of the lever guarding the neutral and reverse positions.

The Bentley has an old-style bonnet, levers on each side of the scuttle interior releasing the appropriate, centrally-hinged bonnet panel. The roof lamp is operated automatically as the doors are opened, with an overriding finger-switch on the near-side door pillar.

Visibility through the big curved windscreen is excellent, thanks to slim screen pillars, and both front wings are just visible to a driver of average height. In-built headlamps are naturally fitted, the semi-inbuilt sidelamps above having tiny red inserts to show them to be alight. The headlamps provide good but not exceptional light for fast night driving and are dipped by a big rubber knob on the floor. Lucas fog-lamps are effective in mist and fog.

The doors of the modern Bentley do not close with quite the "coachbuilt" action of pre-war models and we have to report some minor body rattles and an irritating creak from the region of the near-side screen pillar.

Nevertheless, one of the first impressions is of the wonderfully quiet functioning of the car. The 4.9-litre engine is inaudible and with all the windows closed wind-noise dies away to less than a whisper, so that it is possible to converse in low voices while cruising at a speed of 100 m.p.h. To enable the full benefit of this charming absence of effort to be enjoyed, very thorough air-conditioning and heating is provided, each system separate, so that the car can be driven with all windows shut as a matter of course. A separate duct ventilates the rear compartment. Demisting is looked after by cold or hot air feeds to the windscreen, and by an electrically-warmed rear window.

A central, scuttle-mounted mirror provides the driver with a good rear view, at the expense of slightly impaired left-forward vision. The hand-brake is a pull-out toggle on the right under the scuttle; on the car tested it did not hold very well and came out so far that the driver was in danger of hurting his knees on it when vacating the driving seat. The doors are amply wide, for dignified entry and egress, although, as the seats are high, one steps down quite a long way.

The body lines of the S-series Bentley are beautifully proportioned, rendering this big car handsome as well as imposing. The luggage boot is of very generous capacity, with a flat floor broken only by the fuel filler-pipe, as spare wheel and tools, the small tools in a tray, the large ones in clips, are stored beneath it. The lid has over-centre hinges, and locks. The radiator is, of course, a dummy, the actual cooling element being located some way to the rear of it. Twin screen vizors are fitted, arranged to swivel sideways when detached from clips, the passenger's having a vanity mirror. The steady white-tipped needle of the speedometer provides for easy reading of the Bentley's speed but the matching rev. counter of earlier cars was missed, albeit with automatic transmission the driver has little control of engine speed towards maximum r.p.m. The beautiful hand-throttle control, too, is missing on the modern Bentley. There is no choke, starting being automatic.

THE ESSENCE OF LUXURY allied to high performance places the modern Bentley in a class of its own.

ENGINE OF THE 4.9-LITRE BENTLEY.—Note the typical Bentley high-grade finish, twin S.U. carburetters with their huge air cleaner and neat external pipe-runs, etc.

Road Behaviour

Having thus set down the technical specification of the S-series Bentley and examined its complete and luxurious detail appointments we set about discovering how the present-day representative of this famous make performs on the road.

Following a preliminary canter of some 250 miles we set out to drive to Scotland and back, and, being gluttons for punishment and having to compress the test into as short a space of pre-Motor Show time as possible, we decided to return without an overnight stop, thus gaining experience of the modern Bentley in strenuous round-the-clock motoring. We had anticipated without thought of road conditions in Britain. From London that Sunday until "over the border" we had to contend with an almost continual stream of week-end traffic, nosing along at 20-30 m.p.h., as well as with congested towns (Doncaster!) and long hold-ups at one-way road blocks. That the Bentley was able, under these conditions, to average 47.6 m.p.h. to beyond Abington, including getting clear of London and two stops, respectively for refuelling and screen-cleaning, is a clear measure of its excellent roadholding and high performance.

A very big car to contemplate at the kerbside—it has a track of 5 ft. and is 17 ft. 8 in. long—it has that elusive but desirable quality of seeming quite small when gaps in tight-packed traffic have to be negotiated. It goes up to 100 m.p.h. in a very short distance and is arrested from such high speeds very easily by the servo brakes, in spite of its considerable weight. Because of these characteristics we made comparative light of the appalling traffic conditions, making Stamford from East London in 1 hr. 59½ min., Grantham in 2 hr. 23 min., Doncaster in 3 hr. 21½ min., Boroughbridge in 4 hr. 25½ min., and arriving at Scotch Corner in 5 hr. 2 min. total time. Carlisle was reached in 6 hr. 16 min., Gretna in 6 hr. 33 min. Abinton, where we branched off A74 for A73, was accomplished in 7 hr. 28 min., after which in rain and mist, over incredibly slippery roads which had resulted in one unfortunate in a Ford overturning into a field, we crossed to Edinburgh and, after about an hour's sleep at daybreak, were in Cambridge, looking at the new F. II Lister-Climax, soon after 9 a.m. on the Monday.

This journey had not been devoid of interest. Our best hour's average while hop-scotching the mimsers was 53 m.p.h. Just before Newark a police Austin did a lurid dive across the oncoming traffic, presumably to make sure of "escorting" the Bentley through the town, and at Grantham we noticed the "Antone" Rolls-Royce van rolling steadily South on its way from Oulton Park. Just before Boroughbridge a Type 44 Bugatti tourer was encountered, and towards the end of Bowes Moor a Cooper, bearing racing No. 50, was spotted on its trailer outside a garage. A forlorn traction engine with tattered tarpaulin was noted by the Editor as being on the left of the road out of Locherbie. The object of the run, however, was to check the behaviour and habits of the Bentley. Its automatic gearbox provides a brisk step off, after which progressive upward changes occur at 6, 11 and 20 m.p.h. using a light throttle opening or at 18, 31 and 65 m.p.h. if full-throttle is employed. Similarly,

111

when the accelerator is lifted, automatic downward gear changes occur at 14 (top to third), 8 (third to second) and 4 (second to bottom) m.p.h. The gear-lever, however, enables second and third gears to be held up to the normal maxima when desired, and by dropping into the third position excellent acceleration can be obtained, although one never quite loses one's desire for a normal gearbox. That acceleration isn't stifled by the automatic transmission is indicated by times of 10 sec. from 0-50 m.p.h. in either fully automatic or third gear and of 18.4 sec. from 0-70 m.p.h., holding third gear, which increased to 19.5 sec. using fully-automatic transmission. A steady 50 to 70 m.p.h. in top gear occupied 9 sec. and 10-30 m.p.h. in third gear took 6.5 sec., actually improved to 4.6 sec. in automatic control.

This excellent acceleration, accomplished in complete silence without any suggestion that the big six-cylinder engine is even running, carries one past the worst of the hold-ups, and along the double-track piece of road before Scotch Corner the speedometer went to the stop at 110 m.p.h., as it was to do again on several occasions. This performance is matched by the aforesaid exceedingly powerful, mechanical-servo, hydrastatic brakes, which are certainly a feature of the Bentley. With practically no effort on the driver's part it is possible to pull the car up on dry roads in a straight line, and so powerful is the action, with a slight lag as the servo comes in, that some practice is needed to make smooth, progressive non-emergency stops. A slight squeak from the servo was occasionally noticed, otherwise the brakes are vice-free—in the dry. On exceptionally slippery roads care is necessary, as the back wheels lock first and the tail of the car slides all too easily.

Roadholding and cornering are very good for a large and heavy luxury car of this sort, and rapid cornering calls forth only mild protest from the Dunlop Fort tubeless tyres. Moreover, the tail-wag we recall on the Mark V has been completely eliminated. There is neither appreciable over- or understeer and roll is subdued, especially with the ride-control set to "hard," although the difference in suspension characteristics seems less pronounced than on pre-war Bentleys, the ride being fairly hard on both settings. This transmits some slight shake to the steering wheel, but return action through the front wheels is conveyed only over really bad surfaces and then not as pronounced kick-back.

The steering asks four turns from one to other of a generous lock (turning circle, 41 ft. 8 in.) and is exceptionally smooth and light at all times save for very low-speed manoeuvring, although it is essentially spongy steering. There is little sense of low-gearing at speed, but the driver is called upon to do considerable wheel twirling when making pronounced changes of direction, in which he is aided by very strong castor-action. In future, on overseas' models, power steering will be available and it seems probable that this will be provided more from a desire to raise the ratio of the steering than from any need for nicer or lighter action at speed. The automatic transmission does not jerk excessively, although once or twice it juddered in moving the car away from rest and on slippery roads automatic upward gear changes tend to promote momentary wheelspin and loss of adhesion. Incidentally, it seems a pity that this transmission is of American origin, especially as once upon a time Rolls-Royce made most of their own equipment, even to electrical components.

There is no gainsaying the ease of control provided by the Bentley in its present form. As one eases up to a traffic obstruction and stops the car merely by depressing the brake pedal, the nose with its winged-B badge dipping faintly in gentle acknowledgement of the superb braking power, the car ready to glide away in bottom gear, engine inaudible, with no movement of the owner's hands from the wheel, appreciation of first-class engineering is unrivalled, especially in view of the very high performance, in terms of speed, acceleration and road-ability, combined in this comfortable and elegant motor car.

We made a careful check of the essential fluids on our fast run to Scotland and back. Petrol consumption was 13.6 m.p.g. Unfortunately, this represents a fuel range of only 245 miles, which, on Continental roads, could be covered in some four hours. A larger tank seems to be called for. No water and less than half a pint of oil was consumed in the hard 1,000 miles, which is eminently satisfactory, while the automatic chassis-lubrication virtually obviates chassis maintenance. On long runs the front seats, for all their depth of cushion, feel hard and a headrest for the front-seat passenger would be a worthwhile addition. In the back compartment, however, sheer luxury and comfort prevail, while in all seats there is ample leg-room and an entire absence of fumes.

Other cars can equal the Bentley's top speed and its vivid acceleration, but it is the astonishing mechanical silence and absence of wind-noise when the windows are shut, allied to the manner in which the

---

THE BENTLEY S-SERIES SALOON

Engine : Six cylinders, 95 by 114 mm. (4,887 c.c.). Push-rod operated o.h. inlet valves, side exhaust valves. 6.6 to 1 compression-ratio.

Gear ratios : Hydramatic transmission with over-ride and "kick-down" control. First, 13.06 to 1; second, 9.00 to 1; third, 4.96 to 1; top, 3.42 to 1.

Tyres : 8.20 by 15 Dunlop Fort tubeless on bolt-on steel disc wheels.

Weight : 1 ton 19 cwt. 1 qtr. 7 lb. (without occupants, but ready for the road, with approximately 14 gallons of petrol).

Fuel capacity : 18 gallons (range approximately 245 miles).

Wheelbase : 10 ft. 3 in.

Track : Front, 4 ft. 10 in.; rear, 5 ft. 0 in.

Dimensions : 17 ft. 8 in. by 6 ft. 2½ in. by 5 ft. 4¼ in. (high).

Price : £3,495 (£5,243 17s., inclusive of purchase tax).

Makers : Bentley Motors (1931), Ltd., Crewe, England.

---

vehicle can be driven in and out of traffic obstructions, steering accurately with but a light touch on the wheel, those outstandingly powerful brakes in reverse, that render it the supreme high-performance luxury car. At its price of £5,243 17s. inclusive of p.t. not many can afford it, but for those who can the genuine quality of its interior, appointments, no less than the splendid finish of the engine and the air of security imparted by the deep veneered dash and broad bonnet, will be a source of constant pleasure and inspiration. In the course of our round-the-clock drive we counted many Bentleys and at least seven of the latest S-series, in which hydramatic transmission coupled with the over-riding gear-lever represents the ultimate in automatic gearboxes.

The British lion may spend much of its time lying down these days but it is still capable of getting up and facing the world in a bold and dignified manner, and such productions as the Bentley S-series remind us that this is so. This is the Company Director's motor car *par excellence* and it is fitting that men who control British destiny should drive these fine cars from Crewe rather than chromium-draped floating drawing-rooms of other than British origin.—W. B.

THE BEST CAR IN THE WORLD

Map reproduced by courtesy of Hallwag, Bern

3-FIGURE CRUISER

November roads are clear and colourful from Le Touquet to Tossa

IF England does enjoy sunshine it is likely to be during the summer months, so why not make the most of it at home and then chase its warmth south towards the end of the year? Well, it makes a good retrospective argument—not that anyone would need to argue about whisking a Continental Bentley to Spain in any season.

Summer last year had been too full to take holidays, and for me France and Spain spell good food and wine, picturesque towns and villages, clear roads, and smiles even if they accompany a bill. Warm weather and blue skies help, but they are not essentials.

Ferryfield and Le Touquet were bright with November sunshine and unseasonably mild. The air and the sea below had been calm. A company of three, we set a southerly course from the Pas de Calais coast over dry roads. Our plans were happily few; time and details bothered us little. We had determined to visit a few places of special interest and the prospect of many fast miles in the Bentley pleased us all. The Costa Brava in its almost deserted beauty was to be our principal destination, and on the home journey we had in mind to visit some of the historic places between Perpignan and Limoges over a route taking in Carcassonne and Albi, along the western slopes of the Massif Central.

We had no desire to make marathon day journeys nor to set up records for this was a holiday drive, during which we expected to form some

Tossa will seem strangely deserted to those who have seen its picturesque bays only in the summer season of sunshine

impressions of the Bentley when used in its true element.

Returning two years ago from Le Mans, a night stop had been made at the Lion d'Or at Bernay: now, over a late lunch at Le Touquet, as we considered where to stop on our first evening out—and preferably with Rouen behind us—this hotel came to mind again.

On arrival the signs were right. The yard through the archway into this unpretentious coaching inn was well filled with French cars; the kitchen and dining room were already active. No Ritz except perhaps for the *cuisine*, this was our sort of French hotel —unchanged in decades, the smell, colour, and staff all as they should be in a town like Bernay. We were made very welcome, fed well and charged moderately.

In the miles that followed we grew hourly fonder of the Bentley, for it seemed to offer the best of several worlds. In particular it has all the performance one could desire, outstanding comfort of upholstery and equipment, and a substantial feel which dispels anxiety and infiltrates a glow of wellbeing.

We found that on the straighter and better-surfaced roads we felt relaxed at a cruising speed of just over 100 m.p.h. The engine was very sweet at 3,500 r.p.m., which gives 105 m.p.h. Over the rougher cambered roads we dropped back to about 80 m.p.h. (2,600 r.p.m.).

The Continental Bentley's size and firm suspension served it well when, to pass, we were compelled to leave the road-crown and take to the unmade verges. It remained stable and comfortable though the suspension gives none of the float sometimes deemed suitable for big luxury cars. To express a personal opinion, a floating ride on a car which is designed to cruise at 100 m.p.h. or over makes more for apprehension than for comfort. One does not always feel completely at home in a large car at speed; this Mk 2 Continental gives confidence, and is the best compromise of any to date. And its brakes demonstrated their ability to stop its two tons loaded weight whenever called upon.

Not long ago our formal road test of this car was printed (December 21, 1956). The maximum speed attained was 120 m.p.h. For these high-speed tests, tyre pressures were raised to 35 lb front and 40 lb rear. For our Spanish visit we ran at 27 and 30, for the very high pressures desirable for sustained speed were uncomfortable for ordinary road work. A two-position ride control is a Bentley feature, and on the Continental it stiffens the ride and roll slightly with, it seemed to us, particular advantage on mountain bends.

We have spent a dull night at Souillac, after finding Limoges disappointing and confusingly signposted in the dark. Quillan and its gorge lie behind us. Tonight, Perpignan is full of life and lights, and a Mediterranean mildness is apparent. We have just made our second circuit of a one-way river-side block and become hopelessly entangled in some quaint narrow streets in the old town. We know where the hotel is, but how to get there and park outside?

Finally we got the bags out and the car garaged. The bolts and lock on the boot lid had become slightly deranged before we took

Open-car driving was invigorating, but with coats on, it was a squeeze to share the two front seats between three. The brilliant autumn colours were a constant pleasure

The Mk 2 Continental Bentley is in its element on the straight French roads. Winter was late in stripping the trees this year

Near the bridge over the River Vienne of Châtellerault. Women may be seen rinsing clothes which they had brought, steaming, in wheelbarrows from the copper

At home there is quite a different parking problem. This is a section of the promenade at Tossa del Mar in November

Winding coast road near San Feliu de Guixols, lined with dwarf oaks and cork trees. The surface is quite good but the corners may be sharp

3-Figure Cruiser . . .

the car over and the porter, thinking that such a large lid would be heavy, gave a firm tug on the handle, springing the catches without realizing that the key was about to be provided. In fact the lid is very lightly built, lifts easily and has a spring-loaded support.

There was plenty of time for a walk, an *apéritif* and a look round the shops. We fed quite well back at the Hotel de France, where we were staying, having seen nothing else in the way of a restaurant which took our eye. Later, nearby, we thoroughly enjoyed watching the local Pyrenean ring-a-roses dancing, to a most intriguing rhythm and music.

A bedtime check on miles and gallons used to date showed that our best consumption had been in the Tours area with its long, high-speed straights; here we had managed 16½ m.p.g. In the more winding country the figure worked out at 14½ m.p.g. Usually we ordered 40 litres of super and gave a round 3,000 francs to the attendant, the small coins making a tip. Unlike earlier Bentley and Rolls-Royce cars which were expected to consume a little oil, this Continental apparently used none in the first 500 miles. Speed seemed to make no noticeable difference to fuel consumption, but where frequent gear changing occurred, the figure dropped by two or three m.p.g.

Next morning the pink, snow-capped Pyrenean peaks greeted us like a distant mirage above the mist as we made for the border. The Spanish customs officials at La Jonquera were quick and friendly. We all laughed a good deal, but my Spanish is insufficient to know whether it was with or at.

In spite of the size of the car and the off-putting road markings on our maps, we took the coast route from Gerona to Tossa. The lunch at Gerona was interesting and we were much taken with the unusually attractive young ladies returning from church parade. We found it a fascinating little city.

We were very pleased with the Bentley's mountain handling, and its readiness to behave like a much smaller and lighter car. It was happier taking fast or tight bends under power. On rough-surfaced corners quite a lot of road shock came through to the driver's hands, and on occasions there was some axle hop, but this behaviour became apparent only when driving fast; in all other respects the steering was excellent.

Although we favour the automatic transmission, the ability, in addition, to select positively with the lever was a great advantage in mountainous areas or when driving really fast. Between continuous mountain bends the drive in automatic would hunt up and down as you eased up or down on the accelerator, and it was most convenient to be able to hold either third or second according to gradient. The transmission gave no trouble, and the only feature that was not "spot on" was the readiness to kick down into second and third, or to change up from third when the transmission was really hot. In these circumstances there was a slight hesitation or reluctance—no more.

With our first glimpse of the Med. came a little sunshine. The stony road squirmed interminably but attractively round the headlands and inlets and, on arrival, there was enough warmth to tempt us out on the rocks and ruins around the bays at Tossa del Mar (mound by the sea). The water was still warmer than around the English south coast in high summer. We had several restful and economical days and revisited Barcelona—now much cleaner and smarter than in the late 40s, but much more expensive, too.

A night drive back in fine rain was terrifying. Antique lorries, mule carts, mopeds and divers animals weaved about on the muddy, ill-lighted roads. Some had lamps at both ends, some only in the front and several none at all. Mud-caked lorries stationary at the road side were always without lights. The Bentley had brilliant, narrow-beam head lights which dazzled everyone, or dipped beams with such a pronounced cut-off that they left blackness only 30 yards ahead on these unfinished roads. The spot and fog lights had fused. We had 25 miles of eye-straining crawl out of Barcelona, and we were glad eventually to get back to base.

There was one particular kind of pock-marked, wash-board road surface that quite upset the Bentley at speeds from 20 to 60 m.p.h.; above that it skimmed the ridges, but conditions generally demanded slow driving. Fortunately there were not many miles of it, and all were around the Franco-Spanish border.

One tank of Spanish petrol had to be bought in Barcelona. Though of number one branded quality, it did not suit the Bentley at all. The car managed to pull well enough on it, and without pinking, but when it came to stopping there was very severe running on, and before we could do anything about it the induction system was so shaken up that slow running and the choke became badly out of adjustment. The 1,600 r.p.m. tick-over that resulted was embarrassing with an automatic transmission, so next morning we had to get out the "showcase" of plated tools and adjust as necessary. Fortunately the carburettors and linkages are accessible and pleasant to work on. The whole engine remained remarkably clean throughout the trip, and at the end of 2,000 miles had used only half a gallon of oil.

Progress on our home journey started as a leisurely round-up of castles, chateaux and churches, but it continued as a petrol hunt. We greatly enjoyed another look round the *Cité* at Carcassonne. In the pink brick Cathedral at Albi, the bright

Familiar to many but still just as striking as when seen for the first time—a small corner of the fortifications of the ancient Cité at Carcassonne. Inside the walls are the Basilique de St. Nazaire, shops, cottages and a hotel for summer visitors only

interior decorations are at first quite startling; bony relics in tarnished gilt, plush and glass caskets appal me. We had a pleasant, light lunch at the Hotel du Vigan, which looks very unlike a hotel, and contains photos of racing at the nearby Albi circuit.

Our fears about petrol shortage were not lessened at all when we were able to pick up the B.B.C. news on 1,600 metres wavelength not far from Cahors. We felt that we were lucky to get audible reception, 500 miles south of the transmitter on a stormy day, even if the information was cheerless.

Every mile seemed to bring new autumn hues; sunshine highlighted striking reds among the russets, golds and yellows. Deep evergreens provided the contrasts.

The petrol scramble started after leaving Carcassone, but our first life was saved at a pretty little village of Laquepic, where we were able to brim the Bentley with 57 litres of Super. The tank holds 18 gallons but our previous buy, a kilometre or two earlier, had been limited to 10 litres of "ordinary." Unfortunately no positive reserve switch is fitted; the gauge was no more precise than most, and the green warning light used as our check on the way down had proved temperamentally subject to terrain.

A very pleasant day's drive through interesting countryside ended with a miserable crawl through Tulle, picking up the odd 5 litres where we could. We were comfortable and well garaged at the little St. Martin tourist's hotel.

Of the remainder of the journey, suffice it to say that once through the area of panic buying and resulting local shortage, we found petrol easier to get. On the splendid highway between Orleans and Etampes we covered one 25-mile section in 18 minutes without being naughty through the villages.

Having enjoyed seeing again the beautiful rose window in the Basilique St. Nazaire in the *Cité* at Carcassonne, we wondered whether to divert a little to take in Chartres once more. Instead, we settled for the direct road to Paris and another walk round Notre Dame, where most of the stained glass is back again after its removal during the war. After this visit we crossed the river and spent an entertaining hour with the Seine-side *Bouquinistes*. This also had the advantage of taking us on to lunchtime in an area where there are many intriguing restaurants to suit all purses. We walked the meal off in a bright, chilly afternoon.

All that day the car was given a well-deserved rest, for in the evening we engaged in a favourite Parisian pastime, promenading —in this case in the Montmartre and Clichy areas. We retired with full tanks and tummies, knowing that if need be, we could make the Silver City flight out of Le Touquet next day without further replenishment. M. A. S.

Albi's remarkable Cathedral of Ste. Cecile

Unusual aspect from a sidestreet hotel in Paris

116

"Contemporary classic" might be an appropriate description of Rolls-Royce design.

ROAD TEST ROLLS-ROYCE

THE NAME Rolls-Royce has always been synonymous with quality without regard to cost, yet the latest Silver Cloud model costs almost exactly the same as did the Phantom-III of 20 years ago. Thus, in terms of real dollars, the Rolls-Royce is a much better value than it once was, for most cars of today cost four or five times as much as they did two decades ago.

More importantly, Rolls-Royce has kept pace with its lower-priced competitors in the questionably named "prestige" category. In keeping with the trend, the latest "S" series car offers a bigger package, more power and performance, a better ride, etc., as compared to earlier models. The external dimensions and weight of the Rolls-Royce are not so great as those of the Phantom series cars, but all interior dimensions have been increased considerably. The ride is not soft; on the contrary, it feels rather solid (but not harsh) on the boulevard. At speed on the open road, the ride is as near to being perfect as that of any car we have ever driven. It feels soft, yet is so well damped that one never gets the feeling that the body is rolling and bouncing about as if barely attached to the wheels. Ride control of the rear shock absorbers is continued, but is now solenoid operated via a convenient toggle switch on the steering column. The new design is much more effective than formerly, although the "soft" setting is used most of the time. The alternative "firm" position is useful for high-speed cruising over rough or wavy roads, and speeds of 80 miles per hour can be held with impunity and equanimity over surfaces which cannot be traversed safely at 60 mph in comparable automobiles.

As for power and performance, the Rolls-Royce now utilizes the high-performance engine formerly available only in the Bentley Continental models. The compression ratio has been raised from 6.6:1 to 8.0:1, there are larger carburetors (twin SU's) and other minor modifications. Rolls-Royce never give engine output data, but the order of performance achieved indicates about 180 brake horsepower and about 280 pounds-feet, given as "estimated" in the data panel.

Considering the test weight of the machine (4780 pounds) the acceleration data are exceptionally good, though not up to those reported in a weekly British magazine. For example, we got 0 to 70 mph in 16.2 seconds, whereas the British source gave 14.0 seconds. Even so, we

Two-pedal control by virtue of U.S.-designed Hydra-Matic.

Luxuriously appointed in a manner befitting its tradition.

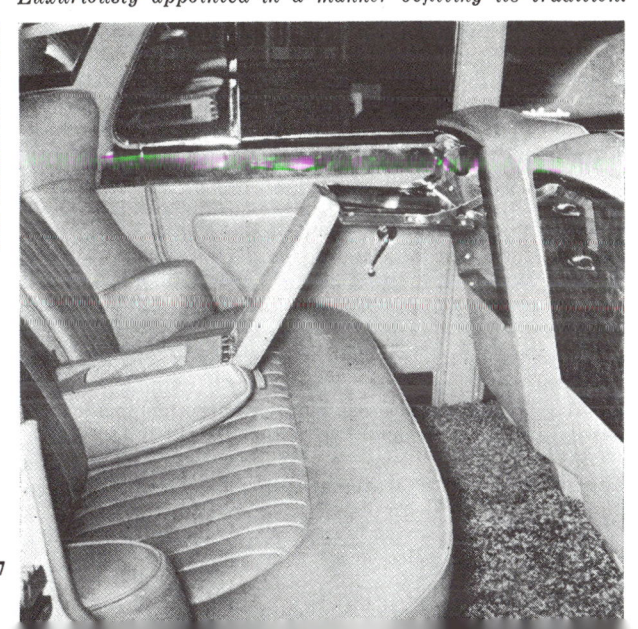

PHOTOGRAPHY: POOLE

Crowded engine compartments are not limited to the U.S.

a lady of quality, unruffled in a crisis

used every trick in the Hydra-Matic book. Starts were made with the brake on, the lever in No. 2 (low) range. With this technique the car starts in 1st gear and shifts up to 2nd at 23 mph. It would then stay in 2nd up to valve float, but we shifted to No. 3 position at 4700 revolutions per minute (45 mph). This over-rules the normal 2-3 automatic upshift at 37 mph and improves the times. Once in 3rd gear, the next automatic upshift occurs at 67 mph when the lever is at No. 4 position, but we left it at No. 3, which puts the 3-4 shift point at 72 mph. In any event, all this "work" saves exactly one second in the 0 to 60 mph test. We obtained 13.1 and 13.2 seconds when starting with the lever at No. 4 or normal "drive" position.

The traditional manual-shift transmission is no longer built, but the Hydra-Matic unit is manufactured entirely by Rolls-Royce. This unit does not employ the small auxiliary fluid coupling developed for the 1957 GM cars. Thus it has the earlier, closer ratios, which are better in our opinion, and with R-R quality control the shifts at part throttle are almost imperceptible. Even at full throttle the shifts are quite smooth, but of course our forced-shift technique produces a noticeable but not too objectionable bump with each upward change.

This famous British firm have long espoused the 6-cylinder engine and the F-head combustion chamber (more properly called i.o.e., meaning intake over exhaust). Certainly the combination is extraordinarily smooth and quiet; in fact, the powerplant sounds and feels no different at 40 or 100 mph. If one is super-critical (which this car and its price invite), it is possible to feel a slight engine tremor. This happens only when slowing down from a higher speed to 20 mph, and then accelerating. The transmission is in 3rd gear and you can just feel the engine under this condition. This is a basic deficiency of the Hydra-Matic transmission for, when accelerating hard from rest, the transmission would normally be in 2nd gear at 20 mph. In other words, a Hydra-Matic does not downshift to 2nd quite when it should, and there is a speed range between about 20 and 36 mph where the acceleration is a little "flat" because you are in 3rd gear. This can be circumvented by shifting to "2," if desired, but the operation produces a fairly severe bump regardless of throttle position.

As for handling qualities, somehow we felt the Silver Cloud is not quite so sports-car-like as was the Silver Dawn, though that test was quite some time ago (Road & Track,

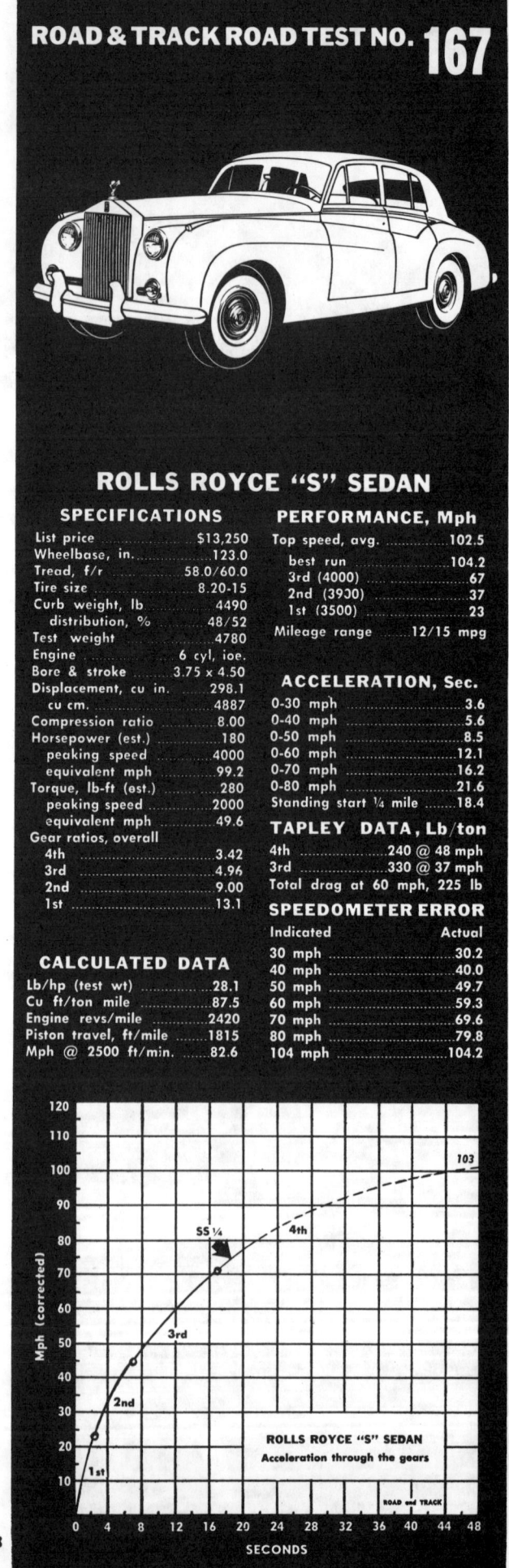

ROAD & TRACK ROAD TEST NO. 167

ROLLS ROYCE "S" SEDAN

SPECIFICATIONS
List price	$13,250
Wheelbase, in.	123.0
Tread, f/r	58.0/60.0
Tire size	8.20-15
Curb weight, lb	4490
distribution, %	48/52
Test weight	4780
Engine	6 cyl, ioe.
Bore & stroke	3.75 x 4.50
Displacement, cu in.	298.1
cu cm.	4887
Compression ratio	8.00
Horsepower (est.)	180
peaking speed	4000
equivalent mph	99.2
Torque, lb-ft (est.)	280
peaking speed	2000
equivalent mph	49.6
Gear ratios, overall	
4th	3.42
3rd	4.96
2nd	9.00
1st	13.1

CALCULATED DATA
Lb/hp (test wt)	28.1
Cu ft/ton mile	87.5
Engine revs/mile	2420
Piston travel, ft/mile	1815
Mph @ 2500 ft/min.	82.6

PERFORMANCE, Mph
Top speed, avg.	102.5
best run	104.2
3rd (4000)	67
2nd (3900)	37
1st (3500)	23
Mileage range	12/15 mpg

ACCELERATION, Sec.
0-30 mph	3.6
0-40 mph	5.6
0-50 mph	8.5
0-60 mph	12.1
0-70 mph	16.2
0-80 mph	21.6
Standing start ¼ mile	18.4

TAPLEY DATA, Lb/ton
4th	240 @ 48 mph
3rd	330 @ 37 mph
Total drag at 60 mph, 225 lb	

SPEEDOMETER ERROR
Indicated	Actual
30 mph	30.2
40 mph	40.0
50 mph	49.7
60 mph	59.3
70 mph	69.6
80 mph	79.8
104 mph	104.2

ROLLS ROYCE "S" SEDAN
Acceleration through the gears

There'll always be a Rolls-Royce radiator grille...

ROLLS

The overall finish of the car is probably the best in the world. Panel smoothness, panel fit, finish and interior are all of the highest standards. The styling may not be modern by U.S. standards, but it certainly is elegant.

August 1953). The Silver Cloud is bigger and heavier—and it feels that way. However, on the credit side, the power steering on our test car may cost quite a lot ($310 extra, not included in list price quoted) but it is positively the best we have ever experienced. It is a progressive type, and it is absolutely impossible to tell it's there. In fact, this is the only power steering to which we would give unqualified recommendation. The Citroen DS-19 and Chrysler types do not have the progressive feature, and the Chrysler, though best in the U.S., still has a deplorable lack of feel.

Brakes are, of course, the firm's traditional mechanical-servo boosted type. There are now two hydraulic master cylinders and a new feature which improves the action and eliminates any chance of a slight lag. This is a brake shoe mounting which allows the linings just to touch the drum at all times, yet causes no undue drag or heating problem. Since there is no clearance to be taken up, the first inch of pedal travel produces braking action. Also, this makes the brakes self-adjusting with no ratchets or other mechanical devices. There is no self-energizing action within the brakes, and Al-Fin drums are no longer used.

The driver sits quite high, and the view which he commands gives great confidence in traffic. We drove the car out for the first time and immediately started down crowded Wilshire Boulevard with only inches to spare on each side, yet felt no trepidation over judging where the right front fender was: a most unusual feeling to experience with a strange machine of such size—and value.

The interiors virtually reek quality and good taste; there are no loud contrasting colors, weird contemporary materials or juke box instruments. The upholstery is soft genuine leather and there are two folding armrests in the middle of the front seat, plus one on each door. The steering wheel is plain black. The instrument panel is natural walnut and the instruments are sheer joy—legible white on black, with real calibrations which tell you exactly what you want to know. A multiplicity of panel-mounted controls take some learning, but this can be excused when you realize some of the unusual features available. To cite just a few, there are a plug for the battery charger, a release for the gas filler, an oil *level* indicator button, and a chassis lubricator pump.

Finally, a word for the die-hard classicist, the man who says "they don't make them the way they used to." Technically, he is right. No longer do we find Rolls-Royce using both a magneto and a distributor for dual ignition, or taper bolts reamed by hand to hold brackets to the frame, or spring leaves fitted to each other with Prussian blue and hand scraping. Unfortunately for the die-hards, a modern coil ignition system is more reliable and gives more accurate timing (spark advance) than the old, expensive system. Modern frames with welded junctions are many times more rigid than those with the traditional tubular crossmembers. Springs function better and last longer when the leaves are separated by modern friction materials. All these items, and many more, save money, of course. But the car is better; the Rolls-Royce would not be acceptable or satisfactory today if it were built as in the Thirties and sold at the 1958 price.

And remember, too, the advent of a standardized body in place of having all bodies built to order has alone accounted for an overall cost saving of nearly $5000 per car. If it were not for these changes in Rolls-Royce policy, the price tags would be at least $25,000 today—and probably more. However, if you want a custom body on a Rolls-Royce, the firm will supply the Silver Cloud chassis with a choice of standard (123-inch) or long (127-inch) wheelbase, or you may have the 133-inch Silver Wraith.

And, if you feel that the traditional R-R radiator is not in sympathy with the rest of the car, you can have the Bentley—identical in every detail except radiator and name plate.

119

MOTORING

The Silver Cloud bears one of the most attractive body designs of all the standard production line Rolls for many years. Its balanced lines belie the fact that this is a large car with overall length of 17 ft. 8½ in

In spite of its dimensions the Silver Cloud has a lithe and eager appearance. Large rear window has heating element wires incorporated in the glass for demisting, de-icing etc

The traditional Rolls-Royce radiator blends with the modern Park Ward standard bodywork. Twin pass lamps also serve as winking indicators and the bumpers, although appearing conventional, effectively protect the wings

NO longer can a Rolls-Royce be lightly dismissed as a stately carriage fit only for elderly aunts or prosperous businessmen. The latest Silver Cloud goes like the wind, although strictly speaking it is not supposed to be a sports car.

With a genuine and useable maximum of 105 m.p.h., the speed potentiality of the current model is apt to administer something of a shock to one's fellow road users. I was highly amused by the antics of a keen type driving a medium-sized saloon beloved by rally enthusiasts, trying to keep up with "my" Silver Cloud. By dint of hard work he managed fairly well in traffic, but when we reached a long straight, down went my right foot, hard; soon his car was a rapidly dwindling speck in the driving mirror, shortly to vanish and not be seen again. Now this is an incredibly good result from nearly two tons of luxury motor car.

Vintage drivers may bemoan passing of the giant pre-war 7-litre engine, traditional wear in Rolls-Royce since we were young, but there is no doubt that the smaller, present-

ON A CLOUD

with Jerry Ames

Touring Trial No. 15
ROLLS ROYCE SILVER CLOUD

day unit of 4.9 litres is much more practical and efficient, and it develops a great deal more power for less petrol. Unfortunately the declared policy of Rolls-Royce is not to disclose the developed power, but one hardly needs to be a genius to reckon that more than 250 b.h.p. must be required to give this weighty machine such a sparkling performance.

Since October last, the compression ratio has been raised from 6.6 to 8 to 1, but present-day fuels and the high degree of turbulence promoted in the light-alloy head permits fully efficient combustion without an accompanying anvil chorus. Valve gear follows post-war Rolls-Royce practice in using side exhaust valves, with overhead inlet valves located immediately above, and the mixture is introduced through two large S.U. carburetters. Despite the high compression ratio, none of the traditional Rolls-Royce smoothness is lacking. No changes have been made to the robust, box-section chassis with cruciform bracing, and the independent front suspension comprising massive helical springs and wishbones remains as before. Semi-elliptic springs and a hypoid axle are used at the rear.

An interesting standardised feature on the modern Rolls-Royce is its fully automatic transmission; enthusiastic drivers will be glad to note that there is an over-riding control. Power steering is available as an optional extra for those who prefer it. The modern Rolls-Royce does not shelter behind tradition, but is right in the forefront of proved design. It retains many well-tried features including one-shot lubrication, operated by a pedal below the facia; a feature that makes an instant appeal to the more seasoned driver.

I collected "my" Silver Cloud from the London service depot at Fulham, once the original workshop of the Hon. Charles Rolls. Taking 17½ ft. of beautiful Rolls-Royce through heavy traffic and along some of London's narrowest streets might appear to call for extreme caution, but my car, equipped with power steering and fully automatic transmission, proved as easy to handle as a modern baby car. These two features alone take all the effort out of driving in congested streets; a driver can forget he is at the wheel of a very large car, because it is so beautifully proportioned and when turning a corner the steering wheel spins so lightly and easily; very soon I found myself threading it through unbelievably tight spaces with the utmost confidence.

When well clear of London I was able to enjoy the satisfying power of the Silver Cloud. Even when motoring quickly a Rolls-Royce never *appears* to hurry, nevertheless it has a deceptively high performance and will quickly settle down to a happy cruising pace of 80 m.p.h. which can be kept up indefinitely. A check on the speedometer proved it to be incredibly accurate throughout its range. Acceleration is outstanding for such a heavy luxury car; 50 m.p.h. could be reached from a standstill in 8.9 sec., whilst 90 m.p.h. could be seen quite often on comparatively short straights, indeed it pays to keep a wary eye on the speedometer, for speed in the Silver Cloud is invariably considerably higher than one realises.

Power steering, of course, is very light at all speeds, indeed some care is advisable at first when handling the car at more than 65 m.p.h., as there is a degree of oversteer, and when negotiating fast curves at high cruising speeds with passengers in the rear seats, some roll oversteer is noticeable. Although power through the steering is considerably reduced in the upper ranges I would have preferred a means of switching it off altogether at speeds in excess of 60 m.p.h. Power steering is only supplied as an extra when specified. I found that the longer wheel-base Silver Cloud I drove recently at Goodwood with power steering has less oversteer at speed than the normal model I have been testing.

The fully automatic transmission, originally imported from America, has been considerably improved by Rolls-Royce engineers and must be without any doubt the smoothest transmission available on any car.

A fluid flywheel coupling is linked with a four-speed epicyclic gearbox, but the driver is provided with an over-riding control which enables him to select second or third gears at will. On most cars with automatic transmission one can kick down and engage a lower gear by jabbing the accelerator hard, but as soon as this pedal is released top gear is automatically re-engaged, usually at the wrong moment. Not so on the Rolls-Royce if one uses the over-riding control. On approaching a really tight main road corner, one can move the gear lever down a notch on the quadrant, third gear is automatically selected and held: One therefore has advantage of braking assistance from the engine. Of course, most of the driving is done with the lever in fourth gear, in fact it is usual for the lever to be left in this position all day for town or country driving.

A great deal of power is required to stop a rapidly moving, heavy car, but the Rolls-Royce hydraulic brakes with servo-assistance are exceptionally good. There is no suspicion of judder or fade, the brakes are wonderfully smooth and even, whether they are used from speeds around 100 m.p.h. or from normal touring gait. One can also make an emergency stop on wet roads with complete confidence. At speeds below 5 m.p.h. the servo motor is not effective and it is better to use the hand-brake, one of the twist and pull variety placed under the right hand side of the facia.

The ride is quite firm and free from pitch. A degree of adjustment can be made to the rear shock absorbers by an electrically operated control on the handsome walnut facia.

Many changes have been made to the sumptuous interior of the latest Silver Cloud and in every respect this is very much a car for the owner-driver, as well as for the professional chauffeur. Two arm-rests are provided in the centre of the divided front seats whilst those on the doors are adjustable for position. Rake of the front seats can be instantly altered by means of a lever on either side which helps also to provide even more leg room and greater comfort for the driver; this is of course in addition to a generous fore and aft adjustment of the seat. The driving position is unusually good, and most comfortable, all the controls are within easy reach, the seating position is high enough to give the driver excellent visibility, both front wings can be seen, and the door pillars do not cause blind spots.

A great deal of care and thought has obviously gone into the arrangement of the driving position. The nicely raked, three-spoke steering wheel closely approaches the ideal; in the centre is a horn button, just above is the lever controlling the automatic transmission. One minor criticism, this lever was somewhat in the way of the facia control for the winking indicators. There is, of course, no clutch pedal and the brake and accelerator are nicely spaced apart, yet leaving ample room to rest the left foot.

For either owner driver or chauffeur comfort is assured. Seats have adjustable backs with centre arm-rest. Instruments are true Rolls fashion with white figures on black dials. Driving is simplified by automatic transmission but an overiding control is provided on the steering column

Test Data: Rolls Royce Silver Cloud
PERFORMANCE
- 0 to 30 m.p.h. .. .. 3.8 sec.
- 0 to 40 m.p.h. .. .. 6.2 sec.
- 0 to 50 m.p.h. .. .. 8.9 sec.
- 0 to 60 m.p.h. .. .. 12.0 sec.
- 0 to 70 m.p.h. .. .. 14.9 sec.
- 0 to 80 m.p.h. .. .. 18.2 sec.
- 0 to 90 m.p.h. .. .. 26.4 sec.

Best timed speed 105.2 m.p.h. Fuel consumption normal 16 to 18 m.p.g. Fuel consumption driven hard 14 m.p.g. Maximum speed third gear 64 m.p.h. Maximum speed second gear 39 m.p.h.

ENGINE
6-cylinder. Overhead inlet, side exhaust valves. Capacity: 4887 c.c. Bore: 95.25 mm. Stroke: 114.3 mm. Compression Ratio: 8 : 1.

TRANSMISSION
Fully automatic with fluid coupling and Rolls-Royce four-speed epicyclic gearbox. Ratios: 3.42, 4.96, 9.0, 13.06. Hypoid bevel final drive.

BRAKES
Hydraulic with Rolls-Royce mechanical servo. Drum diameter: 11 ins.

SUSPENSION
Front: Independent wishbones and coil springs and anti-roll bar. Rear: Semi-elliptic springs. Rolls-Royce double acting hydraulic piston type dampers, adjustable at rear by ride control.

TANK CAPACITY
18 gallons.

STEERING
Cam and roller. Power assistance as on test car, optional extra. Turning circle: 39 ft.

DIMENSIONS
Wheelbase: 10 ft. 3 in. Track: Front 4 ft. 10 in. Rear 5 ft. Overall length: 17 ft. 8½ in. Width: 6 ft. 2¾ in. Height: 5 ft. 4 in. Kerb weight: 38 cwt.

PRICE
Basic £3,795, Purchase Tax £1,898 17s. Total: £5,693 17s. Extra for power steering £165, including purchase tax.

Like all the interior woodwork the facia is in attractive walnut veneer, with sensible and dignified black instruments. No revolution counter is provided, but the very accurate speedometer reads up to 110 m.p.h. Suprisingly enough the oil pressure gauge is merely marked "high" and "low". The water temperature gauge "hot" and "cold", these are grouped in the left hand instrument. By pressing a button the oil level can be read off on the fuel gauge; the petrol filler flap is unlocked by a switch on the dash. Typical of Rolls-Royce thoroughness is the provision of a master key which will turn *all* locks, but a second key concerned only in the running of the car is also provided. Thus an owner can securely put away confidential papers in the locker in the facia and leave the door and ignition key with a garage, safe in the knowledge that only he holds the key to the compartment holding private papers.

The rear seats are every bit as luxurious as those in the

Engine accessibility is limited by bonnet opening—servicing is best left to a Rolls agent. Even a Rolls-Royce now has its radiator filler cap under the bonnet. A large air cleaner surmounts the engine and twin SU carburetters are fitted

front, and allow full leg room and head room. Folding picnic tables are provided and ashtrays and cigar lighters are within reach of all the occupants.

Mention must be made of the de-misting arrangements for the rear window. The device consists of a number of wires within the glass so fine as to be almost invisible to the naked eye. It is brought into use by the de-mister control close by the efficient heater, below the facia.

A luggage boot of gigantic proportions is provided and the spare wheel, tools and battery are accommodated below.

For more than half a century the world has looked up to Rolls-Royce engineers for their skill and fine workmanship. In the latest Silver Cloud they have added Sports performance to the accepted high standards of luxury, making it one of the most desirable 100 m.p.h. cars in the world.

Capacious boot has upward opening lid and spare wheel has its own compartment below, together with tools and battery

ROAD IMPRESSIONS

BEAUTIFUL "S" SERIES BENTLEY

A run in Britain's majestic "silent sports car" convinces us that there are few more impressive motoring experiences.

As for buying one, though—you'd need to win a lottery, and then some!

Shapely tail is one of the nicest yet; indicates to following traffic that the car in front is undoubtedly a Bentley.

SINCE 1931 Bentley has been linked with Rolls Royce, and the name of the "silent sports car" is synonymous with elegance and good breeding.

The latest model to arrive in Australia, the Bentley "S" Series, is no exception. In fact, the 1958 Bentley is probably the most luxurious and best performing stock vehicle ever to leave the English factory.

Credited with a top speed of over 100 m.p.h., it will cruise the highway effortlessly at 85-90 m.p.h. It has also a degree of safety, comfort and excellence that is beyond the experience, and perhaps even the imagination, of the majority of Australian motorists.

It is hard even to realise that it is possible to sit behind the wheel of this car, switch on the ignition, set the gear selector, and by simply pressing the accelerator reach high speeds in a matter of mere seconds. No longer is it necessary — or even possible — to crash gears or jolt passengers as a higher cog is engaged! In this Bentley all that is necessary is to steer, to brake and to accelerate.

Until recently, it was impossible for an Australian motoring journalist to secure either a Rolls Royce or a Bentley in this country for the purpose of gauging the car's capabilities, however, York Motors (Sydney) very kindly made a Series "S" Bentley available to WHEELS, and these were our findings:

Externally, the Bentley displays majestic lines which convey a certain "air", a certain never-out-of-date appearance, which is quite foreign to most other makes of vehicles. Except, of course, Rolls Royce.

From grille to rear bumper the body is perfectly proportioned, yet there is no suggestion of exaggerated streamlining, no flashy layers of chrome, no billowing wings nor sweeping tail fins normally associated with highly priced American or even Continental cars.

In every way the body is that of an aristocrat, and it has modest, cultured, yet cleanly sculptured lines to suit the most conservative and discriminating motorist.

The all-steel body gives no evidence of inexpert workmanship — there are no rough spots anywhere, no ill fitting doors, no badly chromed metalwork. The paint has no orangepeel nor ripples. The Bentley's body is perhaps the nearest approach to perfection that we've yet seen, as regards workmanship.

Inside the Bentley, the finish is equally faultless. The facia panel, window sills and picnic trays are of richly grained, highly polished wood. Door handles, window winders and wind deflector catches all open with the slightest pressure.

But let's climb aboard the Bentley, and head for the open road

True, it is quite a step up into the car, but the doors are so wide that even elderly people will experience little difficulty when either entering or leaving. For some motorists, the seating could be classed as perhaps on the high side, but the superb comfort quickly overcomes

Massive front end is both dignified and impressive — and two tons of motor car follow behind.

any prejudice.

The front bench-type seat is wide enough to accommodate three passengers easily. Flush-fitting arm-rests may be lowered, thus dividing both front and back squabs.

As is the case with many expensive cars, the divided back of the front seat may be controlled individually for tilt. Levers, positioned at either end of the seat, control this movement. Whilst on the subject of front seat passenger comfort, special mention must be made of the adjustable elbow rests fitted to both doors. Gentle pressure on chromed latches adjusts these supports to individual requirements.

Then, too, there's the centrally placed picnic table, which may be eased out from beneath the dashboard when required for use. Large enough to hold a full size dinner plate as well as a cup and saucer, it is fashioned from beautifully grained walnut and is finished with a mirror lustre.

The rear compartment, quite apart from its folding picnic table, ashtrays, arm rests, and spring loaded grip straps, offers each occupant an individual reading light and built-in vanity recess, complete with lighted mirror.

There is ample leg room at both front and rear, while the wide chrome-rimmed windows give excellent vision.

The Bentley we tested was two toned — shell grey over black pearl — and the interior trim was in a delicate shade of blue. However, clients may order colour schemes to suit their tastes.

Close examination of the dashboard and its array of instruments proved interesting. Although it contains more than 20 gauges, dials and switches all told, it still is not "cluttered" nor hard to read. On the contrary, one quick glance gives the driver all information he is likely to require.

In this car there is no need to rely on blinking lights to tell you whether the oil pressure is low or whether the generator isn't functioning — there are gauges which record accurately each individual item.

For example, one of the two circular dials is divided to show the following: oil and fuel capacity, temperature gauge, oil pressure gauge and ammeter. In addition, there is an electric clock having luminous figures and hands.

The speedometer, similar in size and shape to the dial mentioned above, is calibrated to 110 m.p.h.; while there is a trip mileage meter as well as the normal odometer.

Also on the dashboard and within reach of the driver are windscreen wiper and washer switch, cigar lighter, bonnet release, demister and air conditioning controls, instrument and map reading lamps, heater and ventilating control levers, head, side and tail light switches, fog lamp switch, ignition and master key switch, charging plug and petrol cap control. There are other switches for trafficators, ride control and chassis lubrication pump.

Beautifully grained walnut dash is imposing, practical. Only indication that this is a high-speed performance car is grab handle at left.

Note high degree of finish throughout; walnut picnic tables; deep soft seats.

BEAUTIFUL "S" SERIES BENTLEY

As the Bentley has a fully enclosed chassis complete with built-in lubricating system it is unnecessary to have the car greased in the conventional manner — an occasional pump on the automatic lubrication does the job adequately.

Another feature of this magnificent car is the ride control switch, located near the steering column. This controls, electrically, the rear shock absorbers. When driving over rough roads while under a heavy load, the shock absorber tension can be adjusted with a flick of this knob.

But let's get under way!

A half turn of the ignition key— the engine starts — and you are ready to move the gear selector into its first notch. A gentle pressure now on the accelerator and we are off for a dream ride in England's "silent sports car", the Bentley.

Within a few seconds you're amazed at the lightness of the power-assisted steering. It's hard to realise that beneath your hands you are holding the wheel of a two ton motor car — so responsive and sensitive is the wheel. But already we are moving at about 6 m.p.h. and, by listening carefully, we can just detect the motor changing its note as the fully automatic transmission shifts its gearing up one cog. At 12 m.p.h., third gear is engaged, passengers can scarcely catch the sound of the change — and then we go from third to top gear at 24 m.p.h.

The Bentley's automatic transmission, a Rolls adaption of Hydromatic, is the nearest thing yet to being an automatic chauffeur.

With this system, when it is found necessary to suddenly reduce gear ratios, the driver has two alternatives. First there is the accepted "kick down method" with a sudden jab on the accelerator, producing desired results. Or, alternatively, one can shift down by means of the hand lever, since overriding manual control is always available.

In our book there is no transmission to equal that of the Rolls Royce/Bentley system.

Don't think that because the Bentley weighs two tons and measures nearly 18 feet in length that it is a slow moving, overbodied, or in anyway cumbersome vehicle. Plant your foot hard on the gas pedal and the acceleration really amazes. You will beat most of the times recorded by recognised "hot" sports cars; and tests conducted in England show that with its increased compression of 8 to 1, the Bentley "S" is capable of a genuine 104 m.p.h.

When it comes to roadholding, there are few, if any, cars to compare. Even when driven hard into tight bends at high speeds, there is absolutely no indication of body roll or tail slide; and very few Bentleys indeed have been known to turn over.

No doubt much of this stability is due to the solid box section chassis with cruciform bracing, and to the independent suspension with its wide spacing of the lower wishbone pivots. Also, there's the hypoid axle at rear, which, suspended on its semi-elliptic springs, is driven by an open divided propeller shaft and located by a hefty "Z" torque-member.

The Bentley's brakes retain the famous gearbox-driven servo motor, and are most effective — even when applied hard at high speed.

All brakes are fully servo operated, and work on the hydrostatic principle; linings being always in light contact with the drums. Pedal pressure actuates the servo — which in turn operates through the dual master cylinders, supplying hydraulic pressure to the front and rear brakes individually.

Right up until the time re-lining is required the Bentley's brakes never need adjustment. The rear brakes, although fully servo operated, are coupled to the shoes by mechanical linkage.

The huge 4.9 litre motor, with its light alloy head and its 8 to 1 compression ratio, is fitted with two very large S.U. carburettors, each having special automatic starting devices of Rolls' own design. Yet, on a country run the car returns a petrol consumption rate of up to 20 m.p.g.! In our case, however, city driving lowered this figure to just over 14 miles per gallon.

During our run in the Bentley it was necessary to pull into a garage for petrol. We stopped, and merely pulled a switch on the dash-board. Up flew the metal hatch which covers the petrol filling cap in the rear near-side mudguard!

Which brings us to the compartment designed for luggage carrying purposes at the rear of the Bentley. Fully carpeted throughout with matching material, it is roomy enough to hold a veritable horde of suitcases. However, should the load be too large there is provision for extra stowage on the lowered lid.

A most comprehensive array of tools, mounted in a sponge rubber lined tray to obviate rattles, lies beneath the luggage-boot floor; while jack and other heavy items sit snugly in designated places, adjacent to the spare wheel.

Any motor enthusiast, whether his preferences favour glamorous American saloons or smart, nippy sports cars, could not fail to be impressed after a drive in the "S" series Bentley.

The price of the Bentley is every bit in keeping with its grand, even majestic appearance; but by agreement with the distributors, York Motors, we cannot mention the amount in pounds, shillings and pence. We'll say this much though — even if you've won the N.S.W. State Lottery you'll still have to borrow money if you want to buy a Bentley!

The Bentley 'S' series saloon combines silence and luxury with safety and speed

CONTINENTAL LINES. The two-door Mulliner body is both beautiful and efficient, having a smaller frontal area than the normal "S-series" Bentley.

THE BENTLEY CONTINENTAL

The Rolls-Royce and Bentley cars of the current "S-series" combine high performance with luxury in an astonishing manner. These machines now have the high-compression engine that was originally a "Continental" feature, and are capable of speeds around 105 m.p.h., coupled with outstanding acceleration.

Nevertheless, it was found that a demand existed for a Continental version of the new car. This model has now been introduced, and I have recently been able to try one over quite a useful mileage. The theory behind the Continental is that, by reducing the frontal area and the weight, the 4.9-litre engine will pull an axle ratio of 2.92:1, against 3.42:1 for the Standard S-series. This change increases the maximum speed by a full 10 m.p.h. Furthermore, a different governor setting is employed for the automatic gearbox, which enables the driver to run up to a maximum speed of 82 m.p.h. in third gear, as opposed to 67 m.p.h. or so.

Naturally, much of the extra performance is due to the design of the body. The test car was fitted with a Mulliner two-door saloon of light alloy construction. The front seats are special lightweight ones, with adjustable backs which can be folded down for access to the rear seat. This is a bench-type, with folding central arm rest. The luggage boot is of fair size, but the swept tail shape limits its vertical dimension.

The engine is the well-known big six, with inlet valves seating in the head and exhaust valves in the block. It is rubber mounted at three points, and assembled in unit with the four-speed automatic gearbox and fluid clutch. The driver may over-ride the automatic selection by using the hand-lever, with the exception that the governor prevents destructive over-revving of the engine. For instance, if he attempts to change down at 100 m.p.h., the transmission will hold the direct drive until the speed has been reduced to 80 m.p.h. or so. There is a "kick-down" on the accelerator.

There are three notable features about the otherwise conventional chassis. The most important of these is the servo-brake system, which includes the famous Rolls-Royce gearbox-driven servo, and has two master cylinders plus a mechanical hook-up to the rear shoes. Failure is thus impossible. Also notable are the Z-member, which eliminates bouncing of the rear wheels, and the special floating centre bearing of the divided propeller shaft, which avoids vibration and snatch in the transmission.

When one enters the Bentley, a sense of well-being is at once engendered. The sheer quality of the polished wood, the leather upholstery, and the head lining and carpets, is something that no other car can approach. The all-round visibility is excellent, and although this is a very big car, quite a large proportion of it is ahead of the driver. Personally, I would prefer an old-fashioned lever to the pull-out hand brake.

The automatic transmission has the unusual virtue of being entirely free from "creep", even when starting from cold. The big machine may be eased away almost imperceptibly from a standstill, and it will float along at 30 m.p.h. in top gear with the rev. counter in the region of 1,000 r.p.m. Alternatively, firm pressure on the accelerator will produce two black lines on the tarmac, under which conditions the mighty torque of those great cylinders results in really spectacular acceleration.

The high gearing of the Continental makes it a little less lively on top speed than the standard model, but the increased revolution range in third gear more than makes up for this. Quite apart from performance figures, the untiring ease of fast travel with the engine turning relatively slowly is one of those intangibles that render the Continental ideal for long distance touring. The change, either up or down, between top and third gears is barely perceptible, but the wider gap between third and second occasionally makes itself felt. At the timed maximum speed of 115.4 m.p.h., the rev. counter indicated 4,000 r.p.m.

POWER UNIT (left). The well-known big six, with inlet valves seating in the head and exhaust valves in the block, is rubber mounted at three points and assembled with the four-speed automatic gearbox and fluid clutch. The luggage boot (right) is of fair size but the swept tail shape limits its vertical dimension.

126

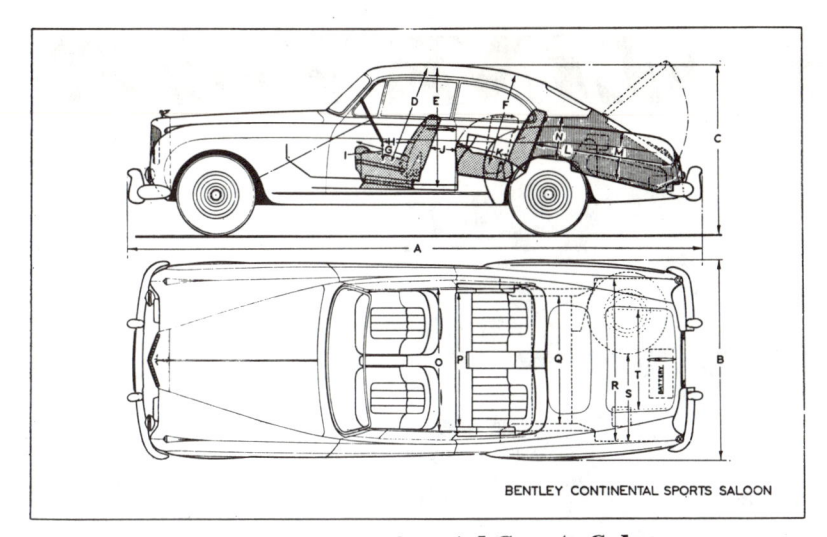

Bentley Continental Sports Saloon

- A Overall length, 17 ft. 8 ins.
- B Overall width, 5 ft. 11½ ins.
- C Overall height, 5 ft. 2 ins.
- D Front seat head room, 3 ft. 1 in.
- E Floor to roof, 3 ft. 7½ ins.
- F Rear seat head room, 2 ft. 10½ ins.
- G Depth of front seat cushion, 1 ft. 8½ ins.
- H Steering wheel to seat squab, 1 ft. 0½ in.
- I Steering wheel to seat cushion, 6 ins.
- J Back of front seat to edge of rear seat cushion, 9½ ins.
- K Depth of rear seat cushion, 1 ft. 8 ins.
- L Depth of luggage compartment, 4 ft. 1 in.
- M Height of luggage compartment, 1 ft. 4 ins.
- N Height at front end of luggage compartment, 1 ft. 3 ins.
- O Width between doors at belt, 4 ft. 4 ins.
- P Width of rear seat between arm rests, 4 ft. 0 in.
- Q Width of luggage compartment between wheel arches, 3 ft. 10 ins.
- R Width of luggage compartment, 4 ft. 10 ins.
- S Width of luggage compartment between side and spare wheel, 2 ft. 6 ins.
- T Width of luggage compartment door opening, 2 ft. 11 ins.

It is a pleasure once again to be able to praise the brakes of the Bentley. The largest American cars may be able to compete with it in sheer speed, but their brakes are not in the same world. There is a smell of hot linings after repeated high speed applications, but no reasonable driver will experience brake fade.

Considering its large size and substantial weight, the car handles well. It remains controllable at high speeds on wet roads, and in this respect it is infinitely better than the Bentleys of a few years ago. If one drives too fast on really bad roads, one becomes conscious that the suspension is hitting on the bump stops, but under all normal conditions the ride is as luxurious as one would expect.

With its good roadholding and brakes, and with really fierce acceleration available at a touch of the pedal, one naturally drives this car fast. Given a clear road, 100 m.p.h. is as good a cruising speed as any other, and the miles, or kilometres, are swallowed up as the passengers relax in comfort. On some of our winding roads, a large car is at a disadvantage, but it is a compliment to say that the Bentley's size is not so apparent as one would expect.

The Bentley Continental is a car of severely classical design which nevertheless achieves a high degree of excellence by sheer quality of construction. In doing so, it develops an individuality, a character, call it what you will, that makes it entirely different from any other *marque*. The man who drives a Continental lives in an enchanted world, for everybody calls him "sir" and he may park where lesser cars can never tread. I have never met so many polite policemen as when I was driving this Bentley!

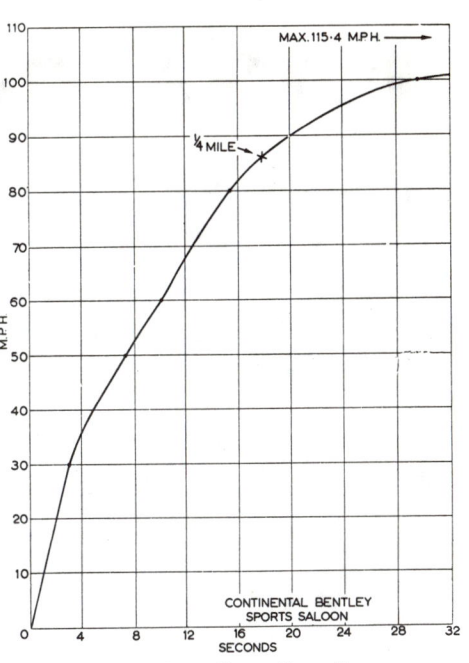

Acceleration Graph

SPECIFICATION AND PERFORMANCE DATA

Car Tested: Bentley Continental two-door Mulliner saloon. Price £5,275 (£7,913 17s. including P.T.).
Engine: Six cylinders 95.25 mm. x 114.30 mm. (4,887 c.c.). Overhead inlet and side exhaust valves. Compression ratio 8 to 1. Twin SU carburetters with automatic choke. Delco-Remy twin contact breaker distributor.
Transmission: Fluid clutch and four-speed epicyclic gearbox with automatic change, plus over-riding hand control and kick-down. Ratios: 2.92, 4.25, 7.69, and 11.17 to 1. Open divided propeller shaft with floating steady bearing. Hypoid rear axle.
Chassis: Box section frame with cruciform bracing. Independent front suspension by wishbones and helical springs. Cam and roller steering box connected by transverse link to three-piece track rod. Rear axle on semi-elliptic springs with combined torque-resisting and anti-roll member. Piston-type dampers all round, with two-position electric control at rear. 8.00-15 ins. tyres on five-stud disc wheels. Hydraulic brakes in front, with hydrostatic non-adjustable shoes, operated by gearbox-driven mechanical servo. Rear brakes 60 per cent. hydraulically operated by servo and 40 per cent. mechanically direct from pedal. Separate master cylinders and reservoirs for front and rear brakes. Pull-out hand brake on rear wheels. 11¼ ins. x 3 ins. finned cast iron drums all round.
Equipment: 12-volt lighting and starting. Speedometer. Rev. counter. Ammeter. Oil pressure, water temperature, petrol and sump level gauges with warning light. Two-speed self-parking windscreen wipers and washers. Flashing direction indicators. Cigar lighter. Built-in heating, demisting, and ventilation system. Clock, radio, spotlights, picnic tables. Ladies' companions.
Dimensions: Wheelbase, 10 ft. 3 ins.; track, front 4 ft. 10 ins., rear 5 ft.; overall length, 17 ft. 8 ins., width, 5 ft. 11½ ins. Turning circle, 41 ft. 8 ins. Weight, 1 ton 15 cwt. (approx.).
Performance: Maximum speed, 115.4 m.p.h. Standing quarter-mile, 17.8 secs. Acceleration. 0-30 m.p.h. 3 secs.; 0-50 m.p.h. 7.2 secs.; 0-60 m.p.h. 10.2 secs.; 0-80 m.p.h. 15.4 secs.; 0-100 m.p.h. 29.6 secs.
Fuel Consumption: 14 m.p.g.

DASHBOARD LAYOUT in the Bentley style. The big steering wheel, large instruments and fine quality panelling are well shown.

FAST and SLINKY—ITS

Town carriage or open high-performance car the 4.9 litre Bentley Continental by Park Ward is clean and graceful. Aerodynamic efficiency is high in spite of an "old-fashioned" radiator. Chromium decoration is restrained and the elegant appearance is produced by body panels of delicate curvature

From the rear the Continental still maintains its dignified, luxurious appearance and blends well with this woodland setting. The rear bodywork is protected from knocks by really substantial bumpers with solid overriders. Beautifully-shaped rear wings have shallow fins and neat rear light clusters

EVEN in this atomic age, 120 m.p.h.-plus machines are by no means commonplace. Perhaps this is just as well judging by the antics of some of our less skilled week-end drivers. It is still possible however to motor safely on the road at speeds of more than 100 m.p.h., provided one picks the right moment and a good car.

One model I would choose for this kind of motoring is the Continental Bentley, one of the fastest, safe, luxury vehicles in the world, in the hands of an experienced driver. It provides the type of motoring most of us dream about, with a maximum speed in excess of 120 m.p.h., good handling characteristics throughout its range and enough acceleration to quietly dispose of most cars as though they were motionless. But to my mind one of its greatest charms is the complete absence of noise and fuss that so often has to be endured in a car possessed of tremendous performance.

This, of course, is in marked contrast to its immortal ancestors, those famous Bentleys which dominated Le Mans and other great races, machines that really established the legend of the British sports car. The latest model offers a ride and comfort undreamed of by that earlier, tough generation, who drove those hard-riding, thunderous Bentleys fast for the sheer hell of it. If only Ettore Bugatti were alive today, no longer would the great man refer to the modern Bentley Speed Model as the fastest lorry in the world.

To enjoy really high speed motoring over deserted roads nowadays, one needs to get up very early in the morning, but this I found no particular hardship with a Continental Bentley awaiting my pleasure. Indeed, what can be more delightful to the keen driver on a spring morning, just as dawn is breaking, than to move swiftly and silently over empty roads with the thrilling anticipation of driving a magnificent thoroughbred really quickly.

Along fast highways 100 m.p.h. comes up very quickly indeed, and gives no more sensation than 70 m.p.h. on the average fast machine. It is pleasant to be able to feel the wheels in contact with the roads, to know what the front end is doing. On this machine a certain amount of feed-back

CONTINENTAL
by JERRY AMES

Touring Trial No. 20
BENTLEY CONTINENTAL

through the steering is noticeable, perhaps slightly more than one expects at high speeds, but the steering is beautifully precise. At lower speeds there is a slight trace of understeer, but this vanishes as one gets into the higher speed ranges.

I found the steering to be without any vices and it is possible to take known bends a great deal more quickly and safely than on the average fast car. Early in the morning several fast curves were taken in the region of 100 m.p.h. without any anxiety at all.

When fast cornering is indulged in, the car is pleasantly free from roll, especially if the switch near the facia controlling the rear dampers is moved over to "hard". The front end always feels in one piece with the rest of the car and this is particularly noticeable coming out of more acute corners under full power when there is no front end patter nor any tyre noise, unless one allows the back end to break away, under these circumstances the car can be easily controlled.

Fastest model in the Bentley range is appreciably lighter than standard, in fact a weight saving of some four cwt. has been achieved on this superb Park Ward coupé. Only a few modifications have been made to the 4.9-litre engine, which otherwise follows current Bentley design, with overhead inlet and side exhaust valves. On the Continental, two huge S.U. carburetters feed the mixture through to larger valves, closed by springs stronger than those used on the touring car. For many years mixture has been compressed at 8 to 1, in combustion chambers noted for promoting an excellent degree of turbulence, although this compression ratio has now been standardized on the slower models, the manufacturers have wisely left well alone on the Continental, preferring not to sacrifice the wonderful smoothness for higher performance. In any case this engine produces ample power (not disclosed in accordance with the firm's policy), for fast motoring on the road.

The Continental under test was fitted with the extremely efficient Rolls-Royce fully automatic transmission, comprising a four-speed gearbox driven by a fluid coupling. Before any of the old school Bentley Drivers Club members choke themselves to death at the very thought of an automatic transmission on their most esteemed motor car, let me hasten to assure them that Bentley engineers like to make their own gear changes too, and so they have provided an over-riding manual control. Therefore, if you wish to make your own gear changes you can do so, but if you feel lazy, place the gear lever in fourth position and this work is all done for you.

Drive is continued by an open propellor shaft divided and supported in the centre by a flexibly mounted ball race, the rear axle is of the semi-floating type and the final drive is through hypoid bevel gears, offset to provide a low body position without sacrificing ground clearance. Large helical springs enclosing hydraulic dampers, together with an anti-roll bar take care of the front suspension, whilst semi-elliptic leaf springs are used at the rear. Some adjustment to the rear dampers by means of a switch below the steering wheel gives the driver a measure of extra control over the back end. The strong box-section chassis is rigidly braced and accounts for a great deal of the solidity of the Continental. Gear ratios are higher that the standard "S" series, but the faster model pulls them with ease.

Another feature new to the Continental is power steering, available only as an optional extra for those who prefer it. Personally I don't care for it on a very fast car, however good it may be, and was therefore delighted to discover this

With the hood erected the coachwork line is still neat and unfussy. A large car, 17 ft. 6½ in. long and 6 ft. wide the Continental is easy to handle at speed. Really large doors permit easy access to rear seat

form of assistance had been omitted on the test coupé. One does not need the strength of a Samson to turn the Bentley wheel, the steering is surprisingly light for such a large and heavy car; after taking a corner the wheel spins easily back with just the right amount of self-centring action. I will concede that ladies might find the steering a mite heavy when parking in a fairly tight space.

For continuous fast motoring, the pressure of the India Super Speed Special tyres should be increased from the normal 22/24 lb. to 30/35; this of course gives a harder but safer ride at speed and brings to light one or two rattles one does not otherwise hear. Incidentally the excellent Park Ward Coupé body was absolutely silent when normal tyre pressures were used.

The last Continental I drove had a manual synchromesh gearbox and I was able to exceed 100 m.p.h. on third, but maximum speed in this gear on the current model proved to be 78 m.p.h. Curiously enough with the present transmission this did not seem to be a real disadvantage, for the acceleration was all that one needed, although I felt that second gear with its maximum of 46 m.p.h. could have been a little higher with advantage.

This fully automatic transmission is one of the smoothest in the world, even when the "kick-down" is employed to engage a lower gear, it is practically free from the customary

The six-cylinder i.o.e. engine propels the Continental at 100 m.p.h. plus, in silence and without effort. Finished in black enamel the engine and engine compartment is a joy to behold, there were no oil leaks. Twin S.U. carburetters are fitted and there are six separate inlet ports

Specification

PERFORMANCE
0 to 30 m.p.h.	.. ..	3.4 sec.
0 to 40 m.p.h.	.. ..	6.1 sec.
0 to 50 m.p.h.	.. ..	8.2 sec.
0 to 60 m.p.h.	.. ..	10.8 sec.
0 to 70 m.p.h.	.. ..	13.6 sec.
0 to 80 m.p.h.	.. ..	17.2 sec.
0 to 90 m.p.h.	.. ..	24.6 sec.
0 to 100 m.p.h.	.. ..	34.1 sec.

Maximum speed:
Top	.. ..	122.4 m.p.h.
3rd	.. ..	78.0 m.p.h.
2nd	.. ..	46.0 m.p.h.
1st	.. ..	23.0 m.p.h.

Fuel consumption: normal, 17 to 18 m.p.g.; driven hard, 15 m.p.g.

ENGINE
6-cylinder. Bore: 95.25. Stroke: 114.3 mm. Capacity: 4,887 c.c. Overhead inlet, side exhaust valves. Compression ratio 8 : 1. Twin S.U. carburetters.

TRANSMISSION
Fully automatic with fluid coupling and Rolls-Royce epicyclic gearbox. Overriding manual control. Ratios: 2.92, 4.25, 7.69, 11.17. Hypoid bevel final drive.

SUSPENSION
Front: independent by coil springs, wishbones and anti-roll bar. Rear: semi-elliptic springs. Double acting piston type dampers, adjustable at rear by switch below steering wheel.

STEERING
Cam and roller steering (power assistance available as optional extra). Turning circle 41 ft. (4½ turns lock to lock).

BRAKES
Hydraulic brakes with mechanical servo.

DIMENSIONS
Wheelbase 10 ft. 3 in. Track (front) 4 ft. 10 in. Track (rear): 5 ft. Overall length: 17 ft. 6½ in. Overall width: 6 ft. Overall height: 5 ft. 2½ in. Ground clearance: 7 in. Kerb weight: 35 cwt.

TANK CAPACITY
Fuel tank capacity 18 gallons.

PRICE
With Park Ward Drop-head coupé as tested. Basic £4,995. Purchase tax £2,498 17s. Total £7493 17s.

---

jerk. No clutch pedal is provided and normally one motors all the time with the gear in fourth position. But a manual over-riding control permits the lower gears to be engaged just by moving the gear lever into one of the other notches on the quadrant, thus the enthusiastic driver trained in the use of the old-fashioned crash box will find the Bentley automatic transmission quite acceptable.

Ignoring the automatic transmission and using the gears in the ordinary way, acceleration can be rather shattering for such a heavy, luxurious car; the Continental Bentley soon showed itself able to whip past production sports cars with an ease that is most deceptive. This is indeed the model *par excellence* for high speed cruising, on fast roads one can keep the speedometer needle continuously between 90 and 100 m.p.h.

With the tyres at normal road pressures the ride is extremely comfortable, if additional damping is required the ride-control below the wheel can be moved over to "H" when the rear dampers will be hardened, but this is not usually needed unless one intends to drive quickly.

Good brakes are of first importance on a fast car and Bentley mechanical servo brakes are as good as you will find on any road machine equipped with drums. They can be used often at speeds of more that 100 m.p.h. without inducing fade, nor do they lose their wonderful smoothness and effective stopping power, only light pedal pressure is needed. A dry disc type servo is employed, driven by the gearbox; the front brakes are operated entirely by hydraulic pressure, but the rear brakes, hydraulically and mechanically by servo simultaneously, in the proportion of 60 per cent hydraulic and 40 per cent mechanical.

The twist and pull handbrake under the facia works on the rear brakes only and uses part of the foot pedal linkage—it is most effective.

A good driving position is essential in a fast car and in this respect the Bentley excells, the soft well-sprung individual leather covered seats give correct support in the right places and the well padded backs can be altered for rake. There is a generous fore and aft adjustment of both seats and between them is a dividing arm-rest, that lifts up to reveal a useful tray below. In the rear seats there is more than average leg room for a close-coupled coupé and of course absolute luxury, armrests are fitted on the sides for each occupant, those in the front being adjustable, whilst ash trays are always within easy reach.

Despite a long bonnet forward visibility is excellent, both front wings can be seen, enabling the car to be aimed accurately at speed, or when manoeuvring in traffic; hind visibility is aided by a fairly large rear window, but of course one has the usual drophead coupé blind spots on the quarters, therefore some caution is needed when turning into a main stream of traffic. I found the very full range of instruments rather well sited, the large 140 m.p.h. speedometer and the rev. counter reading to 5,000 r.p.m. could be quite easily noted when motoring quickly.

An elaborate hot and cold air system enables one to maintain a comfortable temperature. Very efficient demisting is also provided. The large three-spoke steering wheel is nicely raked and all the essential controls are well positioned; by pressing the two-speed wiper control a screen wash is brought into action. A green warning light flashes when the petrol level is down to three gallons, the flap to the filler of the 18 gallon tank is unlocked by a lever on the facia, this did not always work and then one had to resort to a hand attachment in the boot.

At speed the hood remains taut and with the windows closed there are no draughts, the hood can be folded neatly and quickly and covered by a special bag. The unusual shape of the boot interior is deceptive, nevertheless a very large quantity of luggage can be carried; tools and spare wheel are stored in this compartment, but in their own lockers and therefore do not come into contact with any baggage. Although the headlamps are excellent for fast driving at night, when dipped they are a little disappointing and speed needs to be considerably curtailed. Maintenance is kept to

Extremely comfortable seats are adjustable for rake as well as for to and fro movement. Walnut facia is in true Bentley tradition and still retains the white figures on round, black dial instruments. Steering column mounted lever provides overriding control for the automatic transmission

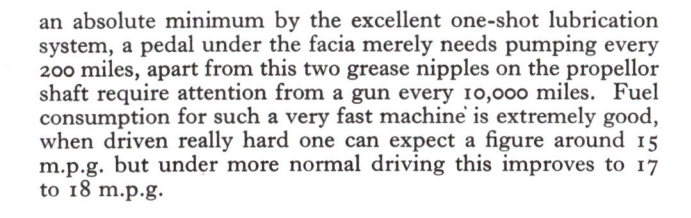

The floor of the enormous boot is covered with carpet. Tools and spare wheel are under covers. Rear lighting is small but adequate, rear lights with separate reflectors. Top lamps are direction indicators

an absolute minimum by the excellent one-shot lubrication system, a pedal under the facia merely needs pumping every 200 miles, apart from this two grease nipples on the propellor shaft require attention from a gun every 10,000 miles. Fuel consumption for such a very fast machine is extremely good, when driven really hard one can expect a figure around 15 m.p.g. but under more normal driving this improves to 17 to 18 m.p.g.

In an expensive motor car one naturally expects standards approaching perfection. So well designed and finished is the Continental Bentley that even after several hundred miles at the wheel one has little but praise for its construction and handling. It is without question one of the most desirable fast machines in the world, but like everything built regardless of cost the price of the finished car unfortunately places it out of the reach of all but the most wealthy drivers. ★

CONTINUED FROM PAGE 101

of fact, on the Oxford By-Pass, the car was cruised at 100 m.p.h., with the passengers carrying on a conversation. So silent was the passage of the vehicle, that none of the occupants had to raise their voices above normal conversational level. This also applies to the radio, which can be heard without interference at these high speeds.

The Bentley "S" does represent a completely new approach to a high-performance machine. It will certainly influence automobile design in general, causing other designers to strive for the uncanny silence which Rolls-Royce engineers have achieved on this superb motor-car. In truth, Bentley has produced the Rolls-Royce of high-performance vehicles—and vice versa!

Specification
Bentley "S" and Rolls-Royce "Silver Cloud"

Engine: Six cylinders, 95 x 114 mm, (3¾ x 4½ ins.), 4,887 c.c. Overhead inlet valves (push-rods), side exhaust valves. Compression ratio, 6.75 to 1. Seven-bearing crankshaft. Pistons with three compression rings and one scraper; top ring chromium-plated. Two SU carburetters; twin electric fuel pumps.

Suspension: Front, independent with wishbones and helical springs; double-piston type Rolls-Royce hydraulic dampers; anti-roll torsion bar. Rear, semi-elliptic with Z-type anti-roll bar; electrically-controlled hydraulic dampers.

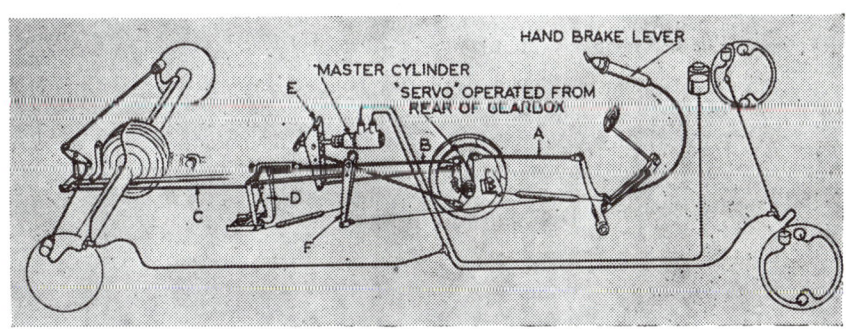

THE BRAKING SYSTEM

A *Connecting rod from pedal to servo.*
B *Extension rod to lever D.*
C *Connecting rod operated by foot or hand brake.*
D *Lever transmitting effort from foot brake or hand brake to rear wheels.*
E *Lever operated by cross rods from servo motor, which draw it forward irrespective of whether a forward or reverse gear is engaged.*
F *Hand brake operating lever.*

Transmission: Four-speed automatic gearbox with over-riding control; ratios, 13.06, 9.00, 4.96 and 3.42 to 1. Divided propeller shaft. Hypoid bevel rear axle.

General: Box-section welded chassis with cruciform centre bracing; 15 ins. disc wheels with 8.20 ins. tyres. Servo-assisted hydraulic brakes; 11 ins. x 3 ins. cast-iron drums. Hydraulic-cum-mechanical operation on rear; handbrake to rear via mechanical linkage. Cam-and-roller steering. 12-volt negative earth electric system.

Dimensions, etc.: Wheelbase, 10 ft. 3 ins. Track, (front), 4 ft. 10 ins., (rear), 5 ft. 0 ins. Overall length, 17 ft. 8 ins. Width, 6 ft. 2½ ins. Height, (unladen), 5 ft. 4¼ ins. Turning circle, 41 ft. 8 ins.

Prices: Rolls-Royce, £3,385 (plus £1,411 10s. 1d. P.T.). Chassis, £2,555 (plus £532 17s. 1d. P.T.). Bentley, £3,295 (plus £1,374 0s. 10d. P.T.). Chassis, £2,465 (plus £514 2s. 1d. P.T.).

Manufacturers: Rolls-Royce, Ltd.; Bentley Motors (1931), Ltd., 14-15 Conduit Street, London, W.1.

IMPOSING as Buckingham Palace, yet retaining a very refined dignity. The Silver Cloud combines superb motoring with tremendous prestige, and the great name still means as much as ever it did.

hydrostatic principle, with the linings in permanent light contact with the drums, and so they never require adjustment, right up to the time when relining becomes necessary. The rear brakes are partly servo operated, too, but the pedal is also coupled to the shoes by a mechanical hook-up, so that some direct "feel" is retained.

Another new feature of the car tested was the power-assisted steering, which is optional. This is a straightforward hydraulic system, which obtains its pressure from a small, engine-drive pump. The operation is so graduated that, although the power assistance does nearly all the work during low speed manœuvring, the driver contributes a fair proportion at other times. This overcomes the "dead" sensation which some power-assisted layouts give, and which might make the driver feel that he was not in full command of his car. With

ROLLS-ROYCE SILVER CLOUD

AMONG the many cars which are submitted for test, the advent of a new Rolls-Royce must always be an outstanding occasion. The magic of the name, of course, has something to do with it, but perhaps even greater interest lies in the sheer continuity of Rolls-Royce development. I have owned seven of the earlier models myself, and have had the privilege of driving nearly every type that has been produced. Most modern cars bear no resemblance to their forbears, but I at once recognized the aluminium name-plate on the bulkhead, because it is absolutely identical to the one that my 1911 Silver Ghost carries. Some knurled nuts on the engine covers would fit the early "Twenty", and other resemblances strike one as the chassis is examined. I almost forgot to mention the radiator.

The Silver Cloud follows the design of recent Rolls-Royce and Bentley cars. It has a box-section chassis with cruciform bracing, and the independent suspension, with very wide spacing of the lower wishbone pivots, is as before. So is the hypoid axle on its semi-elliptic springs, driven by an open divided propeller shaft, and located by a Z-shaped torque member.

The big 4.9-litre engine is still of the inlet-over-exhaust valve layout, but here a change has been made that profoundly affects the character of the car. A very considerable increase in the compression ratio, from 6.6 up to no less than 8 to 1, has improved the performance even more than one would expect. Thanks to the light alloy head and the fundamentally good turbulence of this valve positioning, the engine will still run perfectly smoothly on almost any sort of petrol, and the most expensive grades are not required. The two very large SU carburetters have an automatic starting arrangement of special Rolls-Royce design.

A fluid coupling drives the gearbox, which is of epicyclic principle and has four speeds. It can simply be regarded as a fully automatic transmission, or the driver can take over and make his changes manually. Even then, he is prevented from doing anything destructive. For instance, were he to attempt to change down at 100 m.p.h., the gear would remain in top until the speed was reduced to about 70 m.p.h., when a completely smooth change would take place. There is, as with other "automatics", a kick-down change, achieved by extreme accelerator depression. However, one sometimes wishes to change down with only a light throttle opening, and this the hand lever permits.

I am glad to say that the brakes retain the famous gearbox-driven servo. The front brakes are hydraulic, and fully servo operated. They work on the tyres of 8.2 ins. section, and a car of this size, power-assisted steering must be a worthwhile feature.

The body is of all-steel construction, and is remarkably light in view of its size and luxury, to the benefit of performance and handling qualities. It is large and roomy, with almost every imaginable accessory for passenger comfort. The elaborate built-in heating, demisting and ventilating system includes an electrical demisting element for the rear window.

On taking the wheel, the seating position at first feels rather high, but this gives an impression of command over the big car. Starting is instantaneous, hot or cold, and the machine may be driven off at once in the morning without stalling. Provided that the handbrake is applied, there is no danger of unwanted creeping when a gear is engaged.

The first impression on moving off is

CLASSIC radiator tops a car which gives the appearance of having quality built into every line. This striking view shows the car at its most impressive.

132

DRIVING compartment combines a neat, functional layout of controls and instruments with a "well-bred" air of simplicity and convenience. The driver's position is rather high, and this gives a feeling of complete control over such a large car.

one of immense acceleration. This car really goes, and the higher compression ratio certainly pays dividends when the accelerator is pressed down hard. I would say that the engine is even smoother than the lower-compression unit, and one never has the idea that it is moving about on its flexible mountings. To insulate the unit so completely, without allowing it to shake when it is delivering its massive torque, is a major engineering achievement.

For a car weighing just on two tons, the performance figures are fantastic. The getaway from rest feels terrific, in spite of its smoothness, and the 0 to 30 m.p.h. time of 3.4 secs. can only be equalled by the very "hottest" sports cars. The figure for the standing quarter-mile is 18.1 secs. as a mean of several runs in both directions. On each of these runs, the car was exceeding 80 m.p.h. before the end of the "quarter"! The 0 to 90 m.p.h. time of 25.2 secs. is much better than that of any of our competitively priced sports cars and some of the expensive ones, too. Because these cars are in entirely different classes, it is perhaps not unethical to make such comparisons; or to remark that some famous speed models, of German and Italian origin, would be comfortably out-accelerated.

It is necessary, at this point, to paint the other side of the picture. The Silver Cloud can be moved off almost imperceptibly, and when driven in this fashion it engages its top gear as early as would the most stately chauffeur. It then whispers along in exactly the same way as did the Silver Ghost of immortal memory. The difference is that, if ever it is wanted, that sudden surge of immense power is waiting under the pedal.

It was some 33 years ago that Rolls-Royce first fitted their cars with a gearbox-driven servo for the four wheel brakes. I had one of those early cars, and the brakes were far better than those of most 1958 models. The current type have fully retained that leadership, and they may be applied repeatedly at over 100 m.p.h. without a sign of fade. As most of the world's large cars are woefully short of braking power, the Rolls-Royce is really outstanding, especially in comparison with the biggest American vehicles.

The machine handles well under all conditions, but naturally its size must be taken into account on corners. One cannot, of course, fling it about like a small sports car, but at all reasonable speeds it responds perfectly on dry or wet surfaces. I found that I drove faster in traffic than I have previously driven a Rolls-Royce, both because of the extra acceleration and of the power-assisted steering.

I really like this power-assisted steering. At low and medium speeds, it makes this very large car as light to handle as the smallest vehicle. To manoeuvre such a capacious carriage must demand a fair amount of wheel-twirling, but one finger suffices to spin it from lock to lock. The performance is also excellent at high speeds, but the average driver would be well advised to go gently at first, until he is fully used to power assistance. This is because the steering is so effortless that, in conjunction with the silence and smoothness of

CAPACIOUS boot is unimpeded by the spare wheel which, as can be seen, is carried in a separate compartment.

REAR seats and all interior trim and upholstery maintain the luxurious standard associated with the Rolls-Royce. The rear window is electrically demisted, and picnic tables are mounted on the rear of the front seat squabs.

133

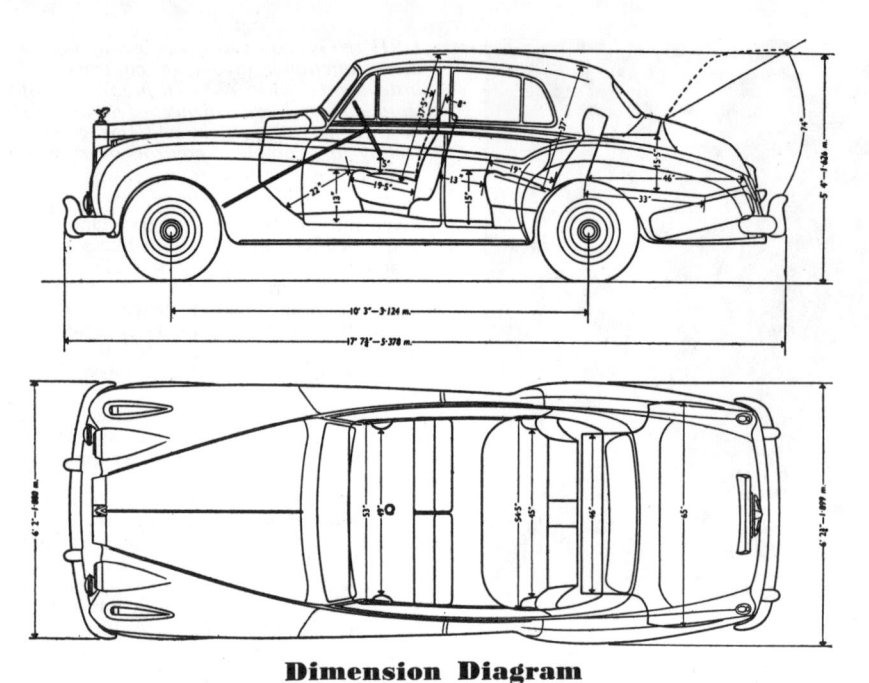

Dimension Diagram

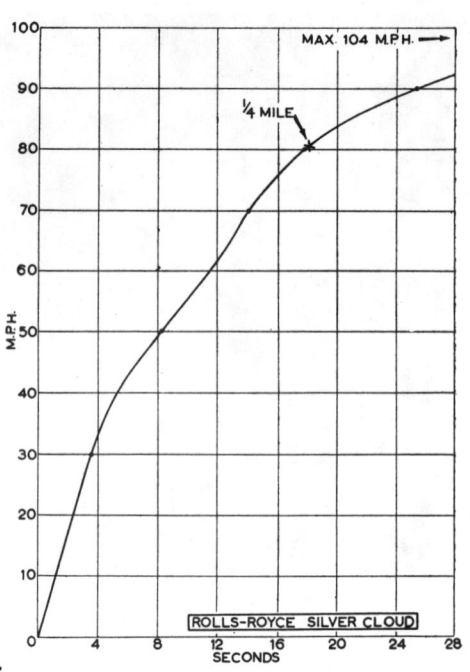

Acceleration Graph

the engine, it is possible through in-attention to enter a corner too fast. It is worth glancing at the speedometer, for it has a way of creeping up to 90 m.p.h. when you thought you were only doing "sixty".

All the usual Rolls-Royce features are found. The rear shock absorbers have a hand control, electrically operated. The suspension is not particularly soft, but gives a welcome freedom from pitch or roll. The chassis components are served by a built-in lubrication system, and it is such a pleasure to return to proper instruments on the dashboard, instead of those nasty little indicator lamps. It goes without saying that the finish is superb, the heater is exceptionally powerful, and even the radio has an unusually good tone.

The pleasure of handling such a car is very great. Its size may be a handicap on certain journeys, but car park attendants, and even the police, give preferential treatment to the driver of a Rolls. The great name still means as much as it ever did, and if the chap who opens the door for you expects half a crown as a matter of course, that is a small price to pay for indulging in this very special sort of motoring.

Sheer quality will never be cheap, and the Rolls-Royce cannot be other than an expensive car. Yet, for the lucky man who can afford it, it is certainly a good investment. As a superb piece of transportation equipment, carrying immense prestige, there is no substitute for this car.

SPECIFICATION AND PERFORMANCE DATA

Car Tested: Rolls-Royce Silver Cloud saloon, price £5,693 17s., including P.T. Extra, power-assisted steering, price £165, including P.T.

Engine: Six cylinders, 95 mm. x 114 mm. (4.887 c.c.). Pushrod-operated overhead inlet valves in light alloy head. Side exhaust valves in block. Compression ratio 8 to 1. Twin SU carburetters. Coil and distributor ignition.

Transmission: Fluid flywheel and four-speed epicyclic gearbox with automatic gearchange, ratios 3.42, 4.96, 9.00 and 13.06 to 1. Open two-piece propeller shaft with steady bearing. Hypoid rear axle.

Chassis: Box-section frame with cruciform bracing. Independent front suspension by unequal length wishbones and helical springs with torsional anti-roll bar. Cam and roller power-assisted steering with three-piece track rod. Rigid rear axle on semi-elliptic springs with Z-type torsional locating member. Piston-type dampers with electrically operated ride control at rear. Hydraulically operated hydrostatic front brakes, hydraulic and mechanical rear brakes, gearbox-driven servo; 8.20-15 ins. tyres on pressed steel wheels.

Equipment: 12-volt lighting and starting. Speedometer. Clock. Oil pressure, water temperature and fuel and oil level gauges. Heater and demister, with electrical demisting of rear window. Self-cancelling wipers and screen washers. Flashing indicators. Radio. Cigar lighters. Picnic tables.

Dimensions: Wheelbase, 10 ft. 3 ins. Track (front), 4 ft. 10 ins., (rear) 5 ft. Overall length, 17 ft. 8 ins. Width, 6 ft. 2½ ins. Turning circle, 42 ft. Weight, 1 ton 19½ cwt.

Performance: Maximum speed 104 m.p.h. Standing quarter-mile 18.1 secs. Acceleration, 0-30 m.p.h., 3.4 secs.; 0-50 m.p.h., 8.4 secs.; 0-60 m.p.h., 11.6 secs.; 0-70 m.p.h., 14 secs.; 0-80 m.p.h., 17.8 secs.; 0-90 m.p.h., 25.2 secs.

Fuel Consumption: 16 m.p.g.

BIG engine of 4.9 litres retains the inlet-over-exhaust valve layout, with its inherently good turbulence. Compression ratio, however, is increased from 6.6 to 8:1 and this improves the performance considerably.

THE engine appears to be even smoother than before, and is perfectly insulated with no shake whatsoever, even when delivering its tremendous torque. Under-bonnet space is well-fitted, but "owner-maintenance" is scarcely intended!

Rolls Royce

THE WORLD'S BEST CAR, and the fastest luxury car, aptly describe the Rolls-Royce and the Bentley, respectively. Except for the name, radiator and price, the cars are identical. Stylewise, they retain the air of elegance associated with England's conservative coach-builders. The interiors vary from polished wood trim to glove-leather padding, and are hand-finished to perfection. Hand-built from the ground up, these cars offer the customer anything he wants, for a price: indirect ceiling lights, bar, sandwich compartment, diamond-studded brocade upholstery . . . and in the case of one Indian Prince, a rolling bathroom! Today's conveniences run along the lines of backseat television, fancy cocktail equipment, and perhaps a lady's vanity cabinet that folds out.

Mechanically the two cars are the same, except for optional gearing. The same six-cylinder engine of 4887cc (298 cubic inches) is used on both cars, along with the same chassis and suspension. An automatic gearbox made from British and American patents is also standard. Two other stock items are the one-shot lubrication system (which, when a button is depressed on the floorboards, shoots metered amounts of grease to all lubrication points), and the ride control. You can make your rear shock absorbers stiff or soft with the flip of a lever. Another standard item is the servo-assisted brake system, which for a passenger car, is nothing short of phenomenal.

Perhaps the most famous feature of the cars is their silence. The engine runs smoothly, without vibration, and the gears are mated and run-in by stethoscope. Optional extras include electrically operated windows, power steering, and air conditioning. P.o.e. prices: $13,750 for the Rolls Silver Cloud sedan, $13,450 for the Bentley series S sedan.

Bentley

ROLLS-ROYCE
BEST CAR IN THE WORLD?

FOUNDED IN 1904, the Rolls-Royce firm introduced its first 6-cyl car in 1910. This model ultimately came to be known as the Silver Ghost and was continued in production with practically no changes for a longer period of time than any other car in the world—even longer than the Ford Model T.

It was the Silver Ghost which set the foundation for the firm's reputation. In an era when cars were generally noisy, unreliable and more than a little undependable, the Rolls-Royce was absolutely unique and unapproached.

The reputation of the marque became so well known, that in 1921 an American plant was set up at Springfield, Mass. There, for eight years, complete chassis were built to standards identical with those of the home office and works in Derby, England.

The Silver Ghost engine was a big one: 6 cyl, 4.50 x 4.75, equivalent to 453 cu in. Cylinders were cast in blocks of three with valves at the side in pockets, since the cylinder heads were not removable. A distinguishing feature was exposed valve springs, and even this early the engine had 7 main bearings, with a diameter of 2.25 in,

ROAD & TRACK
TECHNICAL ANALYSIS

The Silver Ghost engine of 1921.

The latest V-8 is constructed mostly of light alloys.

and a friction driven torsional vibration damper. Aluminum pistons with split skirts came in 1921 and these carried no less than 6 rings. The revolution range was from 200 to 2750 rpm with maximum torque (unspecified) at 1300 rpm and a peak power output of 80 bhp at 2000 rpm. With an axle ratio of 3.25:1 the top speed was given as 80 mph at 2670 rpm. An interesting aside on the performance of this model is the fact that a prominent General Motors engineer once admitted that when the proving grounds at Milford, Mich., were first put into operation, the only car that could lap continuously at 70 mph without trouble was the Rolls-Royce Silver Ghost.

But, by 1925, it was apparent that the 15-year-old engine design had outlived its usefulness. A new engine was introduced for the "New Phantom" series and it had cylinder dimensions of 4.25 x 5.50, equivalent to 468 cu in. The principal new feature was rocker-actuated overhead valves. Power output was increased by over 33%, according to the announcements at the time. In 1929 a Phantom II model was brought out embodying only minor changes in the engine, but with a completely new chassis featuring a lower frame, semi-elliptic rear springs and a hypoid rear axle. Engine power was given as "15% greater."

The new V-8 in cross-section, showing the valve gear.

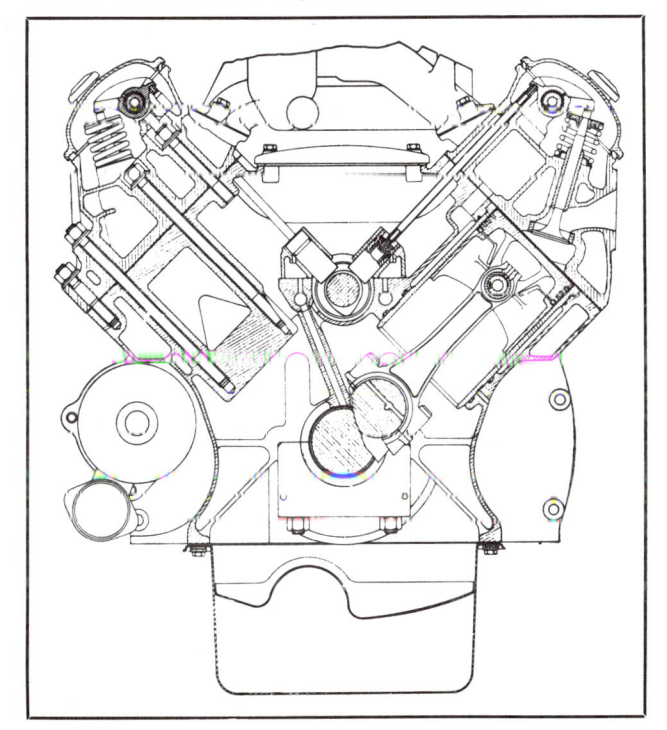

At this point it is not clear (at least to us) whether the New Phantom, or P-I, was actually built at Springfield or not. We think not, and certainly the Phantom II (or P-II) was built only in England. A catalog of the era advertises: "A new Rolls-Royce, built in Derby, England, with left drive and for American road conditions."

Shortly thereafter, in 1933 and 1934, a few cars were built, known as the P-II Continental models. These had a wheelbase of 144 in. (6 in. shorter), minor engine modifications for higher output, and a higher axle ratio of 3.41, instead of 3.73. As a guess, these cars produced about 150 honest bhp and road tests at the time gave the timed top speed as 92/95 mph.

In 1936 the Phantom III was introduced, with a V-12 engine made largely of aluminum alloy. In a manner of speaking, it was somewhat like two of the 25-hp engines (to be described later), for it had a bore and stroke of 3.25 x 4.50. Output was about 150 bhp and the top speed was 95 mph.

Reverting back to 1923, in that year Rolls-Royce acknowledged the demand for a car of more "compact" dimensions and introduced a much smaller car known as the 20-hp model (based on the taxable hp). Making no compromise with traditional R-R standards, the "20" had an entirely new ohv, 6-cyl engine with cylinder dimensions of only 3 x 4.50 (181 cu in.). Through an evolutionary series of changes this powerplant was enlarged to 3.25 x 4.50 and when R-R announced acquisition and resumption of the Bentley name in the fall of 1933, the 3.5-liter "Silent Sports Car" was in fact nothing more than the small Rolls with much lower frame rails.

The original Bentley announcement stated that the 3669-cc engine gave 120 bhp, but some authentic data published in 1938 showed 100 bhp. In 1937 the engine was bored out again, this time to 3.50 in. This gave 260 cu in. or 4255 cc, hence, the term "4¼-liter Bentley." This engine produced 125 bhp at 3750 rpm and was used concurrently in the "25/30 hp" Rolls-Royce, with simpler carburetion.

After the war Rolls redesigned the 20/25/30 hp engine again, this time with a new block which contained the exhaust valves, and a new aluminum head with overhead intake valves. In all this time the original bore centers and bottom end were retained, and the bhp went up from about 40 to approximately 132. This F-head engine was used in both the Rolls-Royce Silver Wraith and the Bentley, for after the war the V-12 P-III was not produced. Ultimately, further modifications were made (higher compression ratio, larger carburetors, still larger bore) and in its final form the many times redesigned "small" Rolls engine attained 4887 cc and an estimated output of 180 bhp. The firm's chief engineer, S. H. Grylls, recently pointed out that, "Since 1923 we continuously developed the 6-cyl engine without altering the length of the crankshaft and without altering the stroke. During those 36 years we greatly increased the reliability of the engine, in spite of multiplying the horsepower by 4.5." This implies an increase from 40 to 180 bhp, which shows the fundamental soundness of the original 7-main bearing crankshaft, which obviously was over-designed at the start.

137

The New V-8

Design work on a new and larger engine, ultimately the V-8, commenced in 1953. The idea of a V-8 was not new to Rolls; the company had built a few V-8 powered cars in 1905, known as the "Legal Limit" model because the law required them to be governed at 25 mph, maximum. Nor was the idea of using aluminum for the cylinder block new to Rolls—the pre-war Phantom III used this material extensively and, of course, the superb Rolls-Royce aircraft engines were also cast largely of aluminum alloys.

The first test V-8 engine was ready in 1955 and prior to production in the late summer of 1959, the V-8's test experience totaled 12,000 hours on the dynamometer and 380,000 miles on the road. Of course, Rolls had an 8-cyl in-line version of the six, but this engine had been used only in the half dozen or so cars supplied to Royalty—the very rare Phantom IV. Concerning it, the firm's chief engineer said, "A straight eight would necessarily have meant increasing the length of the bonnet [hood] and moving the front wheels farther away from the driver. The difficulty of making a nice motorcar free from the sensation of 'jellying' varies with the cube of the distance from the driver to the front wheels." While a military version of the straight eight (with 6516 cc) developed 240 bhp at 4000 rpm, it must undoubtedly have weighed at least 20% more than the six and its length would have created severe thermal expansion problems if aluminum were fully utilized. The V-8 engine, complete with all accessories weighs 713 lb, and with hydramatic transmission the total is 890 lb. Either way, the new assembly saves 10 lb over the corresponding figures for the F-head six.

Power and torque data are not supplied by the manufacturer. Some estimates range as low as 200 bhp, but we put the figure a little higher—at about 220 net bhp at the flywheel, at 4000 rpm. Maximum torque is more closely allied with piston displacement and should be in the neighborhood of 340 lb-ft at 2200 rpm. The maximum design speed was set at 4400 rpm, but a high-torque valve timing, moderate compression ratio and rather small carburetors (two, 1.75 in.) show that there is ample scope for a considerable increase in output—if and when it is needed.

Abandonment of the intake over exhaust valve arrangement, and its replacement by a more conventional wedge-type combustion needs little explanation. Briefly, the F-head chamber is very good up to compression ratios of no more than 8.0:1. Anything above that tends to cause rough-running and also restricts volumetric efficiency. Rolls chose the wedge chamber because it alone offers smooth running at present and future compression ratios. And smoothness of operation has always been one of the marque's greatest virtues.

The choice of oversquare cylinder dimensions was pretty much academic. A V-8 block with offset banks gives plenty of room for big cylinders. Rolls-Royce simply decided that 380 cu in. would be ample for all present and future needs. A stroke of 3.60 in. meant a piston speed of only 2400 fpm at 4000 rpm and a bore of 4.10 in. gave the required displacement with a not unreasonable oversquare ratio of 0.88. The displacement increase of 27% over the six meant that the rear axle ratio could be reduced by 10% and still have a net gain in potential performance equivalent to an engine size increase of 17%. Furthermore, since a very large 7-passenger model was envisioned, the short-stroke engine would allow the necessary higher numerical axle ratio (to offset the extra weight) without running into excessive piston speed.

As a matter of fact, the new Bentley, with an axle ratio of 2.92, has a theoretical cruising speed of 119 mph at 2500 fpm, the Silver Cloud II (3.08 axle) has a theoretical cruising speed of 116 mph, and even the huge Phantom V (3.89 axle) will cruise at 94 mph. Thus the Rolls engineers have cleverly contrived to make the theoretical cruising speed almost exactly equivalent to the true top speed capability of each model.

As is well known, the cylinder block is an aluminum alloy casting. The exact alloy is not specified except by the term "high silicon." We assume this to be one of the commercial alloys containing about 12% silicon, sometimes called "Lo-Ex" for its more favorable low expansion characteristic. The block weighs 85 lb and uses centrifugally cast iron sleeves weighing 4.15 lb each. Considering the fact that aluminum weighs a third as much as iron, the block is surprisingly heavy and indicative of an extremely rigid structure. (A Buick cast-iron V-8 block, for example, weighs 177 lb.) The bore centers are 4.75 in., as required for cylinders of 4.10 in. and the wet sleeve design.

The cylinder sleeves are machined all over and are unique in that the top retaining shoulder, or flange, is narrow to provide a high unit pressure at the gasket sealing surface. This shoulder stands 0.002 to 0.003 in. above the top of the block and the steel head gasket is corrugated to allow for crush when the head is bolted down. Thus R-R engineers have circumvented the problem of cylinder-bore distortion, which can easily happen with wet sleeves if the head bolts are tightened indiscriminately without

The Rolls-Royce chassis; a study in iron-bound and unshakeable conservatism.

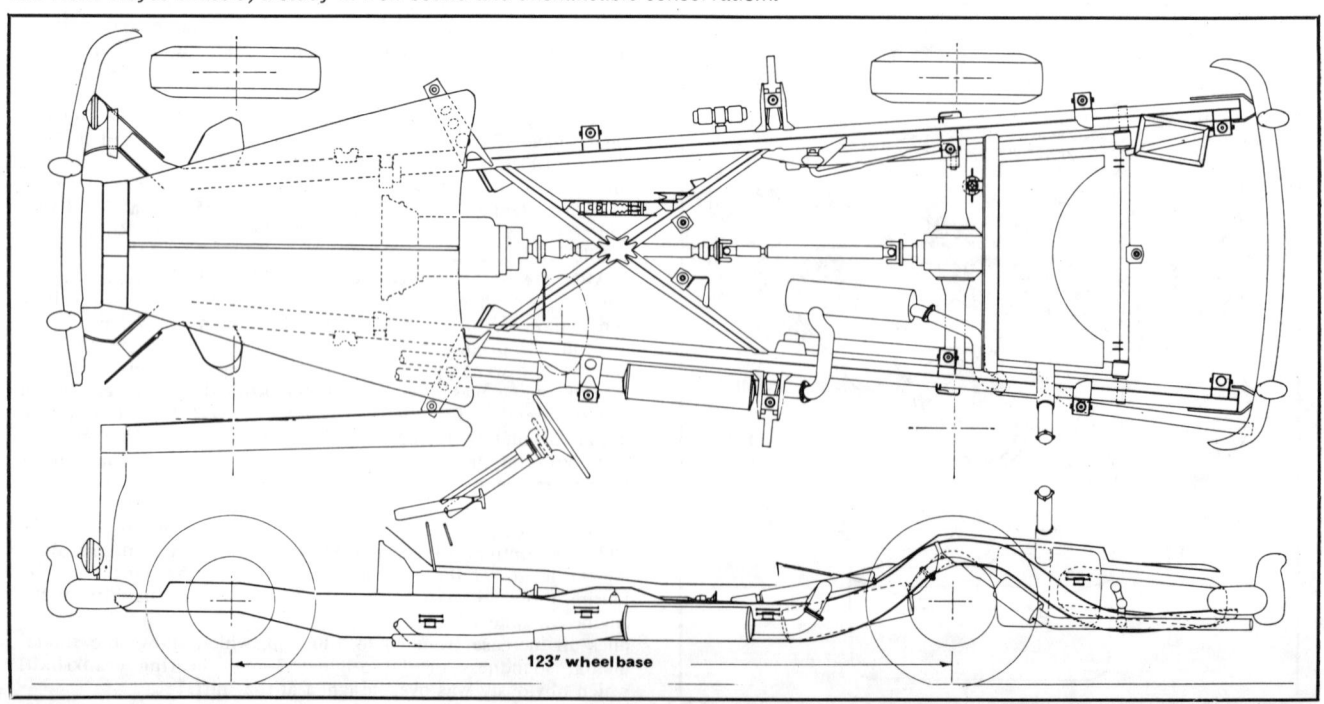

123″ wheelbase

TECHNICAL ANALYSIS

a torque wrench. The bottom of the sleeve is sealed by two synthetic rubber rings with a safety groove and drain holes between. A similar synthetic ring is used near the top of the sleeve, more as a safety precaution than anything else.

The heads are cast of the same alloy as the block, and weigh 28.2 lb each. Protection from corrosion includes anodizing all internal walls in the block and heads, and coating the exposed outside diameter of the iron sleeves with a mix of aluminum pigment and varnish. Recommended coolant is inhibited glycol (such as Prestone), for both summer and winter.

It is interesting to note that the wedge shape of the combustion chambers in these heads is influenced by the choice of a valve angle of 28° from the cylinder bore axis. This is quite a bit more than "conventional" practice which usually lies between 15° and 20°, but is not nearly so extreme as the one important exception of 45°, found in Buicks. Six different forms of combustion chamber were tested and evaluated before a final decision was made. An important factor in their choice of a 28° wedge was the need for keeping the overall width of the engine down so that it could fit into the "old" chassis. The Rolls engine is 22 in. wide, without the exhaust system.

The crankshaft is arranged in the usual way: in two planes, with 5 main bearings and integral counterweights. It is forged from chrome-molybdenum alloy and though the bearing surfaces are not heat-treated they check out at 300 Brinell. The main bearing surfaces are 2.50 in. in diameter and the insert-type bearings are all 1.062 in. long, except for the rear bearing, which is 1.437 in. long. The crankpins measure 2.25 in. in diameter x 2 in. long and the method of cross-drilling for pressure lubrication is noteworthy (see section view). The machined crankshaft weighs 70.9 lb and carries a bonded-rubber torsional vibration damper at the front.

The forged connecting rods are quite conventional and have a length equal to 1.8 times the stroke, or 6.50 in. center to center. Each rod weighs 1.788 lb and the tri-alloy bearing insert has an effective width of 0.79. The wrist pins are hollow with an outside diameter of 0.875.

Lo-Ex alloy pistons are used, weighing 2.09 lb each, complete with wrist pin and rings. These employ the usual slots, slipper skirts, cam grind and tin-plating. There are 3 compression rings and one expander-type oil control ring per piston. Standard skirt to cylinder clearance at room temperature is 0.0012 to 0.0018.

The camshaft is made from alloy cast iron with chilled cam noses. It is carried by 4 babbitt-lined bearings, of 2 in. diameter. The drive is unusual, these days; it is by a steel crankshaft gear and a forged aluminum gear on the camshaft.

Valve gear is also conventional, with hydraulic tappets originally supplied by Chrysler but now made in England. The pushrods are hollow and the rocker arms provide a 1.65 step-up in valve lift. The intake valves measure 1.9 in. and are made of silicon-chrome steel. Exhaust valves (1.535 in.) are rather special, with a stellited seat and tip plus an R-R-developed surface treatment (Brightray) on the exposed head to eliminate scaling. Valve spring pressure is 86 to 174 lb, from closed to open position, and the lift is 0.412. Both valves have larger than usual stems (0.375) and the intake guides are cast iron, while the exhaust guides are bronze. These guides and the austenitic steel seat inserts are shrink-press fitted into the heads.

Engine lubrication is by conventional means at 40 psi, maximum. The pump has helical gears and is located in front of No. 1 main bearing, on the theory that any noise it might create is farther away from the driver and passengers. A full-flow oil filter is used, with a wool felt element.

Twin SU electric fuel pumps supply two side-draft carburetors of the latest SU type HD-6 (1.75 in.). A post-production change has been the incorporation of an anti-percolating device to obviate hot-starting problems. There is an automatic choke and the intake manifold is heated by coolant from the engine circuit.

Nearly all of the length saved by switching from a six to a V-8 is absorbed by the positions of such accessories as the air conditioning compressor and the power steering pump. Frankly, this layout is rather disappointing to us—we had hoped for a nice clean appearance with all these extra gadgets neatly built in and gear driven. However, accessibility is actually better than it might appear at first glance, except for the water pump, which is thoroughly buried. There are two triangular belt drive systems and each uses dual belts in the interest of durability and reliability.

The Frame

The standard wheelbase is 123 in., with an option of 127 in. as before. There are no major changes and the frame rails are still channels welded together to form a box having a maximum depth of 7 in. (below the cowl) and an average width of 3.5 in. These rails, incidentally, are pressure tested for leaks, to insure against moisture and rust. An X-member is continued (see chassis drawing) and a most interesting expedient has been adopted to achieve the 144-in. wheelbase required for the 7-passenger P-V chassis. In effect, the frame is cut across the middle of the junction and a very heavy tube exactly 21 in. long is welded (lengthwise) to the X-junction. Likewise, 21-in. box sections fill out the side rails, giving a frame with a wheelbase of 144 in. Though this may not seem impressive, the P-V frame is 40% stiffer than the 123-in. frame, in torsion.

Torsional rigidity of the 123-in. frame, as used on both the Rolls and Bentley, is 2700 lb-ft/degree. This is not unusually stiff, but additional rigidity is very hard to get on a long wheelbase, even at a heavy cost in added weight. In this connection R-R development excels and the chassis engineers have, in effect, so tuned the body to the frame (via proper placement of mounting points, etc.) that the car takes the roughest roads at back-breaking speed without a sign of body shake, shimmy or stress.

Suspension

With the larger and heavier P-V in mind, an entirely new front suspension has been designed. It is virtually identical to that used previously, except that the wishbones are forgings, and all load-carrying parts are heavier and stronger than before. The 123-in. chassis has a front wheel ride rate of 90 lb/in. and with the normal load and an unsprung weight of 286 lb (2 wheels, brakes, etc.) the front end ride frequency is 55 cycles/min. The front roll center is about 3 in. above the ground. For the P-V the figures are 100 lb/in., with the same ride frequency.

At the rear, the leaf springs are 56 in. long, with the axle on-center. The shackles are arranged to give a variable ride rate: from 122 to 155 lb/in. Unsprung weight is 423 lb and the ride frequency is 77.5 cycles/min. The P-V has longer springs, 65 in., arranged 9 in. asymmetrical. Ride rate is 126 to 145 lb/in. and frequency is 66.5 cycles/min. Roll center at the rear is at the axle centerline in all models. An interesting point arises over the question of understeer. The previous car had definite understeer and the R-R engineers felt that the oversteering effect of much more power available at the rear wheels (via the new V-8 engine) would require considerably more understeer devices in the re-designed chassis, to insure safe handling. To do this they have

The differential—note the bellville springs at arrow.

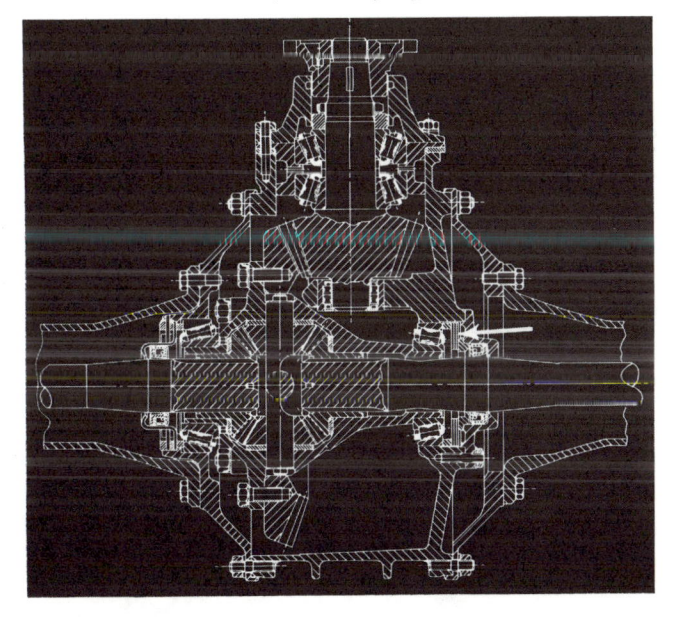

TECHNICAL ANALYSIS

dropped the Z-bar formerly used at the rear to reduce acceleration squat and cornering roll. A single radius rod above the axle and on the right side controls squat. Since the rear anti-roll bar was an oversteering factor, its removal—plus a stiffer bar in front —contributes heavily to more understeer. The actual roll resistance factors are: front, 480 lb-ft/degree; rear, 320 lb-ft/degree (short chassis).

The rear axle (see section view) uses hypoid gears, and the ring gear measures 8.75 in. Pinion offset is a modest 0.65 in and this heavily loaded component has straddle-mounted bearings in approved heavy duty fashion. The main housing is an aluminum alloy casting, open on the left hand side. A pair of steel adapter plates are used on each side, secured by no less than 25 studs each. Axle shaft housings flare out and are attached to each adapter by 24 cap screws (making a total of 98 fasteners, an old and very efficient R-R philosophy). Since differential temperatures often reach 300°F, the problem of maintaining the essential carrier bearing preload at times is, normally, impossible with an aluminum housing. R-R once used a very large coil spring to do the job, but this axle employs three bellville-type springs. These are indicated by an arrow in the drawing and, though they look like flat washers, they are actually slightly dished before assembly. The two axle shafts are of different lengths and are properly turned down for even stress distribution. At the wheel ends the shafts are upset to provide a 5-in. bolt circle for wheel mounting. The rear wheel bearing is an enormous deep-groove ball-type, retained by the press-fit of an extra wide inner race, plus another collar pressed against it. While R-R uses conventional garter spring-type oil seals, as for example on each side of the differential carrier, it is interesting to note that wherever possible slingers and labyrinth-type closures are used because there is no deterioration in efficiency regardless of mileage. The wheel bearings are so protected and the design is excellent, though probably more expensive than conventional oil seals.

Brakes

All brake drums are cast iron and equal in size; 11.25 x 3 in. wide, with a total lining area of 240 sq in. In front there are two trailing shoes per drum, of Girling design, in which the shoes are kept in light contact with the drums at all times. This assures instantaneous availability, allows a higher mechanical advantage for a given pedal travel and provides automatic take-up for wear. At the rear we find opposed shoes, so that some shoe servo effect will be available in either direction. A single master cylinder is used, with a transmission driven clutch assembly designed to augment brake pedal pressure. This booster, or mechanical servo motor, is extremely simple, as shown by the cross section view. The linkage, or hook-up, isn't quite so simple, for in event of servo failure the normal brake pedal travel will still pressurize the hydraulic brake system. Also, the pedal linkage includes a mechanical connection to the rear brake shoes so that 30% of the effort needed is supplied by the driver. Thus, it is virtually impossible to have total brake failure, either through loss of fluid or loss of servo assistance. A third system provides a parking brake via the mechanical links to the rear shoes.

On Bentley models, which are identical to the Silver Cloud II in most details, the brake shoes are further modified by use of a unique 4-shoe brake. No details of this design are available.

Transmission

It is well known that the hydramatic transmission is made by R-R, under license from General Motors. This is the earlier design with a single fluid coupling rather than two, as used more recently by GM. This unit has very low friction losses and provides 4 well chosen gear steps with smooth starts, no possibility of gear clashing and relatively smooth changes either up or down, according to the demands of the driver via his accelerator position and vehicle speed.

The propeller shaft is made in two sections, with a rubber cushioned center bearing. This allows smaller shafts with a critical whirling speed well above the highest possible engine rpm. It also lowers the floor line and lessens vibration problems.

Miscellaneous

The steering gear, linkage and hydraulic booster system is unchanged and it provides a 50% boost for normal driving and an 80% push during parking. However, the V-8 exhaust system's width necessitated moving the column farther to the left (on LHD). This has been accomplished without upsetting steering geometry by adding a small gearbox just above the steering gear itself. This small assembly is beautifully designed, with anti-friction bearings and zero backlash helical cut gears. It is expensive, but was obviously cheaper than a complete redesign of the rest of the system.

The exhaust system was especially designed to eliminate the typical "V-8 beat" noise. There are three mufflers, each tuned to suppress a different range of frequencies.

Manufacture: A Visit to the Factory

Last fall the writer visited the Rolls-Royce automotive division at Crewe, about two hours' train ride northwest of London. Here, before and during the war, the V-12 Merlin engine was built. After the war, car manufacturing operations were removed from Derby to this more modern plant. The area is not very large, but 3600 persons are employed here—though not all on cars.

It undoubtedly would be very interesting to our readers if we could unveil the mystery of the fabulous hand-made Rolls-Royce manufacturing methods. In truth, there is no such thing at all, with a few exceptions. The Rolls-Royce car (and Bentley too, of course) is made rather normally—with proper machinery, tools, jigs and fixtures. The only important exception was found in a corner of one of the shops where a half dozen craftsmen were shaping the traditional "square" R-R radiator shell from sheet steel, by the same simple methods that have been employed for over half a century. Any tinsmith shop can show you the methods and the result is a radiator of the well known shape. Lay a good steel straight-edge on the various "flat" surfaces and you will see that there are no true flat surfaces at all—there are strange waves and curves, as might be expected when no forming dies are employed. But, at least to the eye, the surfaces seem true and flat.

The manufacturing problem at Rolls-Royce is not a simple one. First, there is the one pervading axiom that quality comes before cost. But, space and the number of machines available is limited. Every square foot of space is efficiently utilized, every machine (if at all possible) is designed to do double duty. The problem is complicated by the low production rate of finished cars. There is some secrecy about this, but the output is very close to 10 cars per day, 50 per week and 2500 per year. At the present time, demand is so great that this capacity could, perhaps, be doubled. But the management feels that such an expansion program is not justified. There is no thought of compromising the name which has been synonymous with the ultimate in quality for 55 years.

Reverting back to the manufacturing

The transmission-driven mechanical brake servo unit.

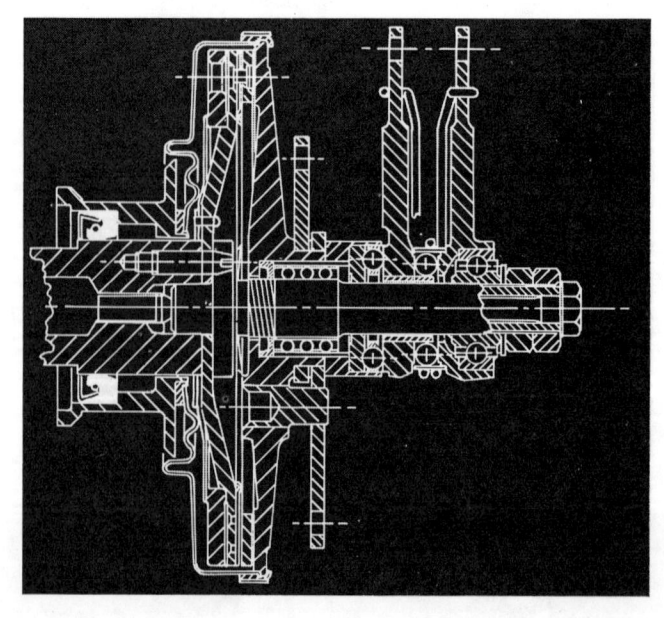

140

R-R TECHNICAL ANALYSIS

processes, most of the operations are straightforward and familiar to any job-shop operation involving similar quantities. A majority of the machine tools are standard types and one machine may be set up for as many as 20 different parts or operations in a single week. However, there *are* some special purpose machines, particularly for the new V-8 engine. For example, both ends of the block are drilled and tapped simultaneously, 38 holes in one end, 29 in the other. Another machine has two spindles for machining the sleeve bores in the block. A clever fixture locates the block accurately at any one of the four locations required to bore 8 cylinders. Each of the two spindles performs a total of 11 machining operations including the seat for the cylinder sleeve and the four sealing ring grooves which are in the block, not in the sleeve.

One of the most interesting operations is boring the main and camshaft bearing holes. This is done twice, first early in the block processing, and finally to finished size after all other machining is completed. Rolls believes this is the only way to insure perfect shaft alignment, because each time metal is removed there is bound to be some stress-relieving and minor distortion. Also, in this case, Rolls uses timing gears and quiet operation of accurately cut gears demands perfect center distance. Incidentally, it is unusual to find a forged aluminum timing gear, and these gears have helical teeth finish ground to super-precision aircraft-quality standards.

Another item of interest is the often heard opinion that R-R makes many more components itself than do most other manufacturers. This isn't quite true; more correctly R-R "buys out" as much or even more than any other company. Rolls does make a lot of components that are traditionally "supplied" (such as steering gears, transmissions, rear axles) but, on the other hand, buys casting, forgings, pistons, valves, bearings, starters, generators, hydraulic pumps, radiators, frames, bodies and the like. Rather unusual for such limited production, R-R makes a very large number of its own screw machine parts. One reason for this may be that Rolls tolerance specifications on screw threads are strictly aircraft quality, plus the fact that the company does use an unusually high percentage of special bolts. For example, all studs that thread into aluminum must be a light-interference fit. Highly stressed threads in places where parts may be removed or worked on more frequently are special "aero" threads. These are not cheap and it's easier to control quality and costs in your own shop.

As a very general statement, production machining tolerances at R-R are no better, no closer, than good American practice. Cylinder bore tolerance is 0.001, pistons are graded in steps of 0.0003, and a few diameters are held to ±0.00025. Nothing unusual or extraordinary here. But the difference comes in material selection and heat-treatment (cost no object) and in inspection and quality control, which takes nearly 15% of the payroll. Finally, there is meticulous assembly by conscientious, skilled workmen—none of the throw-it-together-any-old-way attitude so prevalent over here.

The final assembly line moves very slowly, as required for the production of only 10 cars per day. Bodies are supplied by Pressed Steel Ltd. and all painting, interior trim, seats, etc., are done by R-R. Incidentally, though the bodies are steel, all hoods, trunk lids and doors are panelled in aluminum. Since the appearance, color, etc., of wood trim varies from car to car, each car is coded in such a way that perfectly matched replacements can be made, by ordering from the factory.

Every engine is run for 2.5 hours before being installed, and every completed car is road tested for 200 miles before shipment. Chassis destined for coach-built bodies are also carefully road tested for faults, with an elemental body bolted temporarily in place.

In summary, the R-R way of building cars hasn't changed one iota over the years. At Crewe there is only one way to do the job—the best way, regardless of cost.

Power assisted steering
is now offered on the Bentley 'S' Series
as an optional extra

Take a
BENTLEY
into partnership

BEAUTIFULLY MADE: The most elaborate interior furnishing, which even excels previous models, makes the S2 Bentley a most desirable motor car.

The Bentley S2

THE Rolls-Royce and Bentley cars have for many years been propelled by six-cylinder engines. As the vehicles have grown larger and heavier, with more and more equipment for the comfort of the passengers, the engines have been "stretched" in successive stages to keep up with increasing demands. At five litres, it was decided that the "six" had been developed to the limit, and a new V8 of 6,230 c.c. took its place last year.

If the well-tried design features of the older engine were retained, it was calculated that the big new unit would be so heavy that the roadholding must be adversely affected, not to mention the performance. Accordingly, the V8 was constructed with an aluminium cylinder block, and the triumphant engineers were even able to save weight over the smaller unit. As this engine is "over-square", it was considered advisable to abandon the inlet-over-exhaust valve arrangement and to revert to pushrod-operated valves in the light alloy heads, with more compact combustion chambers.

The new engine is not designed for ultimate power output, but very great pains have been taken to get the last ounce of torque out of it in the middle ranges. This is vitally important because a car with an automatic gearbox tends to seem fussy if the characteristics of the engine are such as to demand frequent changes of gear. The new unit has been specifically designed to suit the automatic transmission which is standard on this car.

The fully counterbalanced chrome molybdenum steel crankshaft runs in five copper-lead-indium lined steel shells. It drives the camshaft by helical gears and the hydraulic tappets eliminate valve adjustment. The valve seats are of austenitic steel in the aluminium heads, and the induction system is fed by twin SU carburetters with an automatic choke for cold starting.

The air cleaner and silencer is normally

AUTOMATIC gearbox combined with power steering and servo assisted brakes make driving this large saloon a very pleasant task.

held down by a wing nut, but it can be quickly attached to a wire which depends from the bonnet. Then, opening the bonnet lifts the air silencer, but it remains connected to the carburetters by its rubber tube. This provision is necessary because it upsets the carburation to remove the silencer, and so one cannot disconnect it when setting the slow running or carrying out other tuning. The exhaust system has three straight-through acoustic-type silencers, each one tuned to a different frequency range.

An automatic gearbox gives four speeds, and is driven through a fluid coupling. It has a "kick-down" change on the accelerator, but one can also change down with the hand lever. The open propeller shaft is divided, with a floating centre bearing, and the rear axle is of hypoid type with a ratio of 3.08:1.

The braking system is very elaborate. In the first place, the pedal supplies part of the rear braking effort directly by mechanical means, and also applies the gearbox-driven servo. This puts on the front brakes hydraulically and also gives a part of the rear braking effort, but both front and rear hydraulic systems are entirely separate with their own master cylinders. Thus, 30 per cent. of the rear braking effort comes from the pedal and 70 per cent. from the servo, while the servo also actuates the front brakes on a separate circuit. It is therefore utterly impossible for a total brake failure to take place, even ignoring the independent hook-up for the hand brake.

The chassis frame is of box section with cruciform bracing. The independent front suspension is by helical springs and wishbones with hydraulic dampers, reinforced by a torsional anti-roll bar. The steering box is of cam and roller type with built-in hydraulic power assistance. The rear axle is on semi-elliptic springs, the hydraulic dampers having a two-position electrical control. There is a radius arm to resist the rear axle torque, and bolt-on wheels carry 8.20-15 ins. tyres.

The steel body is completely insulated from the chassis by rubber. It is beautifully made and the interior furnishing is most elaborate. The sheer quality of the interior finish is hard to put into words —one can only say that it excels that of previous Rolls-Royce and Bentley cars. The very high seat backs give luxurious comfort to the passengers, and there are picnic tables, vanity mirrors, and all the refinements that one expects in this type of motor carriage. The electrically operated windows may be controlled from each individual door switch or by a set of four controls on the driver's door.

The heating and ventilation system is most comprehensive. There are independent recirculatory and fresh-air systems, controlled from knobs on the dashboard. It is worth studying the instruction book before attempting to use the heating system, but once the idea is mastered it is possible to regulate the temperature effectively throughout the car. Refrigeration may be ordered as an extra.

I covered some fairly long journeys on the S2, which I was able to compare with similar runs on the six-cylinder models. As is usual with aluminium engines, it is not quite so silent when

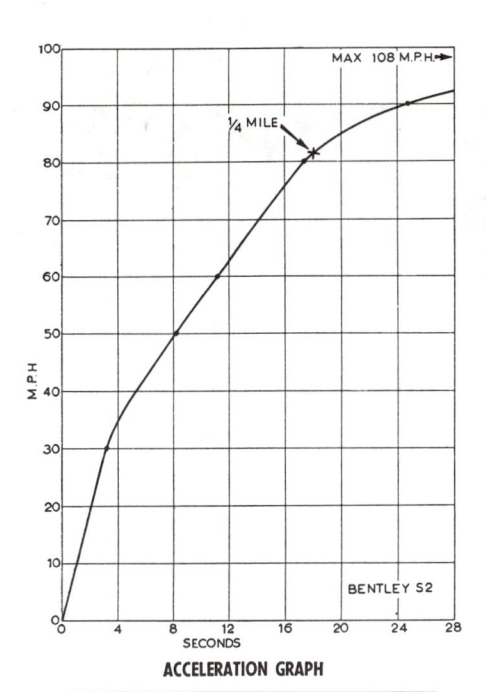

ACCELERATION GRAPH

AMPLE luggage space as well as an accessible spare wheel and cleverly placed tool compartments (to the left of the spare wheel with lined box above) are among the things automatically provided on a car of the quality of the Bentley.

idling as the earlier cast-iron "six". This, however, is only really noticeable from outside the car. Once on the move, the enormous power in the medium ranges makes itself felt. The great car is wonderfully responsive to the accelerator, and after a check it gets back into its stride in a rapid, effortless manner.

Much less gear changing is required than before. The automatic box seems to have been improved in smoothness, too, and one scarcely feels the changes as they go through. The bigger engine certainly pays dividends in providing acceleration which is amazingly rapid yet almost imperceptible.

It is in this respect that the new Bentley excels all other cars. It travels really fast with absolutely no sign of hurry, and the most nervous passenger gains little impression that rapid motoring is in progress. Even a timid driver will cruise at 90 m.p.h., for the car makes it all so easy, and those servo brakes are waiting to "kill" the speed in a few yards. The power-assisted steering is so good that at first I thought that it had not been fitted to my test car! It is literally imperceptible to the driver, but it just gives him that little assistance which makes a big car as easy to drive and park as a little one. Some cars with power-assisted steering have no normal "feel" and are consequently almost dangerous at high speed. During fast driving, the Bentley driver is able to "feel" the road, exactly as with any good car without power assistance.

On fast roads, the effortless acceleration and smooth, silent cruising are a real joy. Naturally, the sheer size of the car prevents it being hurried through thick traffic, and it cannot be cornered in the same way as a lighter machine. Yet, this large carriage is surprisingly controllable, and its very rapid acceleration to some extent permits one to ignore its overall dimensions.

The ultimate maximum speed of the new model is not spectacularly greater than that of its predecessor. Big luxury cars are now expected to be capable of 100 m.p.h., but the margin by which they can exceed it is not of much importance. Curiously enough, the speedometer of the test car tended to read fast to a greater extent than one would perhaps expect.

The Bentley S2, with its sister Rolls-Royce models, gives high-speed travel in silence and luxury, while the driver and passengers enjoy the sense of well being that only the best British craftsmanship can give. The V8 engine, with its flashing acceleration, certainly contributes to this result, and is a definite step forward in Rolls-Royce technique. If its thirst for petrol is decidedly heavier than that of its predecessor, this is scarcely likely to worry the man who can afford this kind of car.

THE NEW V8 engine which replaces the old "six" is a 6,230 c.c. V8 unit with an aluminium cylinder block. It has a fully counterbalanced crankshaft running in five copper-lead-indium lined steel shells.

SPECIFICATION AND PERFORMANCE DATA

Car Tested: Bentley S2 saloon. Price, £5,944 including P.T.

Engine: V8 cylinders with light alloy block and heads. Pushrod-operated overhead valves. Compression ratio 8 to 1. Twin SU carburetters. Coil ignition.

Transmission: Fluid flywheel and four-speed automatic gearbox, ratios 3.08, 4.46, 8.10 and 11.75 to 1. Open divided propeller shaft with floating centre bearing. Hypoid rear axle.

Equipment: Twelve-volt lighting and starting. Speedometer. Fuel gauge with reserve warning light combined with oil level gauge. Oil pressure gauge. Ammeter. Coolant thermometer. Electric clock. Radio. Heating and demisting. Electric door over petrol filler cap. Two-speed windscreen wipers and washers. Cigar lighters. Flashing indicators. Extra: Electric window raising mechanism.

Chassis: Box-section chassis frame with cruciform bracing. Independent front suspension by helical springs, wishbones, and anti-roll torsion bar. Cam and roller steering gear with hydraulic power assistance from engine-driven pump. Rear axle on semi-elliptic springs. Hydraulic dampers all round with electric ride control at rear. Hydraulic drum-type brakes with gearbox-driven servo and reserve mechanical linkage to pedal. 8.20-15 ins. tyres on bolt-on disc wheels.

Dimensions: Wheelbase, 10 ft. 3 ins. Track (front), 4 ft. 10¼ ins.; (rear) 5 ft. Overall length, 17 ft. 7½ ins. Width, 6 ft. 2¼ ins. Turning circle, 41 ft. 8 ins. Weight, 2 tons 1¼ cwt.

Performance: Maximum speed, 108 m.p.h. Standing quarter-mile, 18 secs. Acceleration: 0-30 m.p.h., 3.2 secs.; 0-50 m.p.h., 8.2 secs.; 0-60 m.p.h., 11.2 secs.; 0-80 m.p.h., 17.4 secs.

Fuel Consumption: 10-12 m.p.g.

wheels FULL ROAD TEST

THE CAR

The monumental radiator of the Bentley is topped with a winged B. Fog lamps double as winkers. Bumper bar over-riders are very substantial.

EXCLUSIVE TEST

BENTLEY V8

By PETER HALL

THERE is a Rolls-Royce in every motorist's life.

It might be the one he owns —or will own when his millionaire aunt dies—it might be the old one, the very, very old one his friend bought in rotting condition from a farmer who thought the rusting heap was only an outdated nuisance, and then rebuilt, it might be the one that nearly knocked you down in Collins Street or Martin Place with a monster behind the wheel who had clearly been maltreating the poor and under-privileged all his life.

But most likely it is the Rolls-Royce every motorist would own, if a lot of economic, progenital and horoscopical events had been radically different.

In Australia, actual ownership of a Rolls Royce (or Bentley) car is the preserve of the very few. Only about one new one is sold a week throughout the whole of Australia —and that includes those sold to Rolls-Royce Ltd's best down-under customer — the government.

The cost of a Rolls product here —not much the better side of £9000, when you include registration and insurance—obviously precludes anyone but a very successful company or very wealthy individual from owning one. Recent model used Rolls-Royce and Bentley units still cost the best part of first prize in most Australian lotteries.

A few less wealthy members of the community do occasionally buy Rolls-Royce cars, but the purchasers are usually veteran and vintage enthusiasts and the cars very old and comparatively cheap.

But we all dream about these famous British cars, even motoring writers. Perhaps our close day-to-day association with motor cars of all shapes, sizes and price brackets make us more realistic than most about the possibility of our ever owning one. By the same token it probably makes us even more interested in how the cars which have always been claimed, beyond disputation, to be the best in the world stand up to our road test standards.

In Australia it has been very hard to find out. Because of the comparatively small market for Rolls-Royce and Bentley cars, the local distributors and even Rolls-Royce itself have found it was not worth their while to attempt to maintain any of their own cars as demonstration or representatives' cars.

Very few Rolls-Royce owners have had the slightest interest in having their own cars subjected to road test, and there the matter has lain.

But this year, Rolls-Royce Australia Pty Ltd decided to register a new model for the use of its Australian representative on the motor car side of the business (which is by no means as large these days as

The Silent Sports Car and the Best Car in the World are so closely related that virtually the only difference is the radiator grille.

MOST PERFECT

the aircraft engine division) as what Rolls-Royce quaintly call a "demonstrator car".

It is mainly used to give potential Rolls-Royce buyers a good look at and drive of the type of car Rolls-Royce hope they will buy and for showing old and respected Rolls customers what the new model is like. A new model is never quite sold to them, of course, but occasionally the purchase of one is negotiated.

Good fortune recently came my way when the Rolls-Royce man concerned, Mr John Vidler, offered me a road test of the car.

It proved to be an almost brand new Bentley S2 with the new 6¼ litre Rolls-Royce aluminium V8 engine. I was not actually disappointed that the car to be tested was not a Rolls-Royce itself because, as is now well known, the Bentley S2 and the latest Rolls-Royce Silver Cloud are exactly the same car with only two comparatively minor differences.

Each car retains the distinctive radiator grille that has been a tradition since they both came on the market in the early days of motoring. The Rolls grille, being bigger, more intricate and handmade, makes it imperative that the car costs more than the Bentley— £200 more in Australia.

The other difference is even more unimportant. The instrument dials on the Bentley, identical to those on the Rolls, have the B for Bentley initial on them, the Rolls dials have the famous RR sign.

So, although the name of the car under test was Bentley, I might as well have been driving a Rolls-Royce.

And what a great and difficult-to-describe experience it was to put the Bentley through its paces.

From the outside, apparently an enormous and probably unmanageable car, behind the wheel it proved as easy to handle as a mini-car.

Every aid known for making motoring effortless and rapid was present.

The huge V8 engine developing an amount of horsepower that in the best Rolls-Royce tradition has never been disclosed, but must certainly well exceed 200, gave the 42 cwt car a performance in terms of acceleration and top speed that makes most passenger cars built anywhere in the world look like the doodlings of partly-trained amateurs.

And in the less definable terms of general driver appeal, the Bentley proved itself to be without peer.

And where did that appeal lay? Most of all it was driver comfort and road handling ability.

The steering, brakes and suspension were all of a standard that befitted the best car in the world. There are other magnificent machines that may top the Bentley — and the Rolls-Royce — in any of these three departments, but none that are superior in all three at once.

With typical thoroughness three distinct braking systems are built

---

HOW THE BENTLEY SHAPES UP

GOOD POINTS
- Performance
- Handling
- Appointments
- Workmanship
- Name

NOT SO GOOD
- Price
- Instrument positioning
- Small brake pedal

---

into one. Two are hydraulic, feeding the four shoes on each front wheel and the third is mechanical to the back wheels. Two master cylinders are actuated by the brake pedal, so if one hydraulic system fails, the other remains. In the very remote likelihood of both going, there is still the mechanical system to bring the car to a safe stop.

The whole braking system, of course, is servo assisted with the patented Rolls-Royce system working off the gearbox.

The gearbox itself is fully automatic. It is actually the General Motors Hydra-Matic transmission, as used on the Cadillac and Oldsmobile, but of course, built by Rolls-Royce.

It has a fluid coupling and a four-speed gearbox. A lever on the right of the steering column allows the driver to override the automatic gremlins below and to hold the car in either first, first and second or first, second and third gears. An automatic safety device always operates to make the transmission change up one ratio when the engine's maximum revs are approached.

The lever is typical of the Rolls-Royce thorough approach. It is just in the right place and moves smoothly with luxurious-sounding clicks.

There was a time when the Bentley was the most powerful and successful sports car — they were the days when Ettore Bugatti described it as the best and fastest truck in the world. Later, after Rolls-Royce Ltd acquired the ailing Bentley company in 1931, it became known as the Silent Sports Car.

After the Second World War, and even today, the description sticks, although mainly in advertisements. It is now more widely and quite rightly thought of as the Rolls-Royce with a different face.

But the car I tested showed abilities that would make most sports cars jump up and down in jealous rages.

Top speed on the test was a genuine 115.3 mph. Only one run was made because it was raining at the time and just after the run was completed the rain turned into a cloudburst. So, safety first. However, that one run was made *into* a slight headwind, so it is quite possible that the car's true top speed is even higher.

Even if it isn't, over 115 mph is a very impressive speed by any standards, especially by those that must be applied to a car weighing well over two tons. The fact that at that speed the car was rock steady and gave none of the feeling of

---

How good is the car the maker's modestly claim to be the best in the world? Tester Peter Hall did a full test on the newest Bentley just to find out.

145

THE CAR MOST PERFECT

doubt that is often associated with high speed in a passenger car made it quite outstanding.

Acceleration, too, was a remarkable tribute to the power developed by the new engine and the chassis design which gives it excellent traction.

But, from the performance point of view, it was the Bentley's cruising ability that stood out from everything else.

I did my best to contain its cruising speed to 80 mph, but found that almost completely impossible. I would set the speedo on 80 (and the speedo was pretty well dead accurate throughout its speed range) concentrate on the road and four or five miles later the needle would have crept up to 90 or 95.

And there the pointer would rest for as long as the driver and traffic conditions liked until, I have no doubt, the 18 gallon petrol tank was empty.

With the windows raised there was no sound of air being displaced outside the car at this speed (or at any other) and only an easy, gentle purr from the engine came to the ears.

The bodywork was tight, the power-assisted steering responsive but not so much assisted that feel of the road was lost, and always those magnificent brakes ready for when and if they were needed.

The driving position was superbly comfortable and everything the driver needed just where he wanted it.

And the things he had near his hand included a switch for changing the rear shocker settings from soft to hard. I left it on hard most of the time, because I personally prefer a taut ride, but a more boulevarde-type of ride was there for the flick of a switch whenever I felt like a change.

Nothing man-made is quite perfect and so it was with the Bentley.

But my only serious criticism was the placing of the instrument panel in the centre of the dashboard instead of in front of the driver. I suppose this is a throwback to the days when the vast majority of Rolls owners sat in the back and could see the instruments only if they were in the centre.

It seemed a pity, too, that the footbrake was quite a small pedal that could be operated only by the right foot. It is some years since Rolls-Royce switched Rolls and Bentley cars to automatic transmission exclusively and it is some wonder the firm has not appreciated by now the advantages of a wide brake pedal that can be used by the left foot. Perhaps the directors think it looks vulgar.

But there is nothing vulgar about this regal car.

Some may sneer at its automatic transmission, the snobbery of its name, the enormity of its price.

But they would be forced, along with the many rich men who refuse to own anything else than one or other of the Rolls-Royce-made cars, to admit its performance and excitement to match its dignity and thorough breeding.

They would all agree, I venture, that there are now two best cars in the world.

There is little doubt in my mind that the Bentley is one of them.

The four door bodywork on the test car is what could be described as the standard body, if one can say such a thing about a Bentley.

On the Bentley range, this is the only one with a body actually made by the Bentley Motor Company. It is also the cheapest model, but the one most popular in Australia where import duties and taxes make the price disproportionately high.

As an example, consider that in England a Bentley of the same type as the test car costs just on £6000, but the most expensive one, more or less off the showroom floor, is listed as a spectacular £8700!

The "non-Bentley" bodies are made by the better known coachbuilders such as James Young, Mulliner and Park Ward. Their products are semi-production models usually incorporating many of the customer's own ideas about interior appointments.

If the customer has enough money, he can still have a Bentley (or a Rolls-Royce, for that matter) designed and made specifically to his specifications.

Unfortunately, coachbuilding is almost a forgotten art, but thanks to the demands of Bentley and Rolls-Royce customers, it is still kept going.

Although there are only two in Australia to my knowledge, the Bentley Continental is an even more impressive version of the test car. To start with, it is not fitted with the same body as the normal machine.

Several alternative styles are available — all of them expensive. There are two-door, four/five seaters, drop head coupes and four-door saloons.

Performance has been boosted and there is also a high ratio rear axle available. The cost of landing one of these cars in Australia is huge. Estimates vary depending on the particular car, but I think £14,000 would probably be about right for the drop head, which, of course, has power operated roof.

The first of the post-war Bentley Continentals had a beautifully streamlined body with a right-hand floor gearchange and a maximum of more than 100 mph in third, 120 plus in top.

However, it was not long before this car gave away to an automatic version with a different body. Manual transmission was still available to special order, though.

Finally, the V8 version is now sold in its most expensive form as the Continental.   #

The interior of the Bentley is superbly fitted. The big steering wheel is traditional, but controls for winkers and shocker settings are on column.

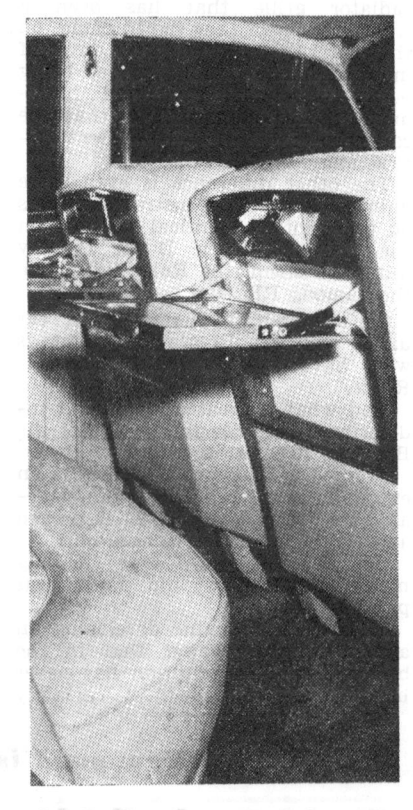

The Bentley passenger lacks nothing. Picnic trays, ash trays, cigar lighters are all there. Upholstery is fine leather. Cappings are wood.

wheels ROAD TEST

TECHNICAL DETAILS OF THE BENTLEY V8

PERFORMANCE

TOP SPEED:
Fastest run .... 115.3 mph (see text)

MAXIMUM SPEED IN GEARS:
Second .... 57 mph
Third .... 79 mph
Fourth .... 115.3 mph

ACCELERATION:
Standing Quarter Mile:
Fastest run .... 18.7 sec
Average .... 18.7 sec
0 to 30 mph .... 3.7 sec
0 to 40 mph .... 6 sec
0 to 50 mph .... 8.6 sec
0 to 60 mph .... 11.8 sec
0 to 70 mph .... 15.3 sec
0 to 80 mph .... 19.9 sec
20 to 40 mph .... 3.7 sec
40 to 60 mph .... 5.6 sec

SPEEDO ERROR:
Indicated   *actual*
To 80 mph .... accurate
101 mph .... 100 mph
117 mph .... 115 mph

GO-TO-WHOA:
0 to 60 mph to 0 .... 16.8 sec

FUEL CONSUMPTION:
Overall for complete test .... 13.4 mpg

SPECIFICATIONS

ENGINE:
Cylinders .... eight, V pattern
Bore and stroke .... 104 by 91.5 mm
Cubic capacity .... 6230 cc
Compression ratio .... 8 to 1
Valves .... pushrod overhead
Carburettors .... two SU
Power at rpm .... na
Maximum torque .... na

TRANSMISSION:
Type .... Hydra-Matic/Rolls-Royce, four speeds
Ratios .... first 11.75, second 8.1, third 4.46, top 3.08
Rear axle .... Hypoid spiral

SUSPENSION:
Front .... independent by coils
Rear .... semi-elliptics
Shockers .... manually controllable hydraulic

STEERING:
Type .... Cam and roller, power assisted
Turns, 1 to 1 .... 4¼
Circle .... 41 ft 8 in

BRAKES:
Type .... drums, hydraulic/mechanical operation

DIMENSIONS:
Wheelbase .... 10 ft 3 in
Track, front .... 4 ft 10½ in
Track, rear .... 5 ft
Length .... 17 ft 7 in
Width .... 6 ft 2 in
Height .... 5 ft 4 in

TYRES:
Size .... 8.20 by 15

WEIGHT:
Dry .... 39 cwt

147

A CAR ROAD TEST

BENTLEY S2 L.W.B. Saloon

CAR *expresses its thanks and appreciation to the Administrator of the Cape, the Hon. J. N. Malan, who kindly granted permission to this journal's staff to conduct this road test of his official car, PA-1.*

(*Right*) "The second best car in the world?" ... the Bentley is identical to the current Rolls-Royce excepting for the shape of the radiator shell, and other minor features.

(*Below*) In spite of the speedometer reading, photographing the instrument panel from the rear seat during the fuel consumption test presented no problem!

SOME 40 YEARS HAVE PASSED since the first Bently became available to the public. In all this time and in spite of its changes of character — whether as a Le Mans-winner of the Twenties, or as "The Silent Sports Car" in the years that followed Rolls-Royce's take-over of the original Bentley concern, or, in more recent times, as the epitome of luxury motoring in the modern manner — the name "Bentley" has remained amongst the most famous in the world. And, by being identical with the Rolls-Royce, but for a different radiator grille, name badge and instrument lettering, the car today shares with R-R the distinction of being the best in the world — a claim, which to our knowledge, has never been challenged.

The car tested was one of the most recent S2 models in long wheelbase form, longer, that is, by four inches by comparison with the standard S2. Features of similarity between the two models include the same 6·23-litre V8 aluminium alloy engine, automatic gearbox, power brakes, power steering and suspension system; features which distinguish this l.w.b. saloon from standard comprise adjustable rear quarter lights, an electrically operated glass division between front and rear passenger compartments, a heating, ventilation and refrigeration system and an intercom. between passengers and driver. The range of instruments and controls (centrally located on the facia) is complete — excepting a rev. counter — the usual range on most high-quality cars being supplemented by such very non-standard items as a switch to open electrically the cap over the rear fuel filler nozzle, an oil level indicator (the level in the sump being read after pushing a button on the facia) and a switch to vary the rear shock absorber settings between "Normal" and "Hard".

The lavish interior appointments are in the best possible taste. Deep pile carpets cover the floor of the rear passenger compartment; the seats are finished in the finest Connolly leather; door and division cappings, facia and picnic tables are of highly polished burr walnut — the whole combining to convey initially a feeling of complete comfort and security. Confronting the rear seat passengers (who have more-than-ample legroom) are the two folding picnic tables surmounted by grab handles. Between these tables is a housing containing the radio loudspeaker-cum-microphone unit. Instructions to the driver are given by speaking into the loudspeaker after turning a control switch on the panel above it to the right; the driver's acknowledgment can be heard with the switch in the central position and when it is turned to the left, the radio can be listened to. On the top of the housing are controls to raise or lower the glass division panel and to operate the radio and refrigeration (air-conditioning) system, a cigar lighter also being provided. Rather surprisingly, the small rear ashtray is somewhat inaccessible being centrally located above the loudspeaker housing at more than-arm's length distance from the passengers.

Externally no quieter than the average large V8 unit, the engine is virtually inaudible from the interior once the car is on the move and can, in fact, only really be heard — and that as a comforting hum — when the accelerator is completely depressed for rapid starts from standstill. The driver enjoys first-class outward visibility with all controls well-positioned; the size of the brake pedal, however, could be increased. Visibility for the passengers at the rear is also excellent to the side but rather less so in a forward direction: such is the comfort of the seats that one sinks down into them to find the top of the division at about eye level.

This feature, however, in no way detracts from the feeling of security which — as already mentioned — is experienced the moment the car is entered. With the glass division raised it was found that at 90 m.p.h. the only sound which could literally be heard in the rear compartment was a gentle rumbling from beneath the car.

Interior silence was to some extent marred when the extremely efficient refrigeration unit was in operation. This unit is operated by a four position rotary switch — "Off", "Low", "Medium" and "High". At "Low" and "Medium" speeds, a gentle whistle emanated from the two outlets from which cooled air enters the interior but at "High" speed a whine was heard. Although not loud enough to become annoying, this whine is out of keeping with the otherwise restfulness of high-speed travel in the Bentley: dare we say it? but American car air-conditioning units which we have sampled are a lot quieter in operation! (In cold weather, warm dry air can be provided by operating the heater in conjunction with the refrigeration system.)

Slight irregularities on the road surface are crossed without being noticed and over rather more pronounced bumps the suspension "gives" and "hardens" without in any way disturbing the level ride. Over one really bad railway crossing and that at speed, the car did however "bottom". The shock absorber setting at speed, we found, was better left in the "Hard" position.

Especial mention must be made of the extraordinarily good braking system: as is known, the brakes are of the R-R drum type with gearbox-driven servo assistance; pressure on the brake pedal actuates the servo to operate both front and rear brakes by hydraulic pressure and, in addition, the pedal is connected to the rear brakes by mechanical linkage. Repeated stops from high speeds could induce no fade whatsoever; in spite of its weight, the car pulled up again and again in a straight line as surely as it had done when the brakes were cold. In this aspect of its performance, the Bentley is streets ahead of any comparable American car.

The S2 is superbly easy to drive both at slow speeds in heavy traffic or when the car is being manoeuvred into a parking space. The steering remains light under the latter conditions and provided the engine is running the steering wheel can readily be turned from lock-to-lock even when the car is halted. Cornering hard is accompanied by a squealing of tyres and the body itself leans but in those conditions under which a car of this nature would normally be driven, turns in the road can nonetheless be quickly negotiated without in any way disturbing the serene ride. The turning circle of 43 feet — although rather large by comparison

A luxurious interior: the appointments include a central ash-tray flanked on either side by passenger grab handles. Below these are folding picnic tables. The central housing incorporates the radio loudspeaker-cum-intercom-microphone, the intercom. operating switch being immediately above. Controls on top of the housing operate the air-conditioning unit, the glass division between rear and front compartments, the cigar lighter, the radio and the heating unit.

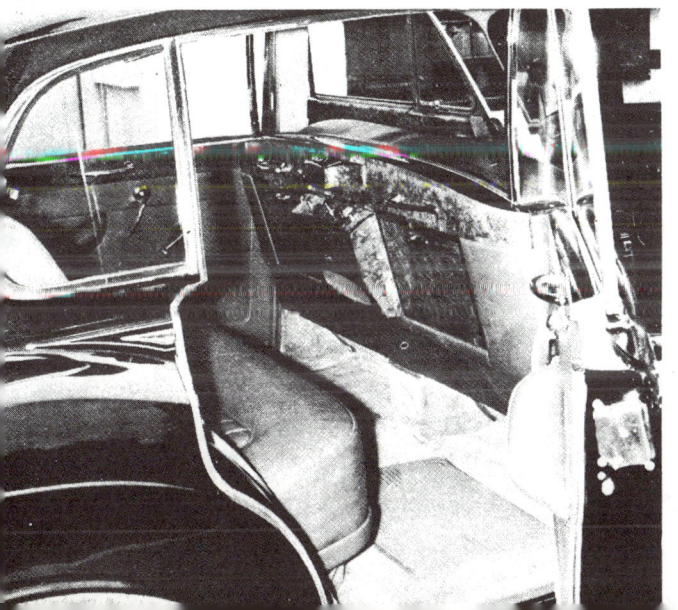

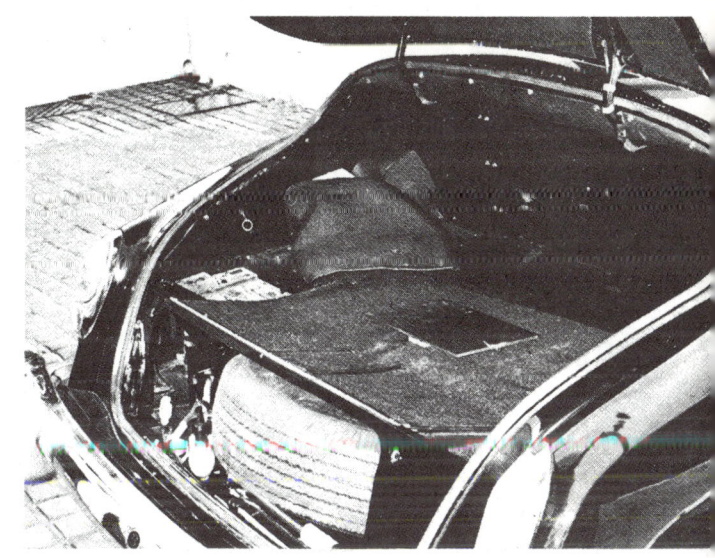

The luggage compartment mat has been folded back in this photograph to reveal the fitted tray of small tools, larger tools being located in the spare wheel compartment. The refrigeration unit is installed behind the rear seats, access to it being gained through the boot. (The ring visible above the tool tray is at the end of a pull cable which releases the locking catch of the fuel filler door should electrical operation of the door, from the instrument panel, fail).

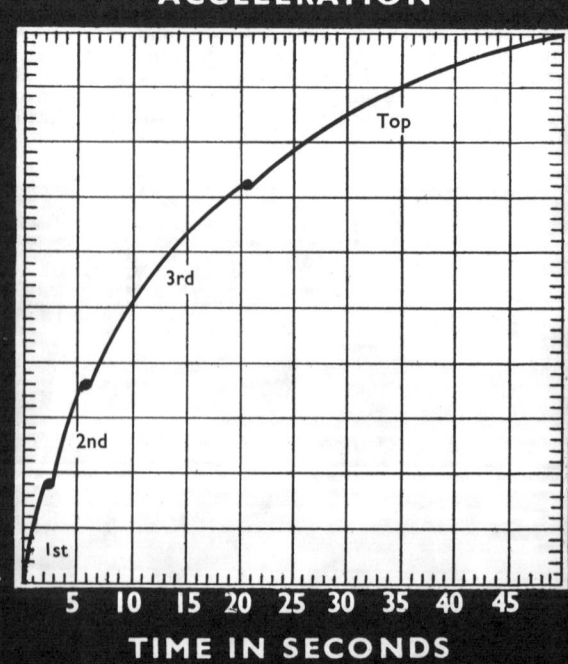

BENTLEY S2 L.W.B. Saloon

SPECIFICATION

ENGINE: V8-cylinder, water-cooled, o.h.v. (push-rod), front-mounted, twin S.U. HD6 carburetters, twin electrical S.U. fuel pumps, compression ratio, 8 to 1.
BORE AND STROKE: 104·14 x 91·44 mm. (4·1 x 3·6 ins.).
CUBIC CAPACITY: 6,230 c.c.
MAXIMUM HORSE-POWER: Not available.
MAXIMUM TORQUE: Not available.
TOP GEAR M.P.H. AT 1,000 R.P.M.: 28·2.
TOP GEAR M.P.H. AT 1,000 FT./MIN. PISTON SPEED: 47
PISTON SPEED AT 100 M.P.H.: 2,127 ft./min.
BRAKES: Power assisted by servo motor, twin hydraulic system with additional mechanical linkage to rear brakes, total lining area, 240 sq. ins.
SUSPENSION: (Front) Independent coil, with hydraulic shock absorbers and anti-roll stabiliser. (Rear) Semi-elliptic leaf springs with controllable hydraulic shock absorbers and axle control rod.
TRANSMISSION: Bentley automatic, incorporating four forward speeds and reverse with over-riding hand and "kick-down" change speed control.
OVERALL GEAR RATIOS: 1st, 11·75; 2nd, 8·1; 3rd, 4·46; Top, 3·08; Reverse, 13·25.
TYPE OF FINAL DRIVE AND RATIO: Hypoid, 3·08 to 1.
TYRE SIZE: 8·20 x 15. **WHEELBASE:** 127 ins. **TRACK:** (Front) 58·5 ins., (Rear) 60 ins.
GROUND CLEARANCE: 7 ins. **LENGTH:** 17 ft. 11·75 ins. **WIDTH:** 6 ft. 2·75 ins. **HEIGHT:** 5 ft. 4 ins.
STEERING: Hydraulic, power-assisted, cam and roller, 4¼ turns, lock-to-lock.
TURNING CIRCLE: 43 ft. **BOOT CAPACITY:** 12·5 cu. ft.
SUMP CAPACITY: 12 pints. **GEARBOX CAPACITY:** 20 pints. **FUEL TANK CAPACITY:** 18 gallons. **DIFFERENTIAL CAPACITY:** 1⅝ pints. **COOLING SYSTEM CAPACITY:** 21 pints.
KERB WEIGHT: 4,816 lbs. **WEIGHT AS TESTED:** 5,720 lbs.
BASIC PRICE IN THE UNITED KINGDOM: £5,140 (R10,280).
INTERIOR DIMENSIONS:
  Width of front seats: 54·5 ins.
  Driver's seat to brake pedal: 17·5 ins.
  *Front seat headroom: 4·5 ins.
  Width of rear seat: 54 ins.
  Rear seat kneeroom: 12·5 ins.
  (*Measurements taken with 6-ft. man seated, no hat.)

CHASSIS LUBRICATION: By grease gun to 21 points, every 10,000 miles.
TYRE PRESSURES: (Front) 22, (Rear) 28.
BATTERY: 12-volt, 67 amp. hr.

PERFORMANCE

(All times given are the mean of two opposite runs).

ACCELERATION THROUGH THE GEARS:

M.P.H.	Secs.	M.P.H.	Secs.
0—30	4·5	0—70	18·8
0—40	6·2	0—80	26·0
0—50	9·8	0—90	35·0
0—60	13·6	0—100	50·0

ACCELERATION IN TOP:

M.P.H.	Secs.	M.P.H.	Secs.
30—50	11·0	60—80	12·5
40—60	10·8	70—90	15·9
50—70	10·6	80—100	24·6

ACCELERATION IN THIRD:

M.P.H.	Secs.	M.P.H.	Secs.
20—40	6·4	40—60	6·8
30—50	6·7	50—70	8·0

MAXIMUM SPEEDS IN GEARS: 1st, 18 m.p.h.; 2nd, 36 m.p.h.; 3rd, 72 m.p.h.
MAXIMUM SPEED IN TOP: 105–110 m.p.h.
BRAKING EFFICIENCY: (Car in neutral at 30 m.p.h.) 95 per cent. equivalent to a stopping distance of 34 ft.
FUEL CONSUMPTION AT STEADY SPEEDS IN TOP:
  30 m.p.h.: 15·6 m.p.g.   70 m.p.h.: 14·7 m.p.g.
  40 m.p.h.: 15·9 m.p.g.   80 m.p.h.: 13·8 m.p.g.
  50 m.p.h.: 16·3 m.p.g.   90 m.p.h.: 11·7 m.p.g.
  60 m.p.h.: 16·0 m.p.g.
TEST CONDITIONS: Barometric pressure, 29·75 ins. of mercury; Altitude, 350 ft. above sea level; Air temperature, 72 deg. F.; Wind velocity and direction, 5 m.p.h. head-on; sunny dry tarmac, 93-octane fuel.
SPEEDOMETER CORRECTION:
  Indicated: 20 30 40 51 62 73 86 100 112
  Actual: 20 30 40 50 60 70 80   90 100
NATIONAL ROAD FUEL CONSUMPTION: (Over a total of 80 miles in opposite directions) 11·3 m.p.g. at an average of 67·75 m.p.h. (12 m.p.g. at 70·2 m.p.h. down and 10·6 m.p.g. at 65·3 m.p.h. up).
OVERALL FUEL CONSUMPTION (including 102 miles for performance figure compilation): 12·34 m.p.g.

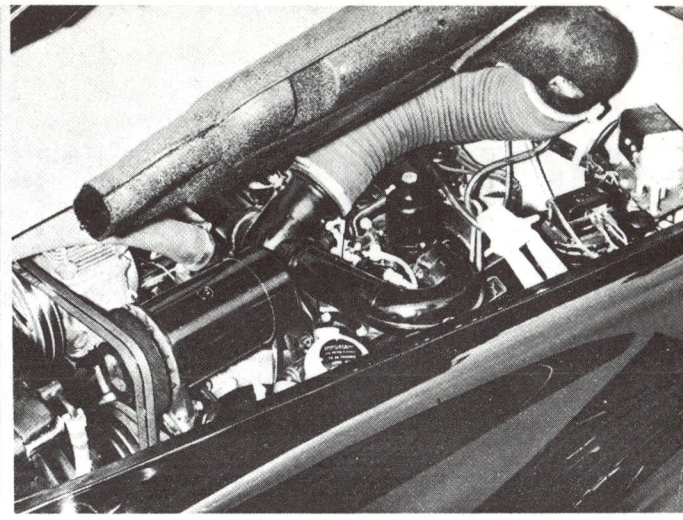

Hardly a cubic inch of space is wasted under the two centrally-hinged panels which are raised to reveal the engine compartment. The engine itself is virtually submerged under a host of ancillary pumps, motors and other devices connected with the air-conditioning, braking and power steering systems.

with the normal saloon car — allows the Bentley to turn full circle within the confines of the verges of a stretch of national road.

The automatic gearbox (a G.M. Hydramatic system built by Rolls-Royce under licence) provides the smoothest of upward changes: so smooth, in fact, that during the compilation of performance figures it was not easy to establish immediately when the changes-up took place. If the car is driven with a light foot on the throttle, the gear shifts take place at about 7 m.p.h. (1st to 2nd), 14 m.p.h. (2nd to 3rd) and 24 m.p.h. (3rd to Top); heavy throttle openings result in changes taking place as indicated in our performance tables. Changes-down, again with light throttle openings, take place at about 16 m.p.h., 10 m.p.h. and 5 m.p.h. Jerkiness in the automatic mechanism can only be induced with the accelerator pedal completely depressed.

The gearbox provides three forward ranges (marked on the control quadrant "4", "3" and "2") and Neutral and Reverse. In "4", 1st, 2nd, 3rd and Top ratios can be obtained; (in "3", 1st, 2nd and 3rd with a safety change to Top), and in "2", 1st gear by kick-down only and 2nd. In this last position, the engine can be over-revved if a speed of more than 50 m.p.h. is attempted, no safety change-up provision being made.

Although everything can be left to the gearbox to make changes at the theoretically correct moments, an over-riding control is provided to enable the driver himself to change gears should traffic conditions so require: it was found, for instance, that very smooth changes between "3" and "4" could be made by moving the selector lever from the 3rd position to the 4th. (The handbook provided encourages drivers to make the fullest use of this gear change as with non-automatic systems.)

In a car built to the highest specification standards, it was surprising to find that the doors did not close with that "click" usually associated with the custom-built product; some force was needed, also, to ensure that they closed securely. (The need for complete dust-proofing has made it impossible to retain the effortless closing of doors associated with the older type of Bentley and R-R coachwork which was far from dust-proof.) Another surprising fault in the car tested was that of a remarkably optimistic speedometer. Luggage accommodation is to some extent limited, part of the space available behind the rear seats being taken up by the refrigeration unit. Although this would not be a drawback in smaller countries of Europe, where distances to be covered are not great, the available space for baggage on this particular l.w.b. S2 has had to be supplemented by a roof rack made for the car by the Bentley Company. Thus, when the car is needed on official business in remote areas of the Cape Province, a full quota of luggage can be carried. The roof rack is, of course, removed when the car is in use in the cities.

That the car is not readily owner-maintained is evident from the need, for instance, to remove both front wheels and valance panels behind them to change sparking plugs: the driver's handbook, however, does provide most detailed and comprehensive instructions to aid the owner, pointing out, inter alia, that engine oil should be changed every 2,500 miles and fluid coupling-, gearbox-, steering transfer box- and rear axle-oil, every 20,000 miles. (It is of interest to record that the official driver, on first taking over the car, wrote to the Bentley Company, to establish the best cruising speed. The answer was "any speed desired".)

The presence of such minor faults as have been mentioned tends to become exaggerated only because they are found in so perfect a car. For there can be no doubt that in terms of sheer luxury of modern road transport, the Bentley (and, of course, the Rolls-Royce) has no equal. Its combination of virtues — silence at speed, comfort, superb engineering and the quite fantastic attention to detail which every corner of the car reveals — certainly confirm the saying that possession of a Rolls-Royce product advertises not the owner's wealth but only his discrimination.

(Readers are asked to note that with so unique a vehicle in this country, and with less than 1,200 miles covered when it was in our hands, no attempt was made to push the car to its limit either in regard to performance or handling. The engine speed graph, also, should not be read literally: as is widely known, maximum output and revs. are not disclosed by Bentley or R-R.) ●

PHOTOS: PETER BIRO

ROLLS ROYCE SILVER CLOUD

THE LAST ROLLS ROYCE WE TESTED was in 1953 — the so-called "baby" Rolls. While an exceptionally nice car, it gave us at least an inkling that the staff of this prestige English firm are, after all, human beings and not miracle workers. Our test this time bore this out. In other words, we didn't approach the test completely enthralled by the clock being the noisiest thing in the car and similar hocum. Matter of fact, the clock can't be heard with everything turned off and your ear next to it. Point; the Rolls is an automobile, not a state of mind, so we encountered no burst bubble and accompanying bitterness, as one writer did in a test for a non-automotive publication.

Conservative, naturally, is the word for the concept in both design and engineering of the Rolls. While they would perhaps like to keep it as classic as a Roman temple, it remains a product offered in a modern, competitive world and, where absolutely necessary, they have made concessions to this effect. Most certainly, the Rolls has been constructed with the solidity of a Roman temple.

"You aren't going to *speed* test the car, are you?" was the sales manager's hopeful question at British Motor Cars in San Francisco. We assured him we weren't interested in its performance. After all, you don't expect an old dowager to be a track star. By the time we'd travelled a few miles, however, it became increasingly apparent that the Silver Cloud, automatic transmission and all, has absolutely

Huge frontal area and louvered radiator have been Rolls trademarks for decades. High quality likewise remains unchanged.

nothing to be ashamed of in this category. There is power on demand and lots of it. Further, it is one of the most stable sedans above 100 mph that we've ever driven. Not only that, it's extremely roadable and can be four-wheel drifted through high-speed corners with amazing ease! And stopping power of the car is well able to cope with the velocity it easily attains. All this proved to be one of the biggest surprises we've ever encountered in a car test and, though most Rolls owners may never take the opportunity to barrel their own machines through a fast, mountainous circuit, they can rest assured that the car is more than capable of doing so. But let's return to the qualities one would expect in a Rolls.

While not appearing overly so, the four-door body is quite tall. We were still surprised to find ourselves sitting a good distance above the ground. The floor level is above the frame and both body and frame construction are as they were in days of yore. Riding in the Silver Cloud produces a very bus-like impression. This is augmented, for the driver, by the large, rather stark steering wheel. The windshield is relatively small by today's standards, as is the remainder of the glasswork (all of flawless quality, incidentally). The doors are quite high, but who cares when they're finished in beautiful leather and polished walnut? The front seat, semi-bucket affairs, are fully adjustable and extra large, making the very average-sized interior appear even smaller. Quite naturally, considerable attention has been paid to assuring rear-seat passengers of adequate footroom and the overall theme of the car is comfort, convenience, and luxury for those within. Examples: lighted vanity mirrors built into each rear quarter, small tables in the seatbacks and one under the right dash, adjustable armrests in both front doors, electrical ride adjustment (solenoid bypasses on the shocks), defroster built into the rear glass, and a blower-operated ventilation system. Optional are a bed, telephone, dictating machine, electric razor, and Expresso-maker. But, being peasants

(continued)

Impressive architecture in foreground and background; Rolls Royce Silver Cloud in front of San Francisco's Court House.

The grain of the walnut dash, above, is noted and kept on file by the factory so that any needed replacement in the future matches the original. Naturally, detail finish is flawless.

Though there's wide access to the luggage compartment, the spare tire well and gas tank make it a bit small in capacity by U. S. standards. Deck lid is locked by double latches. The tail lights are small for car's size but are very powerful.

Detailing some of the luxury appointments in Silver Cloud, the above photos — from top to bottom — show adjustable armrests in front doors, with storage pockets in the lower corners, the lighted vanity mirrors in the rear quarters, and the small tables built into the seatbacks; all standard items.

154

Compartment for big V-8 is quite crowded, and old-fashioned method of access makes it a bit rough to perform tune-ups.

ROLLS ROYCE

by nature, we only tested the "stripped" version. Surprisingly, the trunk space — though beautifully finished — was rather modest. Perhaps Rolls clientele buy new things when they get where they're going.

The mechanical aspects are conventional, well-made, and well-proven. Unequal arms and coil springs comprise the front suspension. The rear axle is live and connects through leaf springs to the chassis in the normal manner. The very large tires (8.20 x 15's) are tubeless, standard-profile Firestones. Brakes and steering are power-assisted and, along with the excellent 4-speed transmission, seem to be based on designs developed by General Motors. Needless to say, it takes a lot of binders to stop a 4700-pound machine, so 14.5-inch drums are assigned this chore.

We didn't run an actual mileage check but would guess that we were getting something around 15 mpg, judging by the way the gauge for the 21-gallon tank was dropping. When the reserve light warned us we were low, we started hunting a station that matched our credit card. Not finding one (there aren't many BP stations in California, Pete), we picked the first that came along, pulled in, pushed the dash button that flicked the gas cap open, and asked the rather suspicious attendant for a buck's worth.

The 6.2-liter V-8, whose output remains unannounced, starts and runs like a jewel. Well-silenced at both intake and exhaust, it's quiet, but not extraordinarily so. It is, most certainly, quite healthy and we'd guesstimate horsepower at 330 and torque to be in the neighborhood of 355 lbs./ft. Fed by two huge SU carburetors, it's conventional-looking, with pushrod/rocker-actuated overhead valves. The eight exhaust ports feed into cast headers and ultimately into one tailpipe. But you have to look through a lot of plumbing to uncover these facts as the engine compartment is very crowded. It's probable that they need every inch of that huge radiator, and accessibility isn't the greatest.

To sum up the Rolls Royce Silver Cloud, it seems worth its price to the buyer concerned with quality, comfort, and prestige. While certainly not sporty, it has first-rate roadability and performance. Its assignment as high-class transportation it fulfills to very near the ultimate and, while they may not be the most modern, Rolls is still establishing automotive standards. — *Jerry Titus*

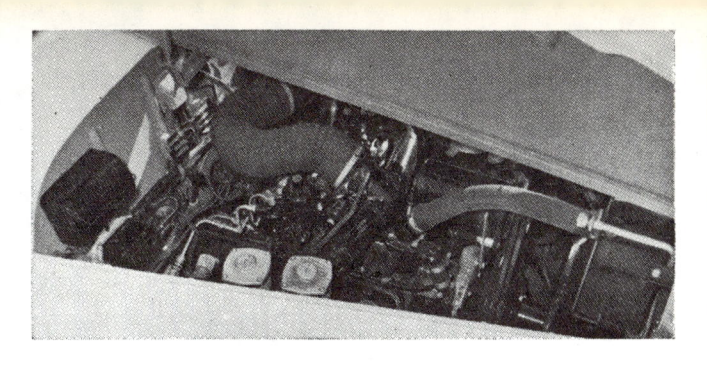

ROAD TEST 25/62
TEST DATA

| VEHICLE | Rolls Royce | PRICE (as tested) | $16,105 |
| MODEL | Silver Cloud | OPTIONS | Just about anything |

ENGINE:
- Type: V-8
- Head: .... Removable
- Valves: .... OHV pushrod, rocker
- Max. bhp .... N.A.
- Max. Torque .... N.A.
- Bore .... 4.1 in. 104 mm.
- Stroke .... 3.6 in. 91.5 mm.
- Displacement .... 380 cu. in. 6230 cc.
- Compression Ratio .... 8.0 to 1
- Induction System: .... 2 SU Carbs
- Exhaust System: .... Cast headers, 8 into 2 into 1
- Electrical System: .... 12V distrib. ign.

TRANSMISSION: 4 speed, Automatic, and manual selection
- Ratios: 1st .... 3.82 to 1
- 2nd .... 2.63 to 1
- 3rd .... 1.45 to 1
- 4th .... 1.0 to 1

DIFFERENTIAL:
- Ratio: .... 3.08 to 1
- Drive Axles (type): .... enclosed

STEERING: Power-assisted sector
- Turns Lock to Lock: 4.5
- Turn Circle: .... 37 ft.

BRAKES: Power-assisted
- Drum or Disc Diameter .... 14.5 in. x 2.5
- Swept Area .... 240 sq. in.

CHASSIS:
- Frame: .... Steel, channel, rails
- Body: .... Conventional steel
- Front Suspension: .... Unequal A's, coil springs
- Rear Suspension: .... Live, leaf springs, arm shocks
- Tire Size & Type: .... 8.20 x 15 Firestone tubeless

WEIGHTS AND MEASURES:
- Wheelbase: .... 123 in.
- Front Track: .... 58.5 in.
- Rear Track: .... 60 in.
- Overall Height .... 64 in.
- Overall Width .... 74.8 in.
- Overall Length .... 212 in.
- Ground Clearance .... 8.5 in.
- Curb Weight .... 4705 lbs.
- Test Weight .... 4910 lbs.
- Crankcase .... n.a. qts.
- Cooling System .... n.a. qts.
- Gas Tank .... 21.6 gals.

PERFORMANCE:
- 0-30 .... 3.4 sec.
- 0-40 .... 5.1 sec.
- 0-50 .... 7.5 sec.
- 0-60 .... 10.0 sec.
- 0-70 .... 14.5 sec.
- 0-80 .... 18.6 sec.
- 0-90 .... 25.1 sec.
- 0-100 .... 32.5 sec.
- Standing ¼ mile 17.4 sec. @ 78 mph
- Top Speed (av. two-way run) 114 mph (over 120 indicated)
- Fuel Consumption:
- Test: .... 14 mpg (estimated)

More power and new frontal treatment for Mk. III models

A lower radiator and more sloping bonnet with modified wings to house four headlamps distinguish the Silver Cloud III (*right*) from its predecessor (*below*).

Rolls-Royce and Bentley

ROLLS-ROYCE are not given to introducing new designs at frequent intervals and the Silver Cloud II and Bentley S2 have been in production since September, 1959 when the 6¼-litre light alloy V-8 engine superseded the 4.9-litre six. The Rolls-Royce Silver Cloud III and the virtually identical Bentley S3, which are now announced, have slightly more powerful engines, modified power steering and a re-styled frontal appearance. The basic specification has altered very little but there are numerous changes in smaller items of equipment.

The height of the radiator has been reduced by about 1½ in. giving a sloping bonnet line which markedly improves forward vision. The previous long range headlights have been replaced by four sealed-beam lamps which have been blended extremely skilfully into the frontal appearance and which give a much wider and more even spread of light. Two foglights are still fitted but these no longer serve also as flashing indicators, the latter being combined with the new sidelights. One further recognition feature is a change to smaller, neater overriders on cars for the home market, export models retaining the old pattern.

More Power

Now that 98 octane petrol is becoming generally available in most parts of Europe, the compression ratio has been raised from 8:1 to 9:1 and 2-in. bore S.U. carburetters replace the former 1¾-in. models. Although the company never reveals the power output of its cars, these alterations have increased it by about 7%. Higher combustion loads have been met by an increase in gudgeon-pin diameter and by the use of a crankshaft with nitride-hardened journals; fuel consumption has also been improved, partly through the addition of a part-throttle suction advance to the centrifugal timing mechanism.

In accordance with changing modern standards, hydraulic power assistance to the Rolls-Royce cam and roller steering has been increased so that the effort required for parking and manoeuvring has been approximately halved.

Entirely separate seats, which replace the previous semi-bench type, have very high backs which are quickly adjustable for rake and give considerably more cornering support.

The body shell remains unchanged but a more upright rear squab has been moved back about 2 in. to increase legroom and modified armrests increase the effective width so that three passengers can be seated more comfortably. Entirely separate front seats with adjustable-rake backrests replace the previous semi-bench type. A number of other internal improvements which were made during the three-year production life of the S2 model have naturally been incorporated in the new cars. These include a revised heating and ventilating installation with separate heater matrices to allow independent control of the

156

fresh-air and re-circulatory systems, a separate fresh-air inlet, a headlamp flasher, a second loudspeaker in the back parcel shelf and footrests for the back seat passengers.

Rolls-Royce are often asked why they have not adopted such modern features as disc brakes and independent rear suspension. Their own complex and highly refined drum braking system uses two trailing shoes and linings which increase in coefficient of friction up to about 400° C. so that fade and unevenness are practically eliminated; even at higher temperatures the effects of a falling coefficient are minimized by the anti self-servo shoe arrangement. Such brakes would be impossibly heavy to use without the unique gearbox-driven mechanical servo which multiplies the driver's efforts by a factor of approximately 7, far more assistance than the usual vacuum device will give. Nearly every part of the brake system is duplicated for safety and the Rolls-Royce view is that when disc brake installations become as good, as quiet and as dependable as their own expensive system, they will use them.

Similarly with independent rear suspension. This 40-cwt. car does not have the same problems in achieving a good sprung/unsprung weight ratio that beset light cars and the rear wheels can be kept firmly on the ground with an orthodox but properly controlled layout of semi-elliptic springs and radius arms.

On the Track

Partly for security reasons which dictated a private venue and partly because they believe that the luxury and refinement of their cars are well known but their sporting qualities are underrated, Rolls-Royce laid on a demonstration of an unusual nature. They hired the Oulton Park circuit and placed at our disposal a new Rolls, a new Bentley and a Silver Cloud II for comparison. A soaking wet track and persistent drizzle made this test more rather than less illuminating and revealed an inherent balance which very soon enabled one to forget both the conditions and the value of the car. Very light steering and brakes need appreciably less effort than most cars of one-third the weight and the new rather higher seats give more lateral support than the old ones and provide a commanding view down the sloping bonnet.

Rolls-Royce engineers have been to considerable trouble to develop neutral steering characteristics in order to make maximum use of the cornering power of all four massive 8.20-15 tyres, a fact which was impressively obvious on the circuit. Even on the faster corners the driver would have liked more lateral support to resist a level of centrifugal force that made the tyres wail furiously on the wet surface; only a gentle progressive unwinding of the wheel was necessary to keep on line as full throttle was used out of bends. It was intriguing to find that one could play tricks with this dignified saloon that one would hesitate to use in many sports cars, like locking over hard with the brakes on before Lodge Corner to break away the front wheels and then using full throttle in third from the apex to bring the back round with barely a wriggle at the transition point. The 100 m.p.h. was indicated just before Knicker Brook on each circuit and the brakes were used hard on lap after lap without any change in characteristics.

Flashing indicator bulbs are now incorporated in the sidelamp housings and smaller overriders are used on all but export cars.

Although engine appearance has changed very little, the two S.U. carburetters are larger and a vacuum advance capsule has been added to the distributor.

Extra leg room and width have been found in the back by modifications to the seat, armrests and side rests.

157

JOHN BOLSTER tests

THE ROLLS-ROYCE SILVER CLOUD

It would be trite to remark that there is no other car like a Rolls-Royce. Sometimes, one dares to wonder if foreign makes have at last surpassed the Rolls, but to drive one is to realize that the great machine is held higher in public esteem than ever it was. Policemen are polite, taxi drivers are charming, and everybody is pleased to see you, beckoning you on as you sail majestically by. Best of all, more beautiful girls give you *that look* than if you were driving the reddest and most rakish Italian G.T. Truly, it is the magic of a name.

Part of the excellence of the Rolls comes from its traditional design, which is daringly old-fashioned. I have owned most of the earlier models and have been privileged to drive all the later ones, but although the machines are now much, much faster, there is otherwise a great similarity that has lasted down the years. No other make boasts such continuity. A Mini does not remind one of a "bull-nose" Morris Cowley, for example, but a modern Rolls-Royce constantly recalls the Twenty, the Phantom 2, and the Phantom 3, in its response to the controls and its way of going and stopping. If the manufacturers have been tempted to launch an ultra-modern design, they have resisted, for perfection in this field comes from decades of development of an existing type.

The Silver Cloud 3 differs from previous models in having a more powerful engine, with larger carburetters and a higher compression ratio. This is an over-square V8 of light-alloy construction with pushrod valve operation, and it is unusual these days in having timing pinions instead of a chain and sprockets. It is mounted in unit with a four-speed epicyclic automatic gearbox, which it drives through a plain fluid coupling without torque multiplication. An old-style chassis frame, with cruciform bracing, has helical springs in front with wishbones and a rigid rear axle on semi-elliptic springs. The latter are reinforced by a torque arm above the axle which also doubles as an anti-roll bar. It is placed to one side to absorb driving torque, no doubt with an anti-tramp function.

The steel four-door body is of traditional shape, a long bonnet being more important than a great deal of seating space or a cavernous luggage boot. The result is a car of commanding appearance and long, sweeping lines. If it had a short, stubby bonnet and a full-width body, the car would carry more people but it would not be a Rolls-Royce. The high seating position gives a splendid all-round view.

The quality of the interior furnishing is quite simply the best. The equipment is absolutely complete—you name it and this car has got it. A superb tool kit is built in, but the engine is not very accessible, particularly for plug-changing.

Ever since Rolls-Royce espoused front brakes in 1923, a gearbox-driven mechanical servo has taken all effort out of their application. This servo now works through a more modern linkage, which includes hydraulic operation and every imaginable safety precaution, but drums are still used. The brakes are as good as those on my late-lamented Phantom 2 of more than 30 years ago, which means that they will pull up a carriage weighing over two tons, and do it repeatedly with the minimum of fading. One doubts that a disc installation could do more.

The steering, too, is devoid of all effort, being power-assisted. The machine understeers at all times, and is consequently stable. Just once in a while, one feels tempted to correct its course when running on a straight road. This, I think, is simply because the control is so light and not due to any handling defect. Through long, open bends the big car takes up its attitude and corners fast with the wheel steady in the driver's hands. The comfort is good on respec-

158

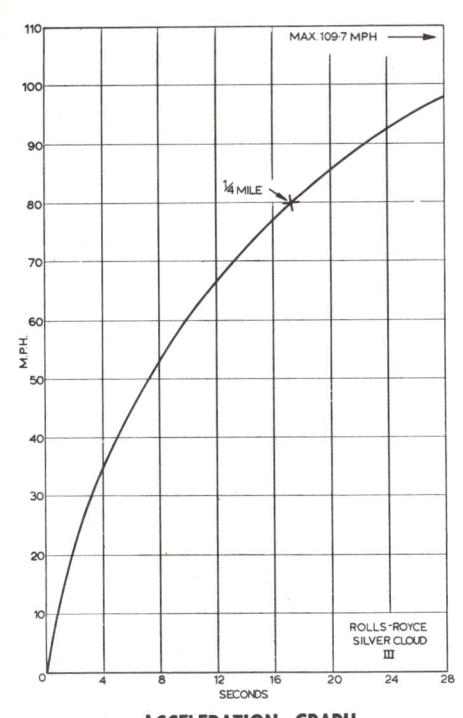

ACCELERATION GRAPH

SPECIFICATION AND PERFORMANCE DATA

Car Tested: Rolls-Royce Silver Cloud 3 saloon, price £5,516 12s. 1d. including P.T. Extras: Refrigeration, £332 5s. 10d.; Electrically operated windows, £84 11s. 8d.; seat belts £16 6s. 4d.; all include P.T.

Engine: Eight cylinders 104 mm. x 91.5 mm. (6,230 c.c.). Pushrod-operated overhead valves. Compression ratio 9 to 1. Twin SU carburetters. Lucas coil and distributor.

Transmission: Four-speed automatic gearbox with fluid coupling, ratios 3.08, 4.46, 8.10 and 11.75 to 1. Open propeller shaft. Hypoid rear axle.

Chassis: Box-section steel chassis frame with cruciform bracing. Independent front suspension by wishbones, helical springs and anti-roll bar. Power assisted cam and roller steering. Rear axle on semi-elliptic springs with single radius rod.

Equipment: Twelve-volt lighting and starting. Speedometer, ammeter, fuel and oil level gauge, water temperature gauge, oil pressure gauge, cigar lighter, clock, two-speed windscreen wipers and washers, heating, demisting, ventilation, and refrigeration system. Flashing direction indicators, electric window operation, radio.

Dimensions: Wheelbase, 10 ft. 3 ins.; track (front), 4 ft. 10¼ ins.; (rear), 5 ft.; overall length, 17 ft. 6¼ ins.; width, 6 ft. 2 ins.; weight, 2 tons 1 cwt.

Performance: Maximum speed: 109.7 m.p.h. (see text). Standing quarter-mile: 17.4 secs. Acceleration: 0-30 m.p.h., 3.2 s.; 0-50 m.p.h., 7.4 s.; 0-60 m.p.h., 9.9 s; 0-80 m.p.h. 17.4 s.

Fuel Consumption: 11 to 15 m.p.g.

table road surfaces for all the occupants, but over potholes the suspension can be heard working though the axle seldom really hops.

Though the engine is, perhaps, more audible to bystanders than was the power unit of the Silver Ghost, it is remarkably silent inside the car. It is just as quiet at maximum speed as at any other rate. I recorded a mean speed of 109.7 m.p.h., but other testers have achieved 115 m.p.h. and it is possible that a longer stretch of road would give this figure. In any case, the Rolls seems determined to cruise at 100 m.p.h. and will work up to this speed on quite a small throttle opening. The acceleration is most satisfactory for a two-ton luxury car.

The gearbox suits the engine well, with the one exception that there is a very large gap between second and third speeds. A kick-down change from third to second can cause a violent jerk, and a driver of sensibility would avoid doing this. The manual change overrides the automatic selection, and a smooth engagement of third or second can always be secured by this means. I would prefer a wider brake pedal, so that I could use either foot, as do most drivers who habitually handle automatics.

Absolutely splendid is the heating, ventilation, and air-conditioning system, with refrigeration. There are separate recirculating and fresh-air systems, which may be used together or alone, according to circumstances. One never needs to open the electrically operated windows, and the refrigerator is powerful enough to keep the car cool inside, even in very hot sunshine. The operation of the controls is logical and simple, so it is easy to turn off the fresh-air ducts when passing a diesel lorry, yet to keep the temperature at exactly the chosen value with the recirculatory arrangement. I am lucky indeed to have driven the Silver Cloud 3 in a heatwave, and I now realize that a vehicle without refrigeration is simply not a luxury car.

The Silver Cloud 3 is a very large car, but it has an excellent steering lock, which is a great help in crowded towns or narrow country lanes. Such a regal carriage may not be everybody's ideal, but it gives the most effortless motoring imaginable and is a real magic carpet. Having regard to the incomparable quality of construction and finish, plus the elaborate equipment, this Rolls-Royce cannot be regarded as an expensive car.

THE EQUIPMENT is absolutely complete (below, left). You name it and the car has got it—unlike many modern cars of today.

In the grand manner

ROLLS ROYCE

"The best car in the world" and Tony Brooks's run to the South of France

Tony Brooks squeezes the bulk of the Silver Cloud through a village street en route for the Cote d'Azur

MORE than a handful of our readers will be aware of the interesting suggestion, bruited of late, that "the best car in the world" today mounts the three-pointed star on its dummy filler cap. The recognised leader used to be made at Derby, of course, but apparently things began to slip with the move to Crewe. Times, with designs, specifications and tastes have changed, it is further argued. It has even been suggested that Rolls-Royce's interest in their car division was waning—a startling suggestion to anyone with any knowledge of the firm and of industry and business generally.

Some months ago an anonymous correspondent posed the question in this journal: "Is there a best car in the world?" And while his leanings, like those of the editor of the journal where most of this wind of change has been blowing rather keenly, were towards the products of Daimler-Benz—principally because of the Mercedes' technical specification—he seemed to doubt if such a thing as a "best" car exists today.

We have had all this before, of course. In the early days of the six-cylinder car, when S. F. Edge was beating the Napier drum so loudly, many good judges thought the Acton cars better than the corresponding Derby-built machines. A decade and more later it was the big Hispano-Suizas that some discriminating motorists put forward as the world's best cars. Certainly on technical specification they had a case; but the hard fact is that Napier, Hispano, Isotta, Duesenberg and Bugatti Royale have all left the stage to Rolls-Royce—and Mercedes-Benz.

The argument of our contemporary is not a very precise one. Mercedes is held up as a best *make*. Their proposition is not a clear-cut case for the so-called Grosser Mercedes 600 as the best, or even the most advanced car in the world today—which we think it is in several respects. They merely submit that the all-round excellence of the cars from Unterturkheim—presumably with design, performance, equipment, workmanship and finish all in mind—add up to a total that beats the corresponding offering of Crewe, or elsewhere. It might also be added that the German firm's range of models is comprehensive, commercially well contrived and of strong appeal. On this score, Crewe would admit that they are not in the hunt, just as they are well behind in terms of sheer output. Mercedes-Benz only lacks cars of 1½-litre and 1000/1200 cc to virtually complete the range; but they now have Auto-Union in the fold and these excellent little two-strokes are worthy representatives in their very competitive field.

Just the same, can anyone say that this or that vehicle is "the best car in the world?" It surely depends on what is required of a motorcar, and what yardstick is employed? Platitudinous maybe, but so often overlooked! There will always be argument, one supposes, and that is not a bad thing.

It happens that at the present time the Editor is running a 220 SE Merc. We have also had one of Crewe's cars—bought new a number of years ago and used for 40,000 pleasurable miles. It turned out to be one of the cheapest cars we had ever owned, too. So maybe we have a fairly sound and, we think, impartial position from which to look at this controversy. It is our honest opinion that there is no *best* car in the world; and the two examples of the two makes put forward as just that by their admirers have given us every satisfaction even if they haven't enabled us to resolve the question.

Putting the Silver Cloud III through its paces

Whether this country, Germany or the United States has the strongest claim to the title of producer of the world's best car might be too involved a matter for most of us; but our personal indecision did not prevent us from greatly admiring a clever piece of selling literature we recently received from Rolls-Royce Ltd. Entitled "A new look at Rolls-Royce motoring" it runs to 48 pages (13½ in by 7 in, unusually) and is produced in fitting style. Before we had opportunity to study it we had read a pretty lengthy notice of it in the "Merc before Rolls" contemporary we have mentioned. Maybe because of this it is more difficult than it otherwise would have been to read and review the booklet with detachment and objectivity. We hope we have succeeded.

The cardinal impression after a careful study of *A new look* is that this decidedly original piece of lush sales literature succeeds because it contains a lot (but not too much) of real information, some of it informed criticism of the car with convincing authoritative replies. In short, the firm has not offered the usual digest of the catalogue supported by reprints of road tests and a few complimentary quotes from motoring journalists and people of more distinction—if, perhaps, less motoring judgment. It is, in short, an intelligent sales aid that is being distributed to intelligent people who are potential customers.

That distinguished racing motorist and contributor to the press Tony Brooks was the central figure in giving Rolls-Royce motoring its new look, and excellently he did it. For a start he took a Silver Cloud III down to Goodwood and put it through its paces on a wet track in a most un-Roycelike fashion. This occupied a couple of hours and further testing on a dry circuit was carried out in the afternoon. The Klemantaski pictures which accompany this section (justifiably entitled "Taking it to the limit") are quite dramatic. After all four-wheel drifts in a Rolls, 100 mph braking of this large and heavy car, and throwing the machine through the chicane are not seen at Goodwood every day. Tony Brooks's comment on the day's testing says that the car ". . . came out of it with flying colours. Very few standard cars, which might handle well on the roads, are quite so at home on a circuit". The pictures gave our contemporary the thought that Brooks had spun the car but we cannot reach that conclusion from the published illustrations. We took the trouble to ask one of the people who watched the day's demonstration and he gave a firm "No" to the suggestion of Brooks "losing it". Standard tyres were used, we were told, but the driver raised pressures a few pounds all round.

739 miles in a day

Main feature of the book is an interesting account of a high-speed Continental journey made in the same car by Mr and Mrs Brooks, Louis Klemantaski and that necessary man of anonymity, an observer—plus fifteen pieces of luggage and equipment.

The heavily laden car left Le Touquet at approximately half-past seven one May morning last year and pulled up at the Grand Hotel du Cap Ferrat half an hour before midnight—739 miles in a day by one driver. Although the Silver Cloud III and its crew were on the road for exactly sixteen hours, almost two hours had been spent over a leisurely lunch somewhere between Troyes and Dijon. On a net time basis we make the average speed an impressive 52.8 mph, and on gross time it is 46.2 mph approximately. Unfortunately, the fuel consumption is not given in the brochure. The party took a well-used route to the Riviera, including passing through Lyons between five and six o'clock in the afternoon. They approached the Riviera by the Estorel-Côte d'Azur Autoroute where, it is recorded, the "speedo reached 120 mph on a downhill section".

In his comment on this remarkably fine performance, Tony Brooks says that they encountered heavy traffic for four hours after lunch and covered three hours after dark. They cruised on really rough roads at 90/100 mph and occasionally touched 105. The car took the bumps "extraordinarily well", and proved for Tony Brooks to be "the most comfortable car for long-distance luxury touring I have ever driven".

It might be said with truth that such a journey is precisely where a car like the Silver Cloud III would show up well. What about its behaviour on mountain passes, where lighter, shorter-wheelbase cars of comparable performance would be in their element? Tony Brooks seems to have thought of that on this trip to the Riviera for the next chapter in his story is a short illustrated account of some stimulated motoring in the Alpes Maritime before the car was turned for home.

Brooks chose a difficult circuit (especially for a big car) that is familiar to Monte Carlo Rally devotees. From Monte Carlo they climbed to La Turbie, thence to Laghet, la Trinité, Drap, L'Escarène, Col de Braus, Sospel, Casillon, Menton, and back to Monte Carlo. With a long-wheelbase heavy car, a fast ascent of the Col de Braus is quite a motoring feat but a fast descent (we know the pass well and have covered it with big cars) is even more of a test of car and driver. There are some thirty hairpins or difficult corners and the straights *are* short! Let the Rolls-Royce booklet now take up the story: "In many of these straights Tony Brooks took the car up to around 70 mph, braked, flung the car round, accelerating hard out of the bends. The two-ton car rushed down the 3,287 ft high Col almost continuously on left- or right-hand lock, taking a corner very fast every few seconds. At the bottom we pulled in to a garage for petrol. Smoke was pouring from the brake drums and the garage lad reached for a bucket of water. No need. In fact, the brakes had never grabbed nor shown any appreciable sign of fade. We motored on over the rest of the course."

The thirty or so words about the braking of the Rolls are exceptionally interesting, remembering that the car has drum brakes—albeit "with a difference", aided by a most effective servomotor. Our contemporary in its "Catalogue review" of the booklet quotes the phrase about the garage lad reaching for a bucket of water but there is no mention of the significant fact that the brakes were not given a cooling suice. Such omission, with what it implies to the journal's big readership, is unfair to the makers of the car and disconcertingly inconsistent with the deserved pat given the Rolls-Brooks combination a line or two earlier in the same notice. The car's handling on the descent of the Col had been very good, it had noted; and probably our contemporary shared with Brooks—and ourselves, we acknowledge—surprise at "the controllability of the car under such very difficult conditions."

After the run to the South of France—to get to and from the Monaco grand prix in this grand style—Brooks

Back home, the Rolls-Royce was put through its paces at Goodwood. These photographs were taken by Louis Klemantaski for 'A new look.

went to Crewe to discuss the performance and design of the Silver Cloud III with Mr S. H. Grylls, MA, the chief engineer of Rolls-Royce's car division. Their two-hour meeting must have been full and mutually satisfying for the *New Look* booklet tells us that its typed record runs to 14,000 words. A most interesting distillation is published, and if most of it deals with Mr Brooks's criticisms and Mr Grylls's answers to them, that can be taken as a fair indication that Rolls-Royce are confident that they "have the answers" for most affluent motorists, including the sporting ones, in their car.

We have not the space to deal adequately with this interesting three pages of the booklet and so we shall say no more about this question-and-answer feature. The contemporary referred to did the Rolls-Royce firm less than justice, however, by listing most of Tony Brooks's criticisms (including some of his minor ones) but omitting to give even a summary of Mr Grylls's reasoned and good replies. Instead, their sin of omission was excused as a desire "not to be thought biased in certain quarters". We think it well to leave the reader, and this very effective new look at Rolls-Royce sales promotion, on that lofty piece of self-deception. We—and perhaps others—must do some 1964 Silver Cloud motoring and report on it . . .

By way of postscript to the above, a few details of Tony Brooks's journey from Le Touquet to Cap Ferrat might be worth recording.

The Silver Cloud III, with four aboard, their luggage and heavy camera equipment, left the French airport at 7.27 one wet, grey-skied morning. "Cruising comfortably at speeds up to 100 mph", in the words of the booklet, Abbeville was reached at 8.04 and Amiens at 8.29, where the traffic was thicker than hitherto. Following the Aisne valley route, Soissons was entered at 9.27, which meant that 112 miles had been covered in two hours. Here a quarter-hour stop was made; and at 10.10 the car pulled up for petrol at Culochy-le-Chateau.

Three hours after leaving the coast, according to the interesting chart in the booklet, 161 miles had been put behind the Rolls-Royce; and we are told that 49 miles had been covered in the last hour. The roads were now dry and cruising at 90/100 mph was the order of things over the straight, undulating roads until the town of Sezanne was reached at 10.54. "Speedo frequently at 105 mph" is here noted on the chart.

Exactly four hours after the start of the trip, Sezanne was left, the attractive road from Troyes to Chatillon, down the Seine valley, being followed. Troyes was reached after five hours of comfortable motoring, with 278 miles on the odometer. The next section involved the climb over the Langres plateau, up to about 2,000 ft; and we are told that on this section Brooks was taking bends "comfortably at 90 mph" in the big car. Precisely at one o'clock, well earned aperitifs were taken in the garden of the Hostellerie de Val Suzon. It is even recorded that they were "vin blanc cassis", a discriminatory choice which suggests excellent Brooks/Klementaski accord. In the words of the car's famous chauffeur, at this first real stopping point: "We've covered some 300 miles before lunch—putting in about 55 miles every hour . . . but it has been a very comfortable run and effortless." The well earned luncheon stop, the chart unashamedly records, took two minutes under the two hours.

At 3 pm a second stop was made for petrol, just outside Dijon; and the next record on the chart is that at 3.58 pm 357 miles had passed under the well-shod wheels of the Silver Cloud. An hour later a further 61 miles had been covered and anyone who knows this well-used route will agree that this was no mean performance for a large, well-loaded luxury car.

The heavy traffic of Lyons was entered at 5.20 in the afternoon, which could hardly have been a worse time for a driver wishing to reach a distant destination as quickly as possible. Valence was the next point recorded on the chart and only 42 miles were covered in this hour, primarily because of the delaying Lyons traffic.

The next hour, including the crossing of the Isere, was better despite quite heavy traffic, for 54 miles were put into it. At ten past seven another stop was made for petrol at Livron and by the end of the next hour 557 miles had been covered. With darkness falling and a great deal of heavy lorry traffic on the road one might have thought that the average speed would fall, but 57 miles were covered in the next hour and 51 in the following one.

Petrol was taken on board at Cagnes and the chart says that 726 miles had been covered at two minutes short of 11 pm—61 of them in the last hour. About a quarter of an hour later Nice was passed through, and precisely 16 hours after leaving Le Touquet the Rolls-Royce pulled up at the Grand Hotel du Cap Ferrat. In 16 hours' gross time, therefore, 739 miles had been covered, a remarkable performance even by one of the world's acknowledged great drivers. On gross time the average speed worked out at 46.2 mph and on net time we make Brooks's speed to be 52.8 mph.

CONTINUED FROM PAGE 109

m.p.h. the machine ran straight without conscious direction from the driver, and the engine still refrained from joining in the conversation. A speedometer reading of 110 m.p.h. was occasionally seen. Perhaps the most endearing characteristic of the car is its remarkable liveliness, in top gear, between 65 and 80 m.p.h. This facilitates overtaking, and is another potent safety feature.

For many years, Rolls Royce and Bentley models have rejoiced in a superb braking system. In spite of adopting the automatic system of transmission, the makers have been able to retain their traditional gearbox-driven servo motor. This controls the whole of the braking force on the front wheels and just over half the effort on the rear wheels, by hydraulic means. About 40 per cent. of the rear braking is still taken from a mechanical hook-up direct to the pedal. In this way, "feel" is retained, and two fully independent braking systems, either of which is complete in itself, are in operation at all times. As there are also tandem master cylinders, with two separate fluid bottles, there is still further duplication for safety.

This method of application really works, and the heavy car can be pulled up from high speeds with never a thought of brake fade. It is in this respect that the Bentley completely dominates American cars of comparable size and speed. The brakes are just as powerful as those on last year's model, but a slight fierceness at very low speeds has now been completely eliminated.

The Bentley, "S" Series, is a very attractive big car with superb lines, appearance, and finish. It costs a lot of money, but to the connoisseur who must have the best, it is worth every penny of it. It is schemed to require the very minimum of attention over large mileages, but it has the famed Rolls-Royce service behind it when any attention is required. A little on the large side in crowded city streets, it becomes a remarkably easy car to handle when the open country is reached. Only a few of us can aspire to own such a magic carpet, but to drive it is to advance another step in one's motoring education. This will still be a glorious car in 45 years' time, just as my own 1911 Rolls-Royce is today.

The Bentley "S" – Specification and Performance

Car Tested: Bentley "S" Series 4-door sports saloon. Price £3,295 (£4,943 17s. with P.T.).

Engine: Six cylinders 95.25 mm. x 114.30 mm. (4,887 c.c.). Pushrod operated overhead inlet valves and side exhaust valves. Compression ratio 6.6 to 1. Twin SU carburetters. Lucas coil and distributor.

Transmission: Four-speed automatic gearbox, ratios 3.42, 4.96, 9.00 and 13.06 to 1. Open propeller shaft. Hypoid rear axle.

Chassis: Box section frame with cruciform bracing. Independent front suspension by wishbones and helical springs. Cam and roller steering box connected by transverse link to 3-piece track rod. Rear axle on semi-elliptic springs with combined torque resisting and anti-roll member. Piston-type dampers all round, with two-position electric control at rear. 8.20 x 15 ins. India tyres on five-stud disc wheels. Mechanical servo motor on gearbox operating front brakes entirely and rear brakes 60 per cent. hydraulically, in addition to direct mechanical 40 per cent. application of rear shoes, in 11 ins. x 3 ins. cast iron ribbed drums.

Dimensions: Wheelbase, 10 ft. 3 ins. Track, front, 4 ft. 10 ins; rear, 5 ft. Overall length, 17 ft. 7¼ ins.; width, 6 ft. 2¾ ins. Height, 5 ft. 4 ins. Turning circle, 41 ft. 8 ins. Weight, 1 ton 19 cwt.

Equipment: 12-volt lighting and starting. Speedometer. Ammeter. Oil pressure, water temperature, petrol and sump level gauges with warning light. Two-speed self-parking wipers. Flashing direction indicators. Heater and demister. Radio. Clock. Spotlights. Picnic tables. Ladies' companions.

Performance: Maximum speed, 102.3 m.p.h. Speeds in gears (approx.), 3rd 67 m.p.h., 2nd 31 m.p.h., 1st 20 m.p.h. Standing quarter-mile, 18.8 secs. Acceleration: 0-30 m.p.h., 3.8 secs.; 0-50 m.p.h., 8.4 secs.; 0-60 m.p.h., 11.6 secs.; 0-80 m.p.h., 20 secs.

Fuel Consumption: Driven hard, 15 m.p.g.

Silver Cloud III

by TONY BROOKS

TONY Brooks, former dentist who drove in turn for Fraser-Nash, Aston Martin, Connaught, B.R.M., Vanwall, Maserati and Ferrari, is now the owner of a garage in Britain specializing in quality cars, and has started a career as motoring journalist and commentator.

Recently Rolls-Royce invited him to do a test of the V8, 6,230-c.c. Silver Cloud III to assess it at his own discretion. He chose to try it on track, cross-country and in the mountains, and his findings were printed in the form of an attractive brochure by Rolls-Royce.

Here is a summary of his impressions as he told them:

TEST 1: Taking it "to the limit" on Goodwood Circuit:

My first impression was one of the car being extremely light to handle in relation to its size, and I was very surprised at its controllability. You could do more or less what you wanted with it and there was no

ROAD IMPRESSIONS

tendency to oversteer or understeer. It had absolutely neutral characteristics . . . It always slid bodily, and it could actually be drifted quite easily in the wet if you wanted to. The car felt extremely safe at all times. I drove the car through the corners as hard as I could and the roll for this class of car was very low indeed.

I was particularly surprised at the way the car could be thrown around and yet behave in such a gentlemanly way. The impression was not of a large heavy car at all.

The braking was very even, particularly in the wet. With four up we stopped at from 90 to 100 m.p.h. as often as we possibly could on two successive circuits of the track. A very exacting and hard brake test, and the brakes stood up extremely well, with only a very slight tendency to fade. Otherwise they were first-class.

The steering of the car was a little too low-geared for my taste but this didn't present any real problem at Goodwood other than at the chicane, where it was necessary to cross my arms to get through in a really fast manner.

Acceleration is extremely good. You can leave black marks on the road from a standing start, and with automatic transmission that is really something. The acceleration is good up through the gears. This is noticeable even at 90 m.p.h.: when you throw your foot right down, the acceleration is still impressive.

The track test was a very hard test for a car of this class, and the Rolls-Royce came out of it with flying colours. Very few standard cars, which might handle well on the roads, are quite so at home on a circuit.

TEST 2: 739 miles across France in one day.

Nearly 14 hours of motoring. We covered roughly 740 miles, and I can honestly say that I was good for many more miles of motoring yet. I've never arrived at the end of a similar journey less tired — even a journey 200 miles shorter — and we haven't had an easy run. We had rough and then fairly twisty roads early in the journey; heavy traffic for nearly four hours after lunch and finally about three hours of night driving.

We are still pretty fresh. It has been effortless high-speed travel, and this aspect of the car impresses me enormously. On some very rough roads we've been able to cruise at 90 and 100 m.p.h. — and even at 105 here and there — and the car has taken the bumps extraordinarily well considering how severe they were.

It is the most comfortable car for long-distance luxury touring that I have ever driven.

The steering is very impressive. It is extremely light at all times, with no variation in the amount of effort that has to be used, and yet it retains plenty of feel. Many power steering systems are effortless, but you just lose all sense of contact with the road, which of course can be very dangerous in slippery conditions. This can't happen with a Rolls-Royce. You always place the car precisely, and it will never do anything unexpected. It is certainly the best power steering I have ever tried. The car really digs in on the corners. I have not sampled better roadholding in a luxury saloon.

The Rolls-Royce is as effortless to drive in traffic as a luxury saloon of this size can possibly be.

What impressed me was the way the car achieved the higher speeds — it was so quiet and effortless. It went up to 100 m.p.h. very quickly, and if you gave it half a chance it soon went up to 115 m.p.h. With any car, however, particularly on the Continent, the absolute top speed is less important than that the cruising speed should be effortless. During the whole of the journey down our cruising speed was governed purely by the conditions on the road, not by the cruising limit of the car. We never at any time felt the car didn't have sufficient performance.

At one point, we reached 120 m.p.h. on a slightly downhill section.

The night section was impressive in that the absence of other cars on the roads in the still of the night emphasized the properties of silence in the car. You felt you were floating along as if on a magic carpet. A little wind and tyre noise was all that could be heard. We covered a lot of ground very quickly and easily.

TEST 3: To the limit in the Alps Maritimes.

The response of the steering was extremely good between left and right handers which often followed in quick succession within a matter of seconds of each other, but I would prefer higher-geared steering for this sort of motoring. Despite the tight hairpins, it was quite easy to get the car round even if you approached them very fast and got a little bit of understeer. On many cars of this size one would have found it very difficult.

The overriding manual control of the gearbox was extremely useful under these conditions. I could play between 3rd and 2nd all the way up the cols; I could keep it in third and change down to 2nd for the really tight hairpins.

On these tight hairpins the car handled very well. Despite its size and weight it was possible to throw it round the corners without any difficulty.

The car gave the driver every confidence with never any impression that it was going to do something unpredictable or let one down. The way it dug in on those turns was most impressive.

My overall impression was one of a vehicle which almost thrived on such conditions. The harder one threw it around, the more it seemed to enjoy it. I was surprised by the controllability of the car under such very difficult conditions.

TONY BROOKS SUMMED UP HIS IMPRESSIONS IN THESE WORDS:

The pleasure that I found in driving the Silver Cloud came from a balance of qualities that is unique in my experience. The result is that the car gives you high-performance motoring as near effortless as it can be under all the varied road conditions experienced on test.

The surprising thing to me is how well adapted the car is to modern traffic conditions. Indeed it tends to solve many contemporary motoring problems. Motoring of this kind cannot be anything but expensive, but I would say that in terms of sheer motoring pleasure, safety and durability, the Rolls-Royce is excellent vale for £5,500.

As a big, luxurious car that can nevertheless be driven in a highly sporting manner, there is nothing quite like it.

(FOOTNOTE: At the end of 2,700 miles and two weeks of enthusiastic treatment by this former Grand Prix driver, the Silver Cloud had a routine service which included renewing tyres and relining brakes, and was then driven to Munich to be used as VIP transport before being given to various motoring journals for road testing.)

Not So Much a Motor Car More a Way Of Life

The Editor's Account of Driving a Rolls-Royce Silver Cloud III

A PLEASURE that had been denied me for a long time materialised last July, when I was able to road test the Rolls-Royce Silver Cloud III. It was not possible, as I had originally hoped, to take this much-discussed car on a Continental journey but it is possible to report on a long test carried out in this country with this £5,632, 4,640 lb., V8 Rolls-Royce.

In General

Some journalists work on the assumption that well-deserved superlatives are necessary when writing of the modern Rolls-Royce and concern themselves with how the public react to the car, how lorry drivers cram on their brakes to give it a clear run at roundabouts, how hotel-porters raise their arms in salute and hotel-managers their tariffs as the classic radiator bows to rest outside their portals, and so on. It is true that a recent-model Rolls-Royce is a status symbol but, in my opinion, such comments are out of place in a report of this kind, if only for the fact that any other costly object, be it suit or mansion, camera or private 'plane or whatever, is apt to have this effect, even in a country under the control of Socialists. I prefer to remark that possession of such a car eases many of life's anxieties, and let it go at that.

The fact, nevertheless, remains that ownership of a Rolls-Royce is a way of life and the price one has to pay for it and its mode of travelling and performance should be related to this fact. I will confess here and now that there exist cars which I would prefer to drive when in a hurry, but, equally, I can say in all sincerity that there are few, if any, other makes which match the tireless, dignified, fast travel offered by the current car from Crewe.

In the first place, there is the extremely restful and very impressive quietness of Rolls-Royce motoring. It has been suggested that certain American automobiles approach closely to the ideal of completely silent running. I think the truth is that the *mechanical* quietness of the Silver Cloud is such that slight sounds like the rumble of the big Dunlops, something loose in the cubby-hole, the faintest creak from the bodywork or a slight air disturbance round a ventilator intake intrude, noises which in lesser cars are absorbed by the hiss of air through the carburetters, a greater degree of tyre noise, and so on, which merely make them *seem* quieter, when in fact their occupants endure a higher level of sound than those in a Rolls-Royce. For the Rolls-Royce is *generally* quiet; if 3rd gear is held in there is a slight increase in noise from the power unit as engine speed rises; acceleration, even under kick-down, passes virtually unnoticed from the aspect of an increase in sound level, providing the windows are kept closed, as they are intended to be, in the refrigerated version of the car. Let us be honest, however, and admit that when starting up from cold the tappets protest mildly for a while and that, if when idling the big engine is so unobtrusive you can hear the ticking of the

The Rolls-Royce luggage boot is not very high but its capacity is deceptive and lots of luggage can be got in. Moreover, the spare wheel can be removed without disturbing the contents of the compartment above.

clock, it does transmit a certain amount of vibration through the steering-wheel, and that in action there is quite a whirr as the automatic transmission gets out of low gear. Unexpectedly, too, the well-sealed doors, not easy to close, require to be slammed, which results in a loud metallic noise. On the other hand, the steel saloon body of this hard-used specimen of Silver Cloud III had but one minor rattle, audible over bad road surfaces, and possibly emanating from the n/s. safety-belt stowage.

Then the brakes! Deceptively powerful yet very light to apply, they have the additional merit of being notably progressive and they are entirely devoid of scream or screech. In normal driving the slight lag when they are applied at a crawl is scarcely noticeable, being less apparent than that experienced with certain vacuum-servo layouts. But these brakes do get quite surprisingly hot, producing an obnoxious smell of near-charred linings after a spell of normal driving, for the ribbed cast-iron drums have to dissipate a great deal of heat to bring a couple of tons smoothly to rest and are almost completely shielded from the air stream by the wheel discs and the adequate mudguarding. Nevertheless, these are brakes, the smooth purposeful fade-free action of which is entirely in keeping with the character of the car they retard and they gave me complete assurance when driving fast, as well they might, for their duplicated hydraulic system with two master cylinders and combined hydraulic and mechanical rear brake actuation was designed with safety in mind.

The cam-and-roller hydraulic power-steering is about the best I have experienced, sensitive, accurate, acceptably light, although not possessing quite that one-finger lightness of those too-sensitive power steering systems which emanate from the other side of the Atlantic. If any criticism is merited it is that the gearing is too low, 4¼ turns lock-to-lock calling for considerable arm work in traffic or when taking right-angled corners, although it has to be conceded that the turning circle of 41 ft. 8 in. is conveniently small. The steering wheel transmits shake-back only when held on right lock and there is useful castor return action. On the test car one or two inches of free play at the wheel was apparent. The Rolls-Royce was unexpectedly easy to park—and to drive.

The suspension naturally provides a pretty good ride and has the traditional hard/soft control of the rear damping. The latter is worth having, although I did not find the variation in stiffness of the rear shock-absorbers very pronounced, and missed the former progressive action of this control. This is, however, far from scientific suspension, for some quite emphatic movements reach the Silver Cloud's occupants on bad roads; there is even some shake transmitted to the steering wheel and scuttle.

The Silver Cloud does not handle like a GT car, nor would one expect it to. I found the suspension unduly supple and when cornering fast strong understeer changes to quite considerable roll as the heavy tail comes round. However, when the initial disappointment had given way to more mature experience, I found that the modern Rolls-Royce can, in fact, be cornered rapidly, with considerable enjoyment. More skilled drivers than I have forced the Rolls-Royce round a race circuit and proclaimed it surprisingly controllable and driving more cautiously (after all, I was on a public road) I found the Silver Cloud predictable and quite an inspiring car in which to press on. This notwithstanding, the suspension system—rigid back axle on semi-elliptic leaf springs with a single radius rod, i.f.s. by unequal length wishbones and coil springs with torsional anti-roll bar—by no means completely disguises the weight and comfort of a hurrying Rolls-Royce. I have at times excused cars with supple suspension by observing that providing they possess good acceleration and effective brakes there is no necessity to rush one's corners. This applies to the Rolls-Royce, although if driven thus a little more low-speed pick-up, even in kick-down, would be welcome, or so it seemed, without putting a watch on the speedometer.

Having said this it is only fair to remark on the fluid manner in which the Silver Cloud III deals with twists and turns and traffic hazards, which derives from a smooth, inaudible, flow of (unpublished) power from the light-alloy 90° 104.14 × 91.4 mm. (6,230 c.c.) V8 engine and the nonchalant arresting power of the now unique (although once well-known) gearbox-driven servo braking system (but for how much longer will it survive, even at Crewe?). Indeed, I rate the Rolls-Royce Silver Cloud III as at its best in heavy traffic, when visibility from the lofty driving seat contributes to the pleasure one enjoys in conducting this superb motor car.

The Rolls-Royce is a large car, measuring 17 ft. 6¼ in. × 6 ft. 2¾ in. × 5 ft. 6 in. high, but to the person fortunate enough to be behind its wheel it feels of medium size—one of the greatest compliments I can pay it and qualifying the foregoing comments on traffic negotiation. The luggage boot, deliberately contrived so that the spare wheel can be withdrawn without disturbing the contents, is deep but low. I cannot think how you would get several large cabin trunks into it, but I do know that five big, heavy suitcases, a sizeable soft-bag, tennis rackets, an oddments box, many coats, a pair of gum-boots and a mass of odds and ends went in eventually, after some re-arrangement. This load made absolutely no difference to the car's handling....

The standard saloon body seats five comfortably, but is not intended to be packed with humanity. Six slim people, perhaps, but seven would be uncomfortably intimate. You don't pack people into the Rolls-Royce as if it were an Underground train...!

I have implied that the Silver Cloud's suspension system is somewhat antiquated and I must confess I have experienced smoother automatic gearboxes, although the Chief Engineer of Rolls-Royce explains the quite vicious kick you experience if open throttles catch the gearbox out before it is in low gear by saying that this is the biggest unsolved problem confronting users of this particular form of automatic transmission. Apart from this peccadillo, it is a smooth, silent gearbox controlled by a lever working in a notably simple quadrant, and the " hold " available in 2nd and 3rd gears makes it nearly the equal, performance-wise, of a manual-change gearbox for enthusiastic drivers. In fact, it becomes customary to select 3rd gear for maximum acceleration. The lever, however, is not best-contrived for operation as an ordinary gear lever. The gap between 2nd and 3rd is too wide, however, causing a falling-off in acceleration if the change-up takes place on a hill.

The 1965 Rolls-Royce gives very adequate but hardly sensational performance. Fully laden with six human beings and their luggage it went to an indicated 85 m.p.h. along the average length of straight road. A stretch of dual-carriageway enabled it to exceed 90 m.p.h., but a Motorway was required before 100 m.p.h. would come up—*but it cannot be too strongly emphasised that at three-figure speeds, to a maximum of 114 m.p.h., the quietness level remains virtually the same as it is at 30 m.p.h., so that the volume of the radio does not need to be increased.* This is an extremely impressive factor which justifies the considerable cost of the Rolls-Royce. Qualifying the theory that Rolls-Royce ownership is a way of life is the fact, normally apparent to any observant person who has seen the car being built at Crewe, that not only is this machinery built to endure for a very long time, but is mechanism which will not quickly get out of adjustment. Continual full-throttle cruising or repeated heavy application of the brakes should have no detrimental effects on the Silver Cloud, and over the years a Rolls-Royce can be expected to behave the same and impart the same " feel " as it did when it was a brand-new car.

STATUS SYMBOLS

This longevity factor was emphasised by the test car itself, because it was a 2½-year-old Series SCX Silver Cloud III which had nearly 50,000 hard miles of demonstration behind it, including many fast journeys to the South of France, to Spain, to Germany, etc., to its credit before I had the chance to drive it. Rolls-Royce expected me to be critical but did not deem it necessary to produce a cosseted new car for the MOTOR SPORT Road Test. This alone makes such a car, which in character but not in timed performance possesses much the same indefinable appeal as the older Rolls-Royce models, seem well worth its high purchase-price to a surprisingly large number of people in most parts of the World. That a front wheel has to be removed before the sparking plugs can be changed and the service charges seem astronomical to family-car owners are items of absolutely no concern to Rolls-Royce's clients.

In Detail

The concept of the Rolls-Royce is decidedly vintage. The driver sits high on a separate seat upholstered in the best English hide, a seat out of the ordinary yet comfortable for hour after-hour, the shoulders well supported by the high squab. Each of these separate front seats has an adjustable squab which can be angled over many degrees (although not falling into a posture anything like so vulgar as a bed) and its own folding armrest, matched by deeper adjustable armrests-cum-pulls, on the front doors. Forward vision is excellent, over the handsome but not unduly long bonnet. Over the classic radiator rides the " Silver Lady," the latter now fairly secure against petty theft. I am of average height and I could see comfortably the n/s front wing. The central mirror gives a good but vanishing rear view and is supplemented by well-placed wing mirrors.

The big 3-spoke steering wheel is pleasantly thin-rimmed, with finger grips below the rim, and is unencumbered, the deep-note horn being sounded by a modest button on the hub. Below the wheel, slightly inaccessible, is a flick switch labelled "N/H" for varying the settings of the rear shock-absorbers—the famous R.-R. ride control.

The facia is a thing of beauty indeed—made of polished French inlaid walnut veneer, and carrying discreet Smiths instruments, all dials bearing the R.-R. initials. These instruments, mounted on a central panel, comprise the combined fuel gauge, water thermometer, oil pressure gauge and ammeter (which while not so impressive as separate dials, is at least a more compact way of accommodating these instruments), a clock and a 120-m.p.h. speedometer, with decimal trip and total odometers. The calibration of the combined dial is not in figures, inscriptions in white sufficing :—High/Low (oil), Hot/Cold (water) 0/½/F (fuel), and 30 minus/30 plus for the Lucas ammeter. The needles of the oil pressure gauge and of the water thermometer normally stand horizontal so that a glance suffices to ensure that all is well in these departments. The accurate clock is matched by a plated cigar lighter, and in the centre is a typically R.-R. circular electrical switchbox, incorporating warning lights for dynamo not charging and low fuel level, the turnswitch bringing in the side lamps, the dual Lucas sealed beam headlamps and finally the Lucas foglamps in conjunction with the sidelamps, when turned in a clockwise direction.

A small ignition key starts the engine, or, turned the other way, enables the radio and wipers, but not the horn, to function when the engine is not running. Matching inset panels carry the air conditioning switches, respectively for upper and lower ventilation, the former switch pulling out to bring in a 2-speed fan, and other switches controlling panel lighting and wipers/washers. Throughout the identification lettering is in white and, needless to say, the plated switches are easy to operate, well positioned and work with precision.

Below the instrument panel are the radio speaker with knob for front/rear selection and the Radiomobile radio which is fitted as standard. A simple radio aerial on the leading edge of the roof can be rather untidily retracted to avoid fouling a garage roof or overhanging trees, etc., by turning a knob on the upper screen rail. Small lidded ashtrays are provided at each end of the non-dazzle capping rail above the facia, this being crash-padded. The door sills are of French walnut veneer, matching the facia and each other. Under the facia there is a pull-out ashtray and polished veneer walnut shelf and on the left the facia contains a deceptively commodious cubby-hole, its lid having a Yale lock and clicking shut with a pleasing suaveness.

The vents of the elaborate Rolls-Royce ventilation system occupy each side of the facia rail and there is another at foot level. These vents each have five hinged vanes, the positions of which can be controlled by plated handles and thus direct the flow of hot or cold air where required. The wipers, which are of the 2-speed type sweep an adequate area of the flat windscreen; pulling out their switch brings in the washers. The panel-light switch rotates to select dim or bright illumination and if pulled out brings in a map light for the convenience of the front-seat passenger. The dual headlamps are dipped by foot button and full beam is indicated by a discreet red light in the speedometer dial. The turn indicators are operated by a well-placed left-hand stalk lever which also provides for daylight headlamps flashing. The turn indicators' warning lamps consist of tiny strips in the speedometer to show which indicator is in use. The rigid anti-dazzle vizors swivel and the near-side one contains a vanity mirror. The accelerator and brake pedals, the latter with the magic R.-R. on its rubber, are of conventional size. A gear lever extends from the right of the steering column in a black quadrant labelled, simply, R, 2, 3, 4, N, a plated press button on the tip of the lever having to be depressed

A Rolls-Royce enhances anyone's estate!

before neutral (necessary before the engine can be started) or reverse can be selected. This lever moves downward for holding third gear and downward again and round the gate to hold second gear. In "hold third" the change-up to fourth speed will take place automatically when peak revs. are reached but in "hold two" the driver is responsible for making the upward change, although if required an automatic change into bottom gear will occur. On light throttle openings upward gear changes occur at 9, 16 and 24 m.p.h., respectively, but kick-down raises these speeds to approximately 26, 40 and 74 m.p.h. Downward changes on light throttle happen at about 16, 10 and 5 m.p.h., respectively. The usual parking position is obviated by having a lock which operates when the lever is in "R" with the engine off. I was surprised to find that there is no hill-hold in any gear position, so the aforesaid slight brake lag can be embarrassing on steep hills.

Reverting to the facia layout, there is a button for releasing the flap over the screw-type fuel filler cap which is concealed in the nearside rear of the car. A flick-switch provides for bringing in the built-in electric rear-window demister, and an inspection lamp socket is provided, while there is a big handbrake-on warning light. Incidentally, the speedometer, calibrated in steps of 10 m.p.h., has its lowest figures on the *right* of the dial, the needle moving steadily in a clockwise direction. Not only does the white needle record speed steadily, but it is sensitive to slight changes of speed as the car is accelerated or decelerated.

The handbrake is a T-type handle affair, twisting and pulling out from its substantial mounting under the facia, for operation by the driver's left hand. It protrudes quite a long way when in the "on" position and, although I became adept at avoiding it, it could be a nasty trap for the elderly or the unwary. However, it holds the car most effectively on gradients.

Much better contrived are the bonnet releases. The bonnet is still sufficiently old-fashioned to have an openable top panel on either side. These panels are released by handles under the scuttle, When the time comes to close them they are drawn neatly down and locked by these same handles, there being no need to slam a Rolls-Royce bonnet.

The doors are far less effective, needing a hearty slam and being noisy to shut, although this is no doubt due to effective dust/air sealing. Two keys are provided, enabling the boot and the cubbyhole to be locked while the car can still be driven. The floor is covered with fitted Wilton pile carpets. There is an additional fresh-air vent on the nearside of the scuttle, opened by a small knob on the right of the handbrake. This air vent "zizzed" faintly, which was irritating in an otherwise so-silent motor car.

The Rolls-Royce is not a vehicle to be littered with oddments, but discreet wells, closed by covers, mounted on plastic slides for quietness, are provided in the front doors, and there is the usual, not very deep, shelf behind the back seat. When the doors are open they disclose token running-boards or steps but these are of little help when dismounting and it is more usual to step directly down on to the road or pavement. The doors can be locked internally with rather old-fashioned but fully effective vertical handles and they have neat press-button external handles, the key action of the external locks being notably precise. The recessed circular roof lamp has courtesy action from any of the four doors, and a recessed plated door-pillar tumbler switch also operates it.

The rear compartment, in which the passengers sit well back from the windows in secluded isolation from the vulgar affairs of the world without, is provided with recessed mirrors in each corner, and adjacent illuminated cigar lighters. The seat has a central folding armrest, and folding picnic tables with inbuilt ashtrays can be pulled out from the backs of the front seats (the o/s table on the test car tended to jam). In addition, there are adjustable footrests permitting the fitting of lambswool rugs to the floor, spring-loaded "pulls," and a re-circulating air vent in the floor. Incidentally, the lamps for the cigar-lighter compartments, which have recessed plated tumbler switches, can also act as reading lamps. The luggage lid is self-supporting and contains a lamp which illuminates the boot.

The test car was provided with neatly-stowed Irvin safety belts for the front seat occupants and a small Bradex fire extinguisher. The car also had two very desirable extras, namely the under-wing refrigerator unit, and the electric window lifts. The former, while not so dramatic or complex as I had anticipated and rather noisy, kept the interior of the car delightfully cool, so that the hot air without slapped you in the face as you alighted. The complete heating and ventilating system enables any desired combination of cool or warm, cold or hot, air to be delivered to those riding in the front or the back of a Rolls-Royce. It costs nearly as much as a Fiat 500 but is certainly worth having if much driving has to be done in a hot climate. It also enables the windows to be kept closed, keeping the dust out of the car and considerably enhancing the low sound level. What this elaborate refrigerator does not do is to give an individual supply of hot or cold air to occupants on different sides of the car, nor does it enable air at a pre-determined temperature to be maintained within the car's interior. The cost of this refrigerating unit includes tinted window glasses and windscreen. The electric window lifts work stolidly but noisily (there are four press-buttons for the driver, over-riding the individual controls provided for the passengers) and are perhaps more immediately desirable than refrigeration, although after driving other cars in which one can keep cool only by opening the windows at the expense of draughts and dust, one appreciates the advantages of possessing a refrigerated Rolls-Royce. In humid weather the refrigerator has a habit of venting on to the road a small quantity of liquid in the vicinity of the o/s front wheel, which is a rather undignified habit for a car of this calibre.

On the road the Silver Cloud III is sheer enjoyment, if you are prepared to drive it as it is intended to be driven. Roll on corners it does, but the roll is consistent, the steering extremely accurate, the brakes superb. I have mentioned previously the desirability of rather more low speed pick-up, but dropping into "Hold 3rd" provides all the acceleration most fast drivers will require, to nearly 75 m.p.h. Incidentally, Rolls-Royce encourage the driver to use the gearbox as a manual one, but operated thus, the lever

The discreet instrument panel and controls of the Rolls-Royce Silver Cloud III, the rather cheap-looking combined-dial being evident on the left—it does, however, look neater than a lot of scattered individual dials. Note the vaned ventilator intakes.

NOT SO MUCH A MOTOR CAR MORE A WAY OF LIFE

is rather stiff and clumsy. Under favourable conditions this Silver Cloud will run up to a maximum speed of 114 m.p.h. or perhaps a little more; at the opposite extreme it can be inched forward mutely at 2 m.p.h. in heavy traffic.

Consumption of the best grade petrol is surprisingly consistent. Determined driving over mainly deserted roads gave 11.7 m.p.g. Traffic work in London and normal country driving improved this fractionally, to 11.8 m.p.g. In a long fast run, heavily laden, together with much country road pottering, it produced a figure of 11.6 m.p.g., so that the overall average came out to 11.7 m.p.g. The fuel gauge gives an accurate indication of how much petrol will go into the tank, except for a slight pessimism towards the empty position, while the low-level light comes on permanently after 164 miles from filling the tank, with scarcely any preliminary flashing. Three gallons then remain, so that the absolute range is approximately 200 miles. This is rather niggardly, but an extra fuel tank can be fitted if required. After 550 miles, warned by the reassuring but actually rather vague press-button oil level gauge which is incorporated in the fuel gauge, an item which has long figured on Rolls-Royce and Bentley cars, I checked the oil level. The engine dip-stick would do service for a sword, and access to the oil filler is obtained by opening the n/s bonnet panel and undoing a wing nut so that the big air cleaner can be swivelled up out of the way and restrained in this position by a short length of cable. Fortunately, this is an operation the owner of a Rolls-Royce is unlikely to perform personally. On the first occasion of checking the oil level I found that a quart of BP Visco Static was required. In all, half a gallon of oil was consumed in a test distance of 1,245 miles. Incidentally, in a fortnight's use the washers' reservoir did not need replenishment.

At the beginning of the test the Dunlop Fort tyres showed 8 mm. tread depth, except in the case of the offside rear tyre, which had a tread depth of 7.2 mm. Checked again after 700 miles, the o/s front tyre had lost 0.9 mm. of tread, the n/s front 1.2 mm., the o/s back tyre 1.0 mm. and the n/s back tyre 1.8 mm. of rubber. There are 21 chassis points to be greased every 12,000 miles. The floor at the n/s front gets warm, which was nice for the motoring dog!

Conclusion

What more is there to say? Going to the Ithon Valley M.C.'s first and very successful grass track meeting at Cross Gates in Wales, the big heavy car with its 8.20-section tyres proved to have unexpectedly good traction uphill on liquid mud. The restrained and dignified appearance of the Rolls-Royce Silver Cloud III grows on one, even if dual headlamps look to be rather an afterthought, and the nose does not dip unduly when making a crash stop. The transmission makes some mild sound when changing up at low speeds and the suspension can be caught out by hump-backs to the discomfort of the rear seat passengers. The brakes work somewhat fiercely when reversing and a faint click can be heard as the linkage takes up again when they are next used going forward. It is hardly necessary to state that in a total distance of nearly 1,300 miles the car never faltered, never put a wheel wrong.

As I have said, the Silver Cloud III is old-fashioned in some respects and for keeping up with the sports cars over winding roads I might be happier, in, say, a Mercedes-Benz 300SE. But the Rolls-Royce is one of the nicest cars in which to drive very fast for hour after hour after hour and *this enjoyment is certainly far more than merely psychological*. The faster the modern Rolls-Royce is driven the more you discover that it can be thrown about with considerable abandon, and it then still corners accurately and feels like quite a small car. By making good use of the gear lever and those very powerful brakes average speeds which would be commented upon in many other cars pass unnoticed and seem like normal unflurried motoring to the more press-on owners of Silver Clouds, yet the car's whole demeanour displays in a subtle manner the characteristics of its forebears. This splendid motor-car from Crewe is indeed one of Britain's best ambassadors and apart from being a status symbol it is a motoring way of life. Regarded thus, its price is quite incidental, even moderate.—W. B.

THE ROLLS-ROYCE SILVER CLOUD III SALOON

Engine: Eight cylinders in 90° vee-formation, 104.14 × 91.44 mm. (6,230 c.c.). Push-rod-operated overhead valves. 9.0-to-1 compression ratio. Horsepower not disclosed.

Gear ratios: (Automatic gearbox). First, 11.75 to 1; 2nd, 8.10 to 1; 3rd, 4.46 to 1; top, 3.08 to 1.

Tyres: 8.20 × 15 Dunlop Fort tubeless, on bolt-on steel disc wheels.

Weight: 2 tons 1 cwt. 3 qtr. (kerb weight).

Steering ratio: (Power steering.) 4¼-turns, lock-to-lock.

Fuel capacity: 18 gallons. (Range approx. 200 miles.)

Wheelbase: 10 ft. 3 in.

Track: Front, 4 ft. 10½ in.; rear, 5 ft. 0 in.

Dimensions: 17 ft. 6¼ in. × 6 ft. 2¾ in. × 5 ft. 4 in. (high).

Price: £4,660 (£5,632, inclusive of purchase tax). With extras as tested: £6,073 13s.

Makers: Rolls-Royce Ltd., Crewe, Cheshire, England.

SOME ROLLS-ROYCE INDIVIDUALITIES

The Rolls-Royce has three independent foot brake systems, so that if one fails the driver still has both front and back brakes. Trailing shoes at the front and equal wearing shoes at the back operate on 11¼ in. × 3 in. drums and are applied *via* a gearbox-driven servo motor. The front/rear ratio is so arranged that the front wheels cannot lock when reversing downhill. Rolls-Royce claim that the car can be stopped once a minute from 70 m.p.h. until the linings are worn out, without fade, and that the servo makes you seven times the man you are.

* * *

There is a fully-fused electrical system for the 12-volt 67-amp.-hr. negative-earthed electrical system.

* * *

Care is taken to avoid gritty switches. Each of the heating and ventilation controls operates its respective tap through an electric servo. The main switchbox is scrupulously engineered and in principle departs very little from the Silver Ghost's counterpart. The starter motor is not connected directly to the driver's switch; Rolls-Royce use a second relay to tell the first relay to tell the starter to engage.

* * *

The pedals are the same distance from the steering wheel as they were on the Silver Ghost.

* * *

A satisfactory tick-over is maintained even after prolonged Motorway driving because the engineering of the throttle linkage to the S.U. HD8 carburetters has received a lot of attention and because the engine will stand several hundred hours at full-throttle without distortion of the exhaust valves.

* * *

The steering geometry is set while the car is moved through its full suspension range, which is the only method of ensuring that errors are negligible in the normal position of the front suspension.

* * *

The screen-wiper blades park tidily, and their blades are then warmed by the demister.

* * *

To contribute to silent running the exhaust system has three separate stainless-steel expansion chambers each tuned to absorb different sound frequencies, and the pressed-steel body is so fully insulated from the chassis that the only direct connection between the two is the speedometer cable.

* * *

Rolls-Royce Ltd. claim that their cars have thicker chromium plate and finer leather than any other car in the World. The body has a dozen coats of paint and primer.

USED CAR TEST

No. 392

1961 BENTLEY CONTINENTAL S2

H. J. Mulliner 'Flying Spur'

PRICES

Car for sale at Maidenhead at	£4,250
Typical trade cash value for same model in average condition	£3,000
Total cost of car when new including tax (excluding extras)	£8,855
Depreciation over 12 years	£5,855
Annual depreciation as proportion of cost new	5½ per cent

DATA

Date first registered	16 January 1961
Number of owners	At least 5
Tax expires	30 November 1973
MOT	23 May 1974
Fuel Consumption	14-16 mpg
Oil Consumption	negligible
Mileometer reading	115,034

PERFORMANCE CHECK

(Figures in brackets are those of the original Road Test of a James Young-bodied Continental S2, published 30 December, 1960.)

0 to 30 mph	3.8 sec (4.0)
0 to 40 mph	6.2 sec (6.3)
0 to 50 mph	8.7 sec (8.9)
0 to 60 mph	11.8 sec (12.1)
0 to 70 mph	15.9 sec (15.9)
0 to 80 mph	21.0 sec (20.5)
0 to 90 mph	28.9 sec (26.9)
0 to 100 mph	41.5 sec (37.1)

Standing ¼ mile 18.5 sec (18.6)

In top gear:

30 to 50 mph	— sec (7.6)
40 to 60 mph	— sec (7.8)
50 to 70 mph	— sec (8.7)
60 to 80 mph	11.3 sec (10.2)
70 to 90 mph	14.1 sec (12.2)
80 to 100 mph	22.5 sec (16.6)

Standing Km 33.9 sec (—)

TYRES

Size: Dunlop Fort cross-ply tubeless 8.20-15in. Approx. cost per tyre £21.90 (tubeless)
Remaining tread depths. 5mm on right, 6mm left front, 4mm left rear, 1½mm spare

TOOLS

The fitted tray in the boot is complete even to spare bulbs and there is a hydraulic jack. Only the tyre pump is missing. A handbook comes with the car.

CAR FOR SALE AT:

Lex Mead Ltd., 34, Market Street, Maidenhead, Berkshire. Tel: Maidenhead 25371.

FROM time to time we have in the past endeavoured to include in our used car test series models which would appeal to a very limited market when new by reason of their price, but which after several years of depreciation could prove an attractive alternative to a new model. Just such a car is the 1961 Bentley Continental S2 with H. J. Mulliner "Flying Spur" coachwork. Rolls-Royce records for this particular car show that the chassis was delivered to the coachbuilders in September 1960, the completed car being handed over to the owner in January 1961. Also confirmed is the fact that the car had three owners in the three-year warranty period. After that the car passed through the hands of Jack Barclay Ltd., the London Rolls-Royce and Bentley distributors, and the most recent owner is Lt. Col. V. Gates, of Cow and Gate fame, during whose ownership the car bore the registration VG23. This period of ownership was from 1969 to 1973.

On getting into the driving seat of the Bentley Continental one is immediately impressed by the good visibility. Once moving on the road the car does not feel at all big to drive. Starting was always first time providing that when cold the accelerator pedal was depressed once to bring in the automatic choke. Manoeuvring when the choke was in operation calls for some care as the vee-8 engine is then ticking over at around 1,500 rpm, at which speed quite a lot of "creep" is available, so that application of the brakes with the left foot was needed to check the car. The 6,230 c.c. aluminium vee-8 unit used in the Bentley Continental S2 is the

Below: The exterior paintwork is in very good condition with only a few minor chips around the doors. There are also one or two very minor bodywork dents in the doors. The front pair of flashing indicators are incorporated in the fog lamps and have yellow bulbs

USED CAR TEST
1961 BENTLEY CONTINENTAL S2...

same as that used in the Bentley S2 saloon and is not as quiet as earlier six-cylinder units, but then one has to remember that the vee-8 delivers about 30 per cent more power. It is certainly true that the engine on this car can be heard both inside and outside at idling speed, but once on the move the engine is as unobtrusive as one would expect up to speeds in the region of 100 mph. It is only when the car is driven hard using full throttle that some commotion from under the bonnet is heard. The four-speed automatic gearbox provides very smooth changes at light throttle openings, but the changes can certainly be felt when accelerating hard. When taking the performance figures we noticed that the car seemed to pause when reaching maximum revs (4,000 rpm) in second gear before changing up into third. Although this is a big heavy saloon it is most impressive in the way it moves off the mark and positively rushes up to around 90 mph. At this point the acceleration is not so rapid and 100 mph seems to take a while to come up. We did not attempt to take a maximum speed figure on the car, which after all is 12 years old and has covered 115,000 miles on its original engine, but there is no reason to suppose that it would not have reached somewhere in the region of 120 mph. When comparing the figures with the James-Young-bodied Continental S2 tested in December 1960, it is important to remember that that model had the normal 3.08 saloon axle ratio and not the higher 2.92 Continental ratio, so this Flying Spur would seem to be well up to standard on performance.

The test car proved so quiet on the road that one was reluctant to spoil the quietness by opening a window. Even at 100 mph it was quite possible to talk without raising one's voice, making this an ideal buy for anyone who has to cover long distances or do a lot of motorway driving. At 70 mph on the motorway the car cruised effortlessly with very little fuss, but on concrete surfaces there was a fair amount of tyre roar and the power assisted steering calls for constant minor correction of the wheel to keep the car running straight. Even so this steering arrangement, which provides 50 per cent assistance when on the move and 80 per cent for parking, has far more feel than the system used on the later Silver Shadow series. Judged by 1973 standards the ride is not as outstanding as it was 12 years ago and one is reminded how much better even low priced cars have become since then. We could not really detect any difference when altering the setting of the rear dampers from normal to hard by means of a switch on the steering column, probably because the suspension is now less taut than it was when new. The brakes have the usual Rolls-Royce mechanical servo and operate quite efficiently although stopping from 100 mph during the acceleration runs produced a pronounced smell of hot linings. This is more a characteristic of the model than this particular example. The test car is on Standard S2 8.20-15 in. tyres as the correct 8.00-15in. tyres for the Continental are difficult to obtain.

In spite of its age the car still looks very good. Bentley Continentals are now appreciating in value, so this would prove a good investment as well as a most enjoyable car to drive. One gets a lot of motor car for the money and it is different enough for its age not to matter so much. The R-Type Bentley Continental is already a much sought after classic and the S-Type Continental will no doubt follow the same upward path in due course.

Below: There is some creasing of the front seats. The picnic table slides away under the facia

Below: Picnic tables and a centre armrest are provided in the back

Above: The centre hinged bonnet allows reasonable access for routine maintenance

Condition summary
Bodywork

According to the log book the test car was originally black. Rolls-Royce's records confirm this and that the original upholstery was Aqua Blue. The car was completely resprayed in its present Regal Red in June 1970 according to the log book, and presumably the upholstery colour was changed at the same time to the present dark red. This colour change appears to have been effected by the use of leather paint. The respray also explains why the windscreen pillars inside the car are blue. The complete colour change has been well executed and there are few signs that this is not original, though a few paint spots on the carpet behind the rear seats and a slight paint run below the rear bumper give the game away on close inspection. Overall the exterior appearance is very good though the bumpers and overriders have a number of dents, and the chrome plating has all but worn through on one corner of the rear bumper. The cover for the number plate bulb is also missing. Both the boot and nearside passenger rear door catches are in need of a little adjustment, the latter tending to bounce back to a half closed position

Above: At 17ft 8in. long the Bentley is a large car. Note the massive bumpers and the twin reversing lamps above the number plate

unless shut very firmly. There are also some minor body creaks and rattles on bad surfaces; these are the more noticeable because of the quietness of the rest of the car.

Inside the car there are quite a few signs that the car has covered a high mileage. As is to be expected the original carpets are now showing signs of wear and are quite grubby with some paint spots behind the front seats. The car appears to have been used a fair bit by smokers who have left their mark in the form of several small cigarette burns on the front passenger seat and one on the rear seat. The grey cloth headlining is also stained, particularly in the front of the car. The front seats are fairly creased as is to be expected in a car of this age and some of the seams are becoming unstitched, particularly on the squab of the passenger seat. The rear bench seat seems to have had comparatively little use as it is in much better condition. The veneered woodwork throughout the car has suffered a fair amount of wear and tear. There are scratches on the "cubby box" lid (as the fine handbook charmingly describes the glove locker) on the facia and on the rear seat picnic tables and their surrounds when in the folded position. In the rear the picnic table surfaces are virtually unmarked but the front picnic table has been slightly scratched. The veneer on the top of the facia has

Below: Fitted tools (right) live above the well containing the jack on the left

crazed and the veneer is peeling on the rear window surround. The door fillets have also suffered and the tops of all four are weathered, more so in the front than at the back. There is a chip out of the veneer on the nearside rear door fillet, and the veneer is lifting around the front quarterlights.

Equipment and Accessories

The speedometer needle wanders suggesting that a new cable is needed and reads 7 mph fast at 100 mph while the mileometer under-reads by four per cent. Neither the central roof lamp nor the swivelling reading lamp (one of the few extras fitted to the car) work. The cigar lighter on the facia does not stay in long enough to glow red. The heating and demisting system works well for hot or warm air, but it is not possible to admit cold air through the heater as the two heater and demister taps under the bonnet have seized in the winter position. The fuel gauge follows the usual Rolls-Royce practice of having alongside a button, which when pressed causes the oil level in the engine sump to be shown in the fuel gauge. Although the fuel gauge itself is working no reading of the oil level is obtainable, suggesting that the float has stuck in the sump. The His Masters Voice push button radio with manually-operated aerial was included in the car's standard specification as were the twin fog lamps. Being a valve set, it takes a little while to warm up, but gives excellent reception. We had some problems with a very high charge rate and came to the conclusion that the battery had a leak. The picnic tables in the back of the front seats, although original equipment on the standard Bentley S2 saloon, were an extra on the Continental as were the badge bar and electrically-operated front windows.

About the Bentley S-Type Continental

The Bentley S-Type was introduced in April 1955 to replace the R-Type and had a larger 4,887 c.c. six-cylinder engine with a new cylinder head and automatic transmission as standard. However it was not until September 1955 that the revised Continental based on the S-Type chassis was announced. Originally the idea of the Bentley Continental had been to provide a long distance touring car capable of maintaining high cruising speeds for long periods. To improve the performance the whole car, including the radiator, was lowered and the coachbuilders went to great lengths to reduce the weight of the aerodynamically styled bodies. Continentals also had a higher axle ratio and increased braking capacity. All the bodies on Continental chassis were coachbuilt using aluminium panels. The S1 Continental continued in production until 1959 when the vee-8 engined S2 version succeeded it. Nearly all the S1 Continentals had automatic transmission, which was standard equipment on the series, although it was possible to specify a manual gearbox up to 1957 and a few manual cars were built. Chassis changes to the standard S-series, often not incorporated into coachbuilt cars until quite a while later, included the provision of twin brake master cylinders in April 1956, and a splayed chassis frame in July of that year. Summer and winter heater taps under the bonnet (mentioned elsewhere in this test) were added in February 1957 and in September that year the compression ratio was raised to 8.0 to 1 and larger 2in. carburettors fitted.

The introduction of the S2 Bentley in September 1959 gave the Bentley Continental a larger 6,230 c.c. vee-8 engine and power-assisted steering, formerly an extra, became standard. Automatic transmission only, using the Rolls-Royce four-speed epicyclic gearbox, was offered and the Continental had 8.00-15in. tyres compared with the standard car's 8.20-15in. tyres. A higher 2.92 final drive ratio was usually fitted to S2 Continentals although it was possible to specify the standard saloon's 3.08 axle ratio. H. J. Mulliner did an exceptionally fine job in producing a four-door version known as the Flying Spur which weighed the same as a two-door Continental. Like all Continentals this car had four-shoe drum brakes giving a lining area of 300 sq.in. and a radiator that was 3¼in. lower than the standard saloon. The S3 Bentley replaced the S2 in October 1962.

171

Rolls-Royce Supplement
TRIPLE SILVER

Three generations of Rolls-Royce's V8 family feel very different to drive. Mark Hughes compares Silver Cloud II, Silver Shadow II and Silver Spur. Photography by Andrew Yeadon

It was such a perfect evening, with the sun sinking from a clear blue sky washed pure by recent rain, that I took the scenic route home from Rolls-Royce at the wheel of RRM 1, the Silver Spur which is the pride of the company's demonstration fleet. South of Shrewsbury, the sight of a ribbon of road curling right over the top of the Long Mynd was irresistible, beckoning me to seek out the way to it through All Stretton. Once up there enjoying the view far into Wales, I had this road completely to myself for half an hour, until an elderly Shropshire farmer driving a tired Escort van came the other way. There we were, face to face in two extremes of modern motoring, with nowhere to pass each other.

I reversed for half a mile to the nearest passing place, and as he drove by the farmer wound down his window to speak. "I just wanted to say how highly honoured I am that a Rolls-Royce should reverse for me," he said, with a mixture of deference and genuine enthusiasm for the car. "Still, mustn't keep you, I'm sure you're a busy man. Good evening, Sir." *Sir!* That is the power of the car you drive...

Apart from being quite the most relaxing car I have ever driven in the way it instils a calming sense of well-being, the Spur, the long-wheelbase version of the Silver Spirit, is hugely memorable for the way people react to it. This encounter on the hills was typical of the courtesy which some people extend to Rolls-Royce drivers, but there are plenty of envious ones who carve you up something rotten because you obviously have lots of money. But almost everybody responds in one way or another, even if it is only to look pointedly the other way, because a Rolls-Royce is still the ultimate automotive symbol of wealth. Most look first at the car, then at the bloke behind the wheel. Since I didn't dig out my only suit for my four days of Rolls-Royce posing – sorry, I meant to write motoring – I guess I was taken for a pop star, a *really* successful yuppie, the heir to a

172

Rolls-Royce Supplement

Rolls-Royce Supplement

fortune, an off-duty chauffeur or perhaps an underworld boss.

The reason for this spell of Rolls-Royce motoring, for indulging in the Spirit of Ecstasy, was to examine one theme of Rolls-Royce's evolution through its Crewe period – the V8 link. Change has always been gradual and considered at Rolls-Royce. For all the strands of modernity possessed by the current range of cars within their hand-crafted character, the crucial features of their design go back many years, and the strongest line of continuity is the mighty V8 engine which all the current cars employ. This engine is shared by the three cars – Cloud II, Shadow II and Spur – compared here.

Rolls-Royce's first thoughts about a new engine to succeed its 4887cc six-cylinder unit, conceived in 1938, go back to the early 1950s. Working under Harry Grylls, an engineering team set about designing an all-new engine which would offer much more development potential than the 'six' without compromising traditional Rolls-Royce qualities of smoothness and reliability. Taking a lead from American manufacturers, the company rejected ideas for a V12 in favour of a simple and sturdy V8.

After experiments with engines of 5.2-litres and 5.4-litres, the definitive 6230cc (from 104.1mm bore and 91.4mm stroke) was adopted for the design which appeared in the Series II version of the Cloud in 1959. Apart from using silicon-aluminium alloy for all its major castings, it was a conventional design with a five-bearing crankshaft, a single gear-driven camshaft operating two valves per cylinder via pushrods, and a pair of SU carburettors.

As installed in the Cloud II, the new V8 offered stronger perfomance and smoothness the equal of the superseded 'six', even if the position of the spark plugs beneath the exhaust manifolds meant that servicing was a nightmare – the left-hand plugs could just be reached from above, but access to those on the right had to be provided through panels in the inner wheelarches. Rolls-Royce have never disclosed figures, but initially power was thought to be around 200bhp and torque at least 325lbs ft.

Apart from some minor suspension, steering and interior modifications, the V8-engined Cloud II followed the existing design. Visually the only change was a tiny one, the front ventilation air intakes becoming black – there were no badges, nor even a second exhaust outlet, to tell the world of the magnificent new engine which lay under the surface. Only with the Cloud III would body revisions be made, notably the adoption of twin headlamps.

Leslie Taylor has owned her Cloud II for 10 years, and says that she will never part with it because it would upset her family too much. The owner of an antiques business in Cirencester, she had little enthusiasm for cars until she won the Cloud in a raffle. She had been given charity tickets at £10 a time to sell from her shop, but the price put off all her customers. In the end she felt obliged to show willing by buying one ticket herself!

"When I collected the prize," she recalls, "I was put in the car and told to drive home. I was very bewildered, and didn't even have a garage to put it in. But I have grown to love the car, and thankfully it hasn't cost too much to run apart from a complete respray two years ago. I don't use it for business as everybody would think I am doing too well, although once when I took it to Scotland I removed the back seat and came home with a gateleg table and some walnut corner cupboards."

It is a lovely, original car which has obviously led a full life, as the 110,000 mileage, leather patina and faded woodwork testify. Its shape exudes elegance and dignity from any aspect, every inch of its voluptuous, billowing curves looking perfect, and the air of sublime quality conveyed by its appearance is matched by the interior's traditional, settling design. The chunky dashboard appears so hewn from solid wood that it could be a piece of furniture, and the veneers continue up the windscreen pillars and back along the tops of the doors. The ergonomics, with instruments and controls scattered on a big, central panel, are from a long-gone era, although some items – the ignition key cluster and the big dial containing four gauges – are identical to those used today.

Mechanically refined though it is, you notice the Cloud II's age by the noise it transmits from the road and the wind rustling past the windows. While these sounds drown the engine at cruising speeds, at standstill the V8 barely intrudes, neither aurally nor in terms of vibration. It is an astonishingly flexible engine which feels as if it will pull from 100rpm, and this low-down willingness is exploited by an automatic transmission set to make upward changes long before you imagine the engine wants them. The four-speed Hydramatic, controlled by a N-4-3-2-R right-hand column selector, will hold top below 20mph, confirming its aversion to using its sprinting ratios. When you do force the V8 to work, it emits a rougher note than its more modern companions, but at gentle throttle openings it feels oily smooth.

Only the wayward cam and roller steering, feeling as loose as a ship's tiller, detracts from the ease of driving the Cloud II. This floppiness and the low gearing (an amazing five turns from lock to lock) force you to work constantly at the three-spoke wheel's slender rim, keeping pressure against the point of resistance in order to steer a straight course. The combination of all-round drum brakes and the heavy pedal pressure of mechanical servo assistance makes slowing down less effortless than on the later cars, but the anchors are good enough if you drive the Cloud II in the gentle manner which it invites. It can bowl along at over 100mph, but driver and car are happiest at much lower speeds.

Only the wayward steering, feeling as loose as a ship's tiller, detracts from the ease of driving the Cloud II

The Silver Shadow began to be sketched on the Crewe drawing boards as soon as the Cloud had been launched. From the start the new car was schemed to be very different in styling and structure, for Rolls-Royce knew that it could not ignore the stiffness and lightness benefits of monocoque construction. Just as the company had been influenced by Detroit in selecting a V8 engine, so it was that the experience of the American giants provided the impulse to reject a separate chassis, even though this raised new problems for the future in evolving coachbuilt derivatives.

Only the V8 engine and GM Hydramatic transmission were carried over from the Cloud, but even these were greatly modified. The engine gained new cylinder head castings and exhaust manifolds to give far better spark plug accessibility, while the four-speed transmission was refined and lightened as a temporary expedient for right-hand drive cars until the new GM400 TurboHydramatic three-speed (which is still used today) could be installed.

Much development work went into isolating road noise and vibration, since a monocoque tends to act as a sounding box. Carefully mounted front and rear subframes carried the suspension, which was now independent at the rear by semi-trailing arms, and a hydraulic self-levelling system was fitted. The brakes also marked a dramatic departure for Rolls-Royce as discs were fitted for the first time.

Two landmarks occurred during the Shadow's life. The first came in 1970 when the V8 was given a long-throw crankshaft to enlarge it from 6270cc to 6750cc, an increase which countered the power losses caused by emissions tuning. Again, no figures were quoted, but power is thought to have remained around the 200bhp mark, although torque was still well over 300lbs ft. The second landmark came with the arrival in 1977 of the Shadow II, which was so thoroughly revised under the surface that it was effectively halfway towards becoming the Spirit.

Among the Shadow II's 2000 detail changes, the most significant were the adoption of split-level air conditioning (already standard on coachbuilt cars), power-assisted rack and pinion steering in place of the old recirculating ball set-up, a new facia and control layout, and revised front suspension geometry to improve stability. The new car was instantly recognisable by its heavy, US-style bumpers and the modest air-dam beneath the nose.

Peter Kimberley, a senior man with an advertising agency, has been a life-long fan of Rolls-Royce cars, but his ambition to own one was realised only when he was able to afford a Shadow II five years ago. He had started looking for a Corniche Convertible, but fell in love with the shape and colour of the Shadow II you see here. Since then he has also acquired a 1924 20hp, and loves Rolls-Royce motoring whether it is vintage or contemporary.

"I suppose my passion for Rolls-Royces is a mixture of a desire to own a beautiful thing and snobbery," he says. "The Shadow is not an easy car to own; it is a constant job to keep on top of the maintenance, and all repairs are very expensive. But the car gives me more pleasure as time goes by; it is wonderful to sit in and relaxing to drive."

Straight away the Shadow II feels very different from the Spur, and rather less intimidating. The view over the bonnet, an expanse of gentle curves despite the car's rectangular proportions, is more appealing, the car feels smaller and the driving position lower and more intimate. No Rolls-Royce is cosy, but this is how it feels after the current giant. The facia, on the other hand, is virtually identical except for the central pair of dials (one a quartz clock which can be heard ticking above a restrained engine note) where the Spur has incongruous digital screens. Many other details – like the style of the electric window lifters and the steering wheel – are also carried through to the current car.

Whereas the Cloud II feels symbolic of its period in its traditional cabin ambience, the Shadow's interior is still with the times. Once moving, too, it is much more closely related to Spur than Cloud II. The only way in which it compares with the older car is in its refinement and laziness – but even the V8 has a quite different, more willing, responsive character. Its steering is a revelation after the Cloud II's sloppiness, the rack and pinion providing perfect accuracy and a degree of weight which surprised me. In this respect, as with the smoothness of the automatic transmission's shifts and the standard of its brakes (apart from lacking anti-lock), the Shadow II is in the same league as the Spur.

In two ways, however, it is inferior to the Spur. Bosch mechanical fuel injection on the current car gives the engine a definite edge on acceleration, which is quite sprightly for a long-wheelbase heavyweight tipping the scales at 2.3 tons. The penalty is that some of the carburettor V8's silkiness has been lost, although this impression is conveyed partly by a more urgent exhaust beat worthy of an American muscle car. The older car, as you might expect, is also less willing to be hustled through bends, wallowing compulsively long before grip begins to give up. The latest Spirits and Spurs have received so much suspension attention that you could almost describe their faithful, sharp handling as sporting.

For all the conservatism with which Rolls-Royce is perceived, the pace of development at Crewe is not as slow as the infrequent changes of body style suggest. These three generations of the V8 family may share an engine bloodline, but the difference between ancient and modern is vast.

Rolls-Royce Supplement

Cloud II: elegant lines, tall grille, plenty of wood and front bench seat

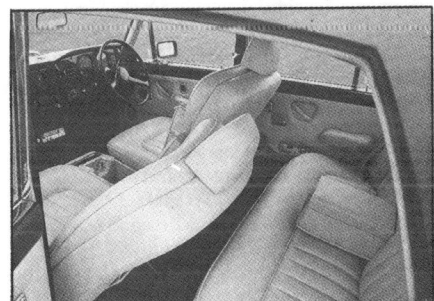

Shadow II: sharper shape, pleasing facia, better steering and air conditioning

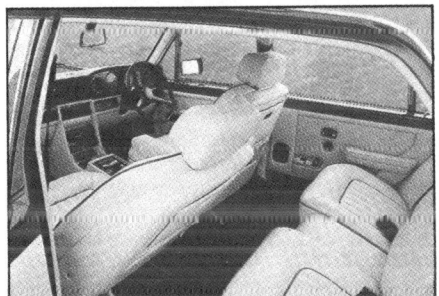

Spur: heavy looks, masses of room inside, precise handling and anti-lock brakes

Crewe or crude?

Mention the MkX and Jaguar enthusiasts instantly split into two camps: love it or hate it. The guys in the hate group have probably never driven one and form secondhand opinions influenced by people who owned a £200 rot-box 10 years ago. MkX lovers usually own one, in good condition, and will insist to their last breath that it was the best saloon car Jaguar ever made.

Me? I've always blown hot and cold about the car, one minute thinking it was the only Jaguar saloon from the sixties worth owning, the next that it was just a vast money-draining barge, not worth the bother. Thirty years have now passed since the biggest Jaguar ever built appeared and it's as controversial as ever. The time was ripe to see how good, or bad, it really is.

A good long drive seemed the best way to get to grips with the car, so we set a weekend aside to take Stuart Holmes' superb 420G (the name given to the last of the Mk Xs built from 1966 to 1970) to Portmeirion in North Wales, using twisting A and B roads, and avoiding motorways. To make things more interesting we decided to bring a rival sixties luxury carriage along, not the obvious Daimler or Mercedes, but a Bentley; in this case a standard S3 saloon dating from 1963. Our angle was value for money: is the £26,000 Bentley worth twice as much as the £13,000 Jaguar?

Saturday

Hampson of Winsford, Cheshire, is lending us the Bentley and it's the morning of its auction when friend Adam Ashmore and I arrive to collect the car. We spend a few minutes looking around the lots with Hampson boss Mark Hamilton. Prices are much more realistic now, reports Mark, pointing out a beautiful left-hooker MGA roadster – one of the nicest we have ever seen – with an estimated selling price of £10-12,000.

Ashmore is more interested in a magnificent 36,000-mile first series Jensen FF. I'm drooling over a Maserati Indy, a big hunk of car priced at £26,000, despite the left-hand drive steering.

Stuart Holmes has already arrived with the Jaguar, stirred from its normal winter hibernation – which usually lasts until April – and fresh from a Granada TV film shoot. He has minor misgivings about the blowing exhaust and a non-functioning offside petrol pump that has put one of the fuel tanks out of commission, but otherwise the car is running beautifully. It looks beautiful too, and Adam and I are hard-pressed to find a single blemish on its glistening and totally original British Racing green bodywork.

Stuart, a tanker driver from Stretford in Manchester and leading light in the local classic car scene (he's the face behind the Tatton Park Classic Car Spectacular), has owned SED 909H for eight years.

"I bought it for £1900 when nobody wanted them" says Stuart, "and now it is on an agreed value policy for £13,000". That's big money for a 420G – you can pick up a perfectly good one for £5000 – but SED 909H is one of the best of its kind in existence. Girlfriend Lesley looks lost in the vast front seat and Stuart reminds us with a grin that the MkX has the widest back seat of any British car.

"Rolls-Royce saw its arse when this car came out" says Stuart. "It had one at the factory at Crewe for six months to weigh it up." Could the Jaguar really be better than the Bentley? I'm itching to get my hands on it.

Mark Hamilton has pulled the Bentley out for us now. It is finished in a lush plum colour that sets off its tall, finely balanced, sweeping lines and there is a nicely preserved powder blue leather interior. It's in good order, with a particularly clean looking engine compartment. This is an S3 with the twin light front and slightly squatter grille, a combination I much prefer.

We throw our bags in the surprisingly mean, shallow boot and I fire the Bentley up. It's a lazy starter, giving Adam and I worries about the resilience of the battery, but the car never lets us down over the whole weekend.

Finally we are in convoy behind the green Jaguar, wafting out along the A54 towards Chester, sitting at Range Rover height on outlandishly thick leather seats and keeping a watchful eye on the fuel gauge, which flickers below the half-way mark with a glowing green warning light. The big alloy V8 warms through after five miles or so and by the time we reach Tarvin, near Chester, it's time to take some fuel on board. The S3 takes a wallet-thinning £28 of 4-star before its big 18-gallon tank is fully satisfied. With an average consumption of 11-14mpg, it won't be the last time we see a garage forecourt.

With Stuart behind this time we set off again, with gathering rain-swelled clouds making it dismal enough for sidelights, although the Bentley's impressively powerful heater makes the cabin feel snug and safe.

We are starting to notice just how powerful the Bentley really is. Opened up on straights it easily keeps pace with the Nova GTE camera car and seems to leave Stuart in the 420G floundering in its spray. A brief stab at the oily smooth throttle pushes the stately bulk forward forcefully. There's a more sporty V8 exhaust crackle than we'd expected too, and the sound of distant thunder emanating from the thickly padded bulkhead is a welcome surprise, but the engine always feels turbine smooth.

Although Crewe never published any power figures we guess the output is well over 200bhp, with perhaps 300lb of torque

Last of the MkXs was called 420G

The Jaguar Mk X was a bargain in the sixties, compared with the Bentley S3. But was it the better car? asks Martin Buckley. Photography by James Mann.

S3 Bentley with twin lights, lower grille

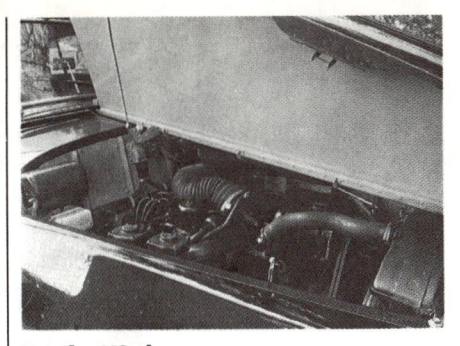

Bentley V8 gives ultra-smooth 200bhp plus, 120mph performance and strong acceleration for a big 44cwt car.

Below: Rear compartment of Bentley cosseting and discreet

Right: Dash is lightly stocked but beautifully finished; old-fashioned compared with Jaguar. Thin wheel needs light touch

peaking low down. We're less impressed by the jerky-feeling, four-speed automatic gearbox though, and we suspect it's a touch low on fluid. "It feels just like a Majestic Major" comments two-time big Daimler man Adam, as I gun the 44cwt hulk out of a turn. Having driven both I have to agree.

By 1.30pm it's time for a photo-call and we turn off the A54 into some thickly wooded parkland. While photographer Mann is jostling the cars into position for a shot we have a chance to compare the styling of the duo. Next to the tall, narrow-windowed Bentley with its sweeping wing-line and bulbous bustle tail the Jaguar looks like some kind of UFO, squat and immensely wide at 6ft 4in.

It is bulky looking from some angles, elegant from others, and although it could never be anything other than a Jaguar, its over-the-top proportions sit uneasily on the traditional Jaguar sports saloon image. Maybe that's why it was the first post-war car Jaguar really had to *sell*.

Still, it's about the only Jaguar that truly matches up to the outlandish artist's impressions of it in the adverts of the day, I snigger to myself. The 420G version of the Mk X (did 'G' stand for 'Grand'?) has a chrome strip to break up the bloated side panels, which we all agree improves the balance of the shape enormously.

We're off again, the rain lashing down now with the double hinderance of a blustery February wind. Stomachs are growling in both cars but there's no time for food if we want to get to Portmeirion in good time and take a few more pictures along the way. Ashmore is sampling the back seat ride now as we dash through Buckley ("must be a dump with a name like that", jokes Adam) Mold, along the twisting A494 to Ruthin and then on to the B5105, a smooth-surfaced

THE BENTLEY IS HARDER WORK, ITS BIG CROSSPLIES NEEDING CONSTANT SMALL CORRECTIONS THROUGH THE LOW-GEARED POWER STEERING

arterial that climbs into the Cambrian Mountains, skirting the Clocaenog forest.

The Bentley is harder work now, its big crossplies needing constant small corrections through the low-geared but quite pleasantly weighty power steering. The tyres dance vividly on cats eyes and every now and again a gust of wind catches the car in the wheel-arches and blows it off course, which makes my hands feel clammy with tension on the big, pencil-thin three-spoke wheel. The Jaguar, in front now, looks much more composed and at home on these roads.

We join the A5 briefly near Cerrigydrudion and then take the B4501 to Bala to investigate a lake for a photographic location. It proves fruitless, so, after the Jaguar has taken on fresh supplies of amber nectar, we decide to head straight for Portmeirion. It's a good, fast drive along the A4212 as my confidence in the S3's handling grows and I discover its surprisingly high road-holding qualities. It's understeer all the way, but long open bends can be taken quickly with surprisingly little lean, while a fast exit out of tighter corners is helped by a manual change down to third. This car is actually *fun* to drive, although both Ashmore and I are surprised by the harshness of the

Above: Classic twin-cam six in potent E-type 3-carb tune. Below: Widest back seat of any British car

Left: Jaguar dash is dramatic but has a skin-deep feel. Driver sits much lower, though cabin is wider and airy

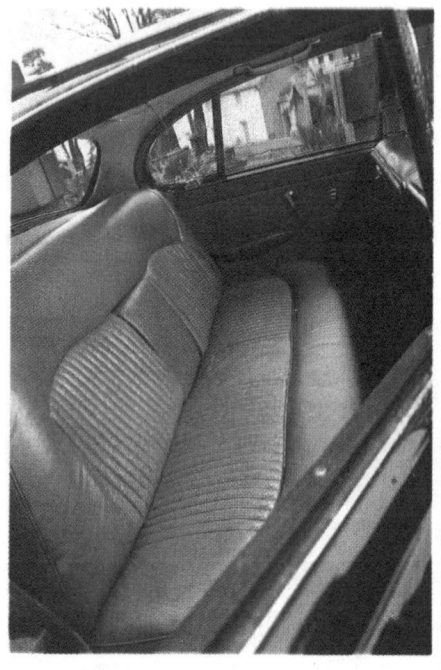

car's ride, the crashy front suspension, and the heavy drum brakes.

By 6pm we're driving through Porthmadog and into the gates of the Portmeirion complex. We have rooms in the Cliffhouse and manage to park the cars right outside the doors. Our rooms are well furnished with colour TVs running episodes of *The Prisoner*.

This is obviously a very popular hotel. We have to wait until 9pm for a table, but it's worth the wait.

Sunday

Sunday dawns much brighter, although the night winds have played havoc and strewn twigs and leaves all over the village.

After breakfast Stuart's out with a bucket of hot water, washing the Jaguar down before the next photo session in the village.

"I love the shape of this car" he says, tenderly rubbing down the shapely side panels. "I couldn't love the Bentley as much. I think the Jaguar styling is much better, more sexy." Stuart's toured Italy in this car and admits to touching 128mph in it on a German autobahn, although he doesn't treat it quite so harshly these days.

After more pictures we're ready to leave the village at around midday and I promise myself a trip back there before too long. It must be fantastic in the summer. In the car park Stuart and I finally swap charges. I warn Stuart about the Bentley's brakes and he shows me how to start the 420G most effectively. (This is the last push-button-start Jaguar). He recommends the 'D1' position on the column selector for the most responsive performance. I'm surprised to learn that Stuart has disconnected the kickdown, mainly to save petrol but also to keep himself in check. "It's fast enough without it", he laughs.

STUART DISCONNECTED THE KICKDOWN TO KEEP HIMSELF IN CHECK. "IT'S FAST ENOUGH WITHOUT IT" HE GRINS

I feel at home in the Jaguar right away, although there's definitely something more ordinary, less special about the whole car, impressive as it is in many ways. You climb over deep sills down into a low-set driving position with a chunky centre console and narrow, dark footwells. Straight away you begin to see where the extra money goes on the Crewe car. The first thing I notice is how stiff the throttle feels compared to the Bentley, the connection between engine and driver immediately less sweet. Although the quality of the Jaguar's veneer has a kind of skin-deep feel we have to agree that the layout is far more impressive, all ergonomic considerations aside. But the switches don't give the same tactile pleasure as in the Bentley, and the optional power windows sound wheezy.

The familiar XK growl is there, more intrusive and sporty sounding than the Bentley, not to mention more vibratory and, with less torque than the Crewe car, it can't get off the mark with the same urgency. Still, it surges forward quickly enough, rushing up to 70mph on the A470 towards Betws-Y-Coed without effort. We are driving quite slowly most of the way though, so photographer

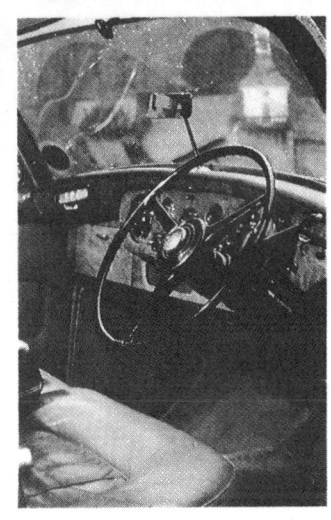

Above: A touch of class... Centre: Bentley S3 corners with surprising poise once you get used to its bulk. Only wandering crossplies spoil the good handling

Mann can take pictures of us at the wheel.

The seats don't hold you like in the Bentley and I miss the commanding driving position and the fantastic view over the tapering snout, but we can see out of the Jaguar more clearly. Glancing over into the back seat I note its immense width but relative lack of legroom for such a big car: it can't match the Bentley. I much prefer the intimacy of the Crewe car's rear compartment with its thicker seats, better quality leather and mirrors in the 'C' posts.

The Borg Warner gearbox, although only a three-speeder to the Bentley's four, changes much more sweetly and the all-round disc brakes inspire total confidence. Thanks to a big oblong pedal I can use left-foot braking in the Jag too, not possible with the S3's regular-sized floor-hinged item.

Before long we are at Blaenau Ffestinog – *Ivor the Engine* country – and then find ourselves driving between vast mountains of slate, an eerie desolate spectacle after the lush green valleys. James Mann tells us later that this is the biggest slate quarry in the world. Driving through the narrow village roads I'm aware of the bulging sides of the 420G and recall Stuart's words about the number of Mk Xs he used to see with scraped sides. "People just misjudged the width and forgot how much they bulge out."

We stop to compare notes on the cars and already Stuart wants to swap back. "It's a lovely car spoilt by horrible tyres, the crossplies make it wander all over the road." He concedes to its extra power though, and admits he could come to love the car but finds the ride and handling very old-fashioned compared with his Jaguar. He even preferred the Jaguar mascot to the winged 'B'.

"I can honestly say", he added, "that there isn't one thing on the Bentley I liked better than the 420G – but don't get me wrong: I would love to own an S3 but it would have to be *as well as* the 420G, not instead of!"

We head further up the road and find a suitable bend for cornering shots. The 420G charges round with familiar Jaguar poise, understeering safely and consistently but with less ultimate tyre scrub, if equally extra-

Above: Thin three-spoke wheel set at odd lorry like angle. Steering column very exposed but leaves lots of flat floor space

Below: S3 in Portmeirion. Narrow build makes the Bentley easy to place, but Drivers tend to forget car's size after a time

Above: The finishing touch...
Centre: 420G feels more modern, more quietly competent than the Bentley when cornering, but possibly less fun. Brakes and gearbox are smoother

vagant roll. It's spoilt only by the curious Varamatic steering, which feels oddly vague in the straight ahead, causing me to wind on more lock than necessary.

"Vague-a-matic would be more like it" sniffs Ashmore, who owned – but never drove – a MkX some years back. What impresses more is the ride: the 420G soaks up bumps superbly thanks to soft springs and effective damping; far better, in fact, than the Bentley and with far less crash from the suspension. It's softer, quieter, more modern feeling, like an XJ6 but not as finely honed.

We swap cars again and head along the A5, chasing the Nova, looking out for streams of icy water coming down from the valleys and covering the road. On one such wet bend the Bentley loses its step for a moment, but it's hardly noticeable and I am glad to be back in the S3, for all its faults.

Driving back to Manchester along the M56 at a steady 70mph, Stuart long gone and the Nova on its way back to London, Adam and I begin to collect our thoughts about the cars. The Bentley is not worth double the price of the Jaguar, and nor was it back in the early sixties when both cars were current.

In cold dynamic terms too the Jaguar is the better car, doing what it was designed to do more scientifically than the Bentley – which is simply a superbly refined version of a design that was old-fashioned 30 years ago: this car had drum brakes all round when even Cortinas had discs at the front, because Crewe didn't trust discs not to squeak.

But personally I'd take the Bentley, faults and all, and Adam comes to the same conclusion, devout Jaguar man though he is. You can see where your money goes in the sheer quality of the Bentley, the way the switches work, pedals operate and the detailing of things that please you every time you use them, and that you *know* are going to carry on that way for a long time.

No Jaguar ever offered that kind of tactile satisfaction. The 420G is a good car – a much better one in later form than it was ever given credit for – but a throw-away item compared to the long-term pleasure you'd get from a Bentley S3.

Above: Traditional Jaguar dash owes little to ergonomic thought; looks impressive all the same. Big, flat semi-bench gives zero sideways support

Below: Bentley and 420G in convoy in North Wales. As high-speed, relaxed saloons, duo is very evenly matched, but bargain Jaguar has the edge on refinement